D0845070

Topics in Applied Physics Volume 20

Topics in Applied Physics Founded by Helmut K. V. Lotsch


Volume 1 **Dye Lasers** 2nd Edition Editor: F. P. Schäfer

Volume 2 **Laser Spectroscopy** of Atoms and Molecules
Editor: H. Walther

Volume 3 **Numerical and Asymptotic Techniques in Electromagnetics**
Editor: R. Mittra

Volume 4 **Interactions on Metal Surfaces** Editor: R. Gomer

Volume 5 **Mössbauer Spectroscopy** Editor: U. Gonser

Volume 6 **Picture Processing and Digital Filtering** Editor: T. S. Huang

Volume 7 **Integrated Optics** Editor: T. Tamir

Volume 8 **Light Scattering in Solids** Editor: M. Cardona

Volume 9 **Laser Speckle** and Related Phenomena Editor: J. C. Dainty

Volume 10 **Transient Electromagnetic Fields** Editor: L. B. Felsen

Volume 11 **Digital Picture Analysis** Editor: A. Rosenfeld

Volume 12 **Turbulence** Editor: P. Bradshaw

Volume 13 **High-Resolution Laser Spectroscopy** Editor: K. Shimoda

Volume 14 **Laser Monitoring of the Atmosphere** Editor: E. D. Hinkley

Volume 15 **Radiationless Processes** in Molecules and Condensed Phases
Editor: F. K. Fong

Volume 16 **Nonlinear Infrared Generation** Editor: Y.-R. Shen

Volume 17 **Electroluminescence** Editor: J. I. Pankove

Volume 18 **Ultrashort Light Pulses.** Picosecond Techniques and Applications Editor: S. L. Shapiro

Volume 19 **Optical and Infrared Detectors** Editor: R. J. Keyes

Volume 20 **Holographic Recording Materials** Editor: H. M. Smith

Volume 21 **Solid Electrolytes** Editor: S. Geller

Volume 22 **X-Ray Optics.** Applications to Solids
Editor: H.-J. Queisser

Volume 23 **Optical Data Processing.** Applications Editor: D. Casasent

Volume 24 **Acoustic Surface Waves** Editor: A. A. Oliner

# Holographic Recording Materials

Edited by H. M. Smith

With Contributions by
R. A. Bartolini K. Biedermann J. Bordogna
R. C. Duncan, Jr. S. A. Keneman D. Meyerhofer
H. M. Smith D. L. Staebler J. C. Urbach

With 96 Figures



Springer-Verlag Berlin Heidelberg New York 1977

*Howard M. Smith*, Ph. D.

Eastman Kodak Company, 343 State Street
Rochester, NY 14650, USA

ISBN 3-540-08293-X Springer-Verlag Berlin Heidelberg New York
ISBN 0-387-08293-X Springer-Verlag New York Heidelberg Berlin

Library of Congress Cataloging in Publication Data. Main entry under title: Holographic recording materials. (Topics in applied physics; v. 20). Includes bibliographical references and index. 1. Holography-Equipment and supplies. I. Smith, Howard Michael, 1938-. II. Bartolini, Robert A. TA 1542.H 64 621.36 77-24503



Monophoto typesetting, offset printing, and bookbinding: Brühlsche Universitätsdruckerei, Lahn-Gießen
2153/3130/543210

# Preface

With most new or advancing technologies, the period of discovery and initial investigation of fundamental principles is immediately followed by a period of intensive search for materials. This is the period when potential applications are being explored and engineering and design begin. The recent discoveries of the transistor and the laser are two prime examples of this type of technological evolution. A third example is the invention and recent renaissance of holography. Following the revival in 1962 to 1964 there came a prolonged (and indeed, ongoing) search for an ideal recording material. The purpose of this volume is to present the up-to-date results of this search.

The timing is right for a book devoted solely to recording materials, not only because so much work has been done, but also because the science and engineering of holography have done a good deal of "settling-in". The needs, aims, and applications have become well defined and the range of useful applications has been narrowed to the extent that whether or not a particular application is viable may well depend on the availability of a suitable recording material. These reasons, and also the fact that, in most cases, each different holographic application has its own unique requirement for a recording material, point up the need for a book on the subject of recording materials.

It is the aim of this volume to satisfy this need. The basic theory and principles of holography have been covered in a summary fashion in the introductory chapter. The purpose here is simply to acquaint the reader with the basic theory and terminology so that the ensuing chapters can be devoted totally to describing recording materials. I have tried to ensure that every known holographic (or laser) recording material is covered. The authors have ensured that each chapter represents the present state-of-the-art, even to the extent that some material is presented here for the first time.

The book is intended to be a monograph, but with so many individual contributors there is bound to be some overlap and inconsistency of notation in spite of the best efforts of the editor. For these I apologize, but I would hope that their effect will be minor and in no way detract from the content of each chapter.

Rochester, New York — *H. M. Smith*
June, 1977

# Contents

**1. Basic Holographic Principles.** By H. M. Smith (With 7 Figures) . . . 1
1.1 History . . . 1
1.2 Basic Description . . . 2
1.3 Classification . . . 3
1.4 Diffraction Efficiency . . . 5
1.4.1 Plane Holograms . . . 6
Amplitude Holograms . . . 6
Phase Holograms . . . 9
1.4.2 Volume Holograms . . . 10
Coupled Wave Theory . . . 10
Transmission Holograms . . . 13
Reflection Holograms . . . 16
Summary . . . 17
1.5 Noise . . . 18
References . . . 19

**2. Silver Halide Photographic Materials.** By K. Biedermann (With 25 Figures) . . . 21
2.1 The Holographic Process . . . 22
2.2 The Silver Halide Emulsion . . . 22
2.3 The Practical Procedure of Making Holograms and Spatial Filters 25
2.3.1 Photographic Materials Available . . . 25
2.3.2 Recording a Hologram . . . 26
Aspects on Material, Size, and Geometry . . . 26
A Deviation on Antireflection Treatment . . . 26
More on Choosing Material, Size, and Geometry . . . 26
Intensity Ratio $K$ and Exposure . . . 27
2.3.3 Processing . . . 28
Standard Procedures . . . 28
Special Developers . . . 31
2.3.4 Reconstruction . . . 31
2.3.5 Real Time Interferometer with in Situ Development . . . 32
2.4 Characteristic Parameters of Thin Amplitude Holograms and Their Evaluation . . . 34
2.4.1 The Two-Step Transfer Process . . . 34

2.4.2 The Macroscopic Characteristic . . . . . . . . . . . . . 34
2.4.3 The Microscopic Characteristic . . . . . . . . . . . . . 36
The Modulation Transfer Function (MTF) . . . . . . . 36
The MTF in Coherent Light . . . . . . . . . . . . . . 37
Methods for Measuring MTF at High Spatial Frequencies 38
The Microdensitometer . . . . . . . . . . . . . . . . 38
The Diffraction Method . . . . . . . . . . . . . . . 39
The Multiple-Sine-Slit Microdensitometer . . . . . . . . 41
Modulation Transfer Data of Silver Halide Materials . . 43
2.4.4 Scattered Flux Spectrum . . . . . . . . . . . . . . . 44
2.4.5 Figure of Merit and Attempts to Improve Photographic Materials . . . . . . . . . . . . . . . . . . . . . . 48
Holographic Exposure Index . . . . . . . . . . . . . 48
Ways of Increasing the Holographic Exposure Index . . . 49
Extending Sensitivity to Infrared Radiation . . . . . . . 49
2.4.6 Holography vs. Photography . . . . . . . . . . . . . 50
2.5 Effects . . . . . . . . . . . . . . . . . . . . . . . . . 50
2.5.1 Macroscopic Nonlinear Transfer . . . . . . . . . . . 51
2.5.2 Microscopic Nonlinear Transfer . . . . . . . . . . . 53
2.5.3 Complex Grating Formation . . . . . . . . . . . . . 55
2.5.4 "Emulsion Stress" . . . . . . . . . . . . . . . . . 56
2.6 Experimental Data on Diffraction Efficiency and Signal-to-Background Ratio of Thin Amplitude Holograms . . . . . . . . . 57
2.7 Phase Holograms . . . . . . . . . . . . . . . . . . . . . 59
2.7.1 Principles of Phase Grating Generation . . . . . . . . 59
2.7.2 Bleaching Procedures . . . . . . . . . . . . . . . . 61
Gelatine Relief Formation . . . . . . . . . . . . . . 61
Bleaching . . . . . . . . . . . . . . . . . . . . . 62
Gaseous Bleaching . . . . . . . . . . . . . . . . . 63
Reversal Processes . . . . . . . . . . . . . . . . . 64
2.7.3 Experimental Data on Diffraction Efficiency and Signal-to-Background Ratio . . . . . . . . . . . . . . . . . . 64
2.8 Thick Holograms . . . . . . . . . . . . . . . . . . . . 66
2.8.1 Characteristics of Thick Photographic Emulsions . . . . 66
2.8.2 Materials Available; Problems . . . . . . . . . . . . 67
2.8.3 Holograms of Extended Thickness . . . . . . . . . . . 68
2.8.4 Experimental Data on Diffraction Efficiency and Signal-to Background Ratio of Thick Holograms . . . . . . . . . 68
2.9 Summary . . . . . . . . . . . . . . . . . . . . . . . . 70
References . . . . . . . . . . . . . . . . . . . . . . . . 71

**3. Dichromated Gelatin.** By D. Meyerhofer (With 5 Figures) . . . . . 75
3.1 Basic Photochemistry . . . . . . . . . . . . . . . . . . 76
3.1.1 Dichromated Colloid Layers . . . . . . . . . . . . . 76

3.1.2 The Properties of Gelatin . . . . . . . . . . . . . . 77
3.1.3 Photosensitivity of the Dichromate Ion . . . . . . . . . . 78
3.2 Dichromated Gelatin as a Photosensitive Medium . . . . . . 79
3.2.1 Dichromated Gelatin as a Photoresist . . . . . . . . . . 79
3.2.2 Photochemistry of Prehardened Dichromated Gelatin Layers 80
3.2.3 The Shankoff Technique of Phase Shift Enhancement . . 82
3.2.4 Variation in Emulsion Thickness During Processing . . . 85
3.3 Preparation and Processing of Hardened Dichromated Gelatin Plates for Thick Holograms . . . . . . . . . . . . . . . . . 87
3.3.1 Preparation of Plates . . . . . . . . . . . . . . . . 87
3.3.2 Development Procedure . . . . . . . . . . . . . . . . 89
3.3.3 Stability of Dichromated Gelatin Holograms . . . . . . 91
3.4 Holographic Properties of Hardened Dichromated Gelatin . . . 91
3.4.1 Phase Shift Available . . . . . . . . . . . . . . . . 91
3.4.2 Sensitivity . . . . . . . . . . . . . . . . . . . . 93
3.4.3 Resolution . . . . . . . . . . . . . . . . . . . . 95
3.4.4 Image Quality and Noise . . . . . . . . . . . . . . . 95
3.4.5 Storage Capacity . . . . . . . . . . . . . . . . . . 96
3.5 Applications . . . . . . . . . . . . . . . . . . . . . 97
3.5.1 Holographic Optical Elements . . . . . . . . . . . . 97
3.5.2 Duplicate Holograms in Dichromated Gelatin . . . . . 98
References . . . . . . . . . . . . . . . . . . . . . . . 98

**4. Ferroelectric Crystals.** By D. L. Staebler (With 13 Figures) . . . . . 101

4.1 Theory of Mechanisms . . . . . . . . . . . . . . . . . . 102
4.1.1 Storage . . . . . . . . . . . . . . . . . . . . . . 102
4.1.2 Erasure . . . . . . . . . . . . . . . . . . . . . . 108
4.1.3 Fixing . . . . . . . . . . . . . . . . . . . . . . 109
4.1.4 Electrical Enhancement . . . . . . . . . . . . . . . 114
4.2 Materials and Preparation . . . . . . . . . . . . . . . . 115
4.2.1 Lithium Niobate ($LiNbO_3$) . . . . . . . . . . . . . 115
4.2.2 Strontium Barium Niobate (SBN) . . . . . . . . . . . 117
4.2.3 Barium Titanate ($BaTiO_3$) . . . . . . . . . . . . . 118
4.2.4 Barium Sodium Niobate . . . . . . . . . . . . . . . 118
4.2.5 Lithium Tantalate ($LiTaO_3$) . . . . . . . . . . . . 118
4.2.6 Potassium Tantalate Niobate (KTN) . . . . . . . . . . 118
4.2.7 PLZT Ceramics . . . . . . . . . . . . . . . . . . . 119
4.3 Methods of Use . . . . . . . . . . . . . . . . . . . . . 119
4.3.1 Read/Write . . . . . . . . . . . . . . . . . . . . 120
4.3.2 Read/Write with Nondestructive Readout (NDRO) . . . 120
4.3.3 Read-Only . . . . . . . . . . . . . . . . . . . . . 121
4.3.4 Multiphoton Absorption . . . . . . . . . . . . . . . 121
4.3.5 Layered Memory . . . . . . . . . . . . . . . . . . 122
4.3.6 Holographic Page Synthesis . . . . . . . . . . . . . 123

4.4 Holographic Properties . . . . . . . . . . . . . . . . . 123
4.4.1 Sensitivity . . . . . . . . . . . . . . . . . . . 123
4.4.2 Dynamic Range . . . . . . . . . . . . . . . . . 126
4.4.3 Resolution . . . . . . . . . . . . . . . . . . . 128
4.4.4 Storage Time . . . . . . . . . . . . . . . . . . 128
4.4.5 Noise and Distortion . . . . . . . . . . . . . . . 129
References . . . . . . . . . . . . . . . . . . . . . . 130

**5. Inorganic Photochromic Materials.** By R. C. Duncan, Jr. and D. L. Staebler (With 20 Figures) . . . . . . . . . . . . . . . 133
5.1 Basic Model . . . . . . . . . . . . . . . . . . . . 134
5.2 Photochromic $CaF_2$, $SrTiO_3$, $CaTiO_3$ . . . . . . . . . . . . 135
5.2.1 Optimum Dopants and Concentrations . . . . . . . . . 135
5.2.2 Photochromic Absorption Spectra . . . . . . . . . . . 136
5.2.3 Optical Recording Modes . . . . . . . . . . . . . . 139
5.2.4 Optical Readout at 514.5 nm . . . . . . . . . . . . . 140
Thickness Effects . . . . . . . . . . . . . . . . . 140
Contrast Ratio . . . . . . . . . . . . . . . . . . 141
5.2.5 Erase Mode Sensitivity at 514.5 nm . . . . . . . . . . 141
5.2.6 Sensitometric Characteristics and Time-Intensity Reciprocity . . . . . . . . . . . . . . . . . . . . 145
$CaF_2$ : La, Na and $CaF_2$ : Ce, Na . . . . . . . . . . . 146
$SrTiO_3$ : Ni, Mo, Al and $CaTiO_3$ : Ni, Mo . . . . . . . . . 148
5.3 Hologram Storage . . . . . . . . . . . . . . . . . . 151
5.3.1 Sensitivity . . . . . . . . . . . . . . . . . . . 152
5.3.2 Maximum Efficiency . . . . . . . . . . . . . . . . 154
5.3.3 Storage Capacity . . . . . . . . . . . . . . . . . 155
5.4 Photodichroics . . . . . . . . . . . . . . . . . . . 156
5.4.1 Model . . . . . . . . . . . . . . . . . . . . . 156
5.4.2 Hologram Storage . . . . . . . . . . . . . . . . . 158
References . . . . . . . . . . . . . . . . . . . . . . 159

**6. Thermoplastic Hologram Recording.** By J. C. Urbach (With 14 Figures) . . . . . . . . . . . . . . . . . . . . . . 161
6.1 Process Description . . . . . . . . . . . . . . . . . 161
6.1.1 Thermoplastic Electron Beam Recording . . . . . . . . 162
6.1.2 Photosensitive Recording . . . . . . . . . . . . . . 163
Photoplastic Devices . . . . . . . . . . . . . . . . 164
Thermoplastic-Overcoated Photoconductor Devices . . . 166
Device and Process Variations . . . . . . . . . . . . 167
6.1.3 Process Models and Theories . . . . . . . . . . . . . 168
Theories of Random Deformation . . . . . . . . . . . 170
6.1.4 Materials . . . . . . . . . . . . . . . . . . . . 171

Substrates and Conductive Coatings . . . . . . . . . . . 171
Photoconductors . . . . . . . . . . . . . . . . . . . . 172
Special Photoconductors . . . . . . . . . . . . . . . 176
Thermoplastic Materials . . . . . . . . . . . . . . . 177
6.1.5 Fabrication Techniques . . . . . . . . . . . . . . . 181
Substrates and Conductive Coatings . . . . . . . . . . . 181
Photoconductor Fabrication . . . . . . . . . . . . . . 182
Thermoplastic Fabrication . . . . . . . . . . . . . . . 184
Overcoating and Matrixing . . . . . . . . . . . . . . . 185
6.2 Holographic Properties . . . . . . . . . . . . . . . . . . 186
6.2.1 Bandwidth and Resolution . . . . . . . . . . . . . . 186
6.2.2 Sensitivity and Diffraction Efficiency . . . . . . . . . . 191
Diffraction Efficiency Considerations . . . . . . . . . . 192
Experimental Exposure Response Results . . . . . . . . 194
6.2.3 Noise . . . . . . . . . . . . . . . . . . . . . . . . 197
Experimental Noise Measurements . . . . . . . . . . . 199
6.2.4 Cyclic Operation . . . . . . . . . . . . . . . . . . . 201
6.2.5 Other Physical Properties . . . . . . . . . . . . . . . 202
Shelf Life . . . . . . . . . . . . . . . . . . . . . . 202
Latent Image Life . . . . . . . . . . . . . . . . . . 203
Archival Life . . . . . . . . . . . . . . . . . . . . 203
Mechanical Durability and Contamination . . . . . . . 204
6.3 Conclusions . . . . . . . . . . . . . . . . . . . . . . . 205
References . . . . . . . . . . . . . . . . . . . . . . . . . 206

**7. Photoresists.** By R.A. Bartolini (With 11 Figures) . . . . . . . . . 209
7.1 Types of Photoresists . . . . . . . . . . . . . . . . . . . 211
7.2 Theory of Mechanism . . . . . . . . . . . . . . . . . . . 211
7.3 Hologram Nonlinearities . . . . . . . . . . . . . . . . . 213
7.4 Diffraction Efficiency and Signal-to-Noise Ratio . . . . . . . 215
7.5 Shipley AZ-1350 Photoresist Characteristics . . . . . . . . . 217
7.5.1 Shipley AZ-1350 Material Nonlinearity Considerations . . 218
7.5.2 Shipley AZ-1350 Resolution Capability . . . . . . . . . 220
7.5.3 Shipley AZ-1350 Relief Phase Hologram Design Procedure 221
References . . . . . . . . . . . . . . . . . . . . . . . . . 226

**8. Other Materials and Devices.** By J. Bordogna and S.A. Keneman (With 1 Figure) . . . . . . . . . . . . . . . . . . . . . . . . 229
8.1 Magneto-Optic Films . . . . . . . . . . . . . . . . . . . 230
8.2 Metal Films . . . . . . . . . . . . . . . . . . . . . . . . 231
8.3 Layered Devices with Photoconductor Write-In . . . . . . . 232
8.3.1 Ferroelectric-Photoconductor Devices . . . . . . . . . . 232
8.3.2 Liquid-Crystal-Photoconductor Devices . . . . . . . . 235
8.3.3 Photoconductor-Pockel's Effect Devices . . . . . . . . 236

8.3.4 Elastomer Devices . . . . . . . . . . . . . . . . . . 237
8.3.5 Advantages and Limitations of Photoconductor-Based Devices . . . . . . . . . . . . . . . . . . . . . . . . 238
8.4 Chalcogenide Glasses . . . . . . . . . . . . . . . . . . . . . 238
8.4.1 Materials Exhibiting an Amorphous-Crystalline Phase Change . . . . . . . . . . . . . . . . . . . . . . . . 239
8.4.2 $As_2S_3$ and Related Materials . . . . . . . . . . . . . 239
8.4.3 Photodoping Devices . . . . . . . . . . . . . . . . 241
8.4.4 Summary . . . . . . . . . . . . . . . . . . . . 241
References . . . . . . . . . . . . . . . . . . . . . . . . 241

**Additional References with Titles** . . . . . . . . . . . . . . . . . 245

**Author Index** . . . . . . . . . . . . . . . . . . . . . . . . 247

**Subject Index** . . . . . . . . . . . . . . . . . . . . . . . . 249

# Contributors

Bartolini, Robert A.
RCA Laboratories, David Sarnoff Research Center, Princeton, NJ 08540, USA

Biedermann, Klaus
The Royal Institute of Technology, Dept. of Physics and Institute of Optical Research, S-100 44 Stockholm, Sweden

Bordogna, Joseph
Moore School of Electrical Engineering, University of Pennsylvania, Philadelphia, PA 19104, USA

Duncan, Robert C., Jr.
RCA Laboratories, David Sarnoff Research Center,
Princeton, NJ 08540, USA

Keneman, Scott A.
RCA Laboratories, David Sarnoff Research Center,
Princeton, NJ 08540, USA

Meyerhofer, Dietrich
RCA Laboratories, David Sarnoff Research Center,
Princeton, NJ 08540, USA

Smith, Howard M.
Research Laboratories, Eastman Kodak Company, 343 State Street, Rochester, NY 14650, USA

Staebler, David L.
RCA Laboratories, David Sarnoff Research Center,
Princeton, NJ 08540, USA

Urbach, John C.
Palo Alto Research Center, Xerox Corp.,
Palo Alto, CA 94304, USA

# 1. Basic Holographic Principles

H. M. Smith

With 7 Figures

With most new or advancing technologies, the period of discovery and initial investigation of fundamental principles is immediately followed by a period of intensive search for materials. The recent discoveries of the transistor and laser are two prime examples of this type of technological evolution. A third example is the invention and recent renaissance of holography. Following the revival in 1962–64, there came a prolonged (and indeed, ongoing) search for an ideal recording material. The purpose of this monograph is to present the results of this search.

## 1.1 History

The history of holography begins with its invention by *Gabor* [1.1] in 1948. His ideas and demonstrations form the foundation of modern holography. Because of the severe limitations imposed by having to work with a totally inadequate light source, among other reasons, few practical uses were found for holography, even though it represented a fundamentally different method of image formation.

However, in the period 1962–64 *Leith* and *Upatnieks* [1.2, 3] published a series of papers describing methods for overcoming two of the major problems associated with the earlier techniques—the limited coherence and intensity of the light and the very problematical mixing of the signal (image forming) and illuminating waves during readout. These papers created a great deal of excitement and the whole field of holography underwent a rebirth. The coherence problem was solved by the invention of the laser. Its use as a source for holography allowed for the diffuse illumination of large, three-dimensional objects and thus the ability to view the reconstruction, complete with parallax and depth effects, with the unaided eye. The use of the laser also permitted great latitude and flexibility in the choice of object to be used and of the physical layout of the recording setup—a degree of freedom undreamed of by the early workers in the 1950's [1.4–11].

The second problem, that of the mixing of the signal and illuminating waves, was solved by *Leith* and *Upatnieks* by the use of an off-axis (noncollinear) reference beam to record the hologram. This allowed reconstruction with a similarly off-axis illuminating beam. Since this light was traveling in a substantially different direction, it no longer obscured the image-forming light.

By this simple technique they were able to solve one of the major problems with the earlier forms of holography.

At about the same time, *Denisyuk* [1.12] in Russia and *Van Heerden* [1.13] in the USA were expanding the basic 1948 concepts to include the recording of waves that could be reconstructed by reflection of white light and waves that could be recorded throughout a large volume of the recording material.

As a result of all of the aforementioned innovations, new life was given to the old idea of holography. Not only was its practicality well demonstrated, but the striking three-dimensional images that could be produced fired the imaginations of many thousands of students and scientists throughout the world. The vast amount of activity in the field during the time span of roughly 1964 through 1970 resulted in many new ideas for application, creating a demand for increasingly better recording materials, a demand that in large part has been satisfied.

## 1.2 Basic Description

If $O$ represents a monochromatic wave from the object, and $R$ a reference wave coherent with $O$, the total field at the recording medium is $O+R$ so that a square-law recording medium responds to the irradiance $|O+R|^2$. After processing, the hologram recording material has a certain complex amplitude transmittance $T_a(x)$ that can be expressed as a function of the exposure $E(x)=t|O+R|^2$ ($t$ is the exposure time):

$$T_a(x)=\beta E(x)=\beta t|O+R|^2$$
$$=\beta t(|O|^2+|R|^2+OR^*+O^*R), \quad (1.1)$$

where the * denotes complex conjugate and $\beta$ depends on the recording material involved. When the hologram with this amplitude transmittance is illuminated with a wave $C$, the transmitted wave at the hologram is

$$\psi(x)=CT_a(x)$$
$$=\beta t(C|O|^2+C|R|^2+CR^*O+CRO^*). \quad (1.2)$$

If the illuminating wave $C$ is sufficiently uniform so that $CR^*$ is approximately constant across the hologram, the third term of (1.2) becomes simply a constant $\times O$. Thus this term represents a reconstructed wave that is identical to the original object wave $O$. The other terms represent the zero-order wave $\beta Ct(|O|^2+|R|^2)$ and the other first-order wave $\beta CtRO^*$. The latter is known as the "conjugate" wave. These terms represent the waves that caused the problems referred to above with the early Gabor holograms. But because of the offset reference beam associated with this type of hologram, these waves are spatially separated from the desired reconstruction wave. This is the main advantage of the off-axis technique and is the principal reason for the resurgence of interest in holography in 1964.

## 1.3 Classification

Holograms are classified as to type with the recording geometry, the type of modulation imposed on the illuminating wave, the thickness of the recording material, and the mode of image formation as factors.

If the two interfering beams are traveling in substantially the same direction, the recording of the interference pattern is said to be a *Gabor* hologram or an *in-line* hologram. If the two interfering beams arrive at the recording medium from substantially different directions, the recording is a *Leith-Upatnieks* or *off-axis* hologram. If the two interfering beams are traveling in essentially opposite directions, the recorded hologram is said to be a *Lippmann* or *reflection* hologram.

These hologram types are further classified, depending on recording geometry, as *Fresnel*, *Fourier Transform*, *Fraunhofer*, or *Lensless Fourier Transform*. Generally speaking, if the object is reasonably close to the recording medium, say 10 or so hologram or object diameters distant, the field at the hologram plane is the Fresnel diffraction pattern of the object, and the hologram thus recorded is a *Fresnel* hologram. If, on the other hand, the object and recording medium are separated by many hologram or object diameters, the field at the recording plane is the Fraunhofer or far field diffraction pattern of the object and a *Fraunhofer* hologram results. If a lens is used to produce the far field pattern at the recording plane, with the lens a focal length distant from both the object and recording plane, then a *Fourier Transform* hologram results, provided the reference wave is plane. Lastly, if the reference beam originates from a point that is the same distance from the recording plane as the object, then a so-called *Lensless Fourier Transform* hologram results.

Any of the foregoing hologram types may also be recorded as a thick or thin hologram. A *thin* (or *plane*) hologram is one for which the thickness of the recording medium is small compared to the fringe spacing. A *thick* (or *volume*) hologram is one for which the thickness is of the order of or greater than the fringe spacing. The distinction between thick and thin holograms is usually made with the aid of the $Q$-parameter defined as

$$Q \equiv 2\pi\lambda d/(n\Lambda^2), \tag{1.3}$$

where $\lambda$ is the illuminating wavelength, $n$ the index of the recording material, $d$ its thickness, and $\Lambda$ the spacing of the recorded fringes. The hologram is considered thick when $Q \gtrsim 10$ and thin otherwise, although it has recently been shown [1.14] that thick hologram theory (coupled wave theory) is quite accurate even for $Q$ values of order 1.

Finally, holograms are also classified by the mechanism by which the illuminating light is diffracted. In an *amplitude* hologram the interference pattern is recorded as a density variation of the recording medium, and the amplitude of the illuminating wave is accordingly modulated. In a *phase* hologram, a phase modulation is imposed on the illuminating wave.

The amplitude transmittance of a hologram is a complex function that describes the change in the amplitude and phase of an illuminating wave upon transmission through the material. If the illuminating field is described by the complex function $C(x)=C_0(x)\exp[i\varphi_0(x)]$ and the transmitted field by $\psi(x)=b_t(x)\exp[i\varphi_t(x)]$ then the amplitude transmittance of the hologram is defined as the ratio of these two quantities:

$$T_a(x)=\frac{\psi(x)}{C(x)}=\left[\frac{b_t}{C_0}\right]e^{i(\varphi_t-\varphi_0)}. \quad (1.4)$$

In most instances, the illuminating wave $C$ is constant or nearly so, so that the complex amplitude transmittance is given by

$$T_a(x)=b_t(x)e^{i\varphi_0(x)}. \quad (1.5)$$

If $b_t(x)$ is not constant but $\varphi_t(x)$ is, it is the amplitude of the illuminating wave that is modulated and the hologram is an *amplitude* hologram. The magnitude of $b_t(x)$ depends on the absorption constant $a$ and the thickness of the recording material according to

$$b_t(x)=C_0(x)e^{-a(x)d}. \quad (1.6)$$

Thus for an amplitude hologram the transmitted amplitude is modulated in accordance with the exposure, since the exposure causes a spatial variation of the absorption constant $a$. The two most common recording materials that can accomplish this are silver halide photographic emulsions (Chap. 2) and photochromic materials (Chap. 5).

If, on the other hand, $b_t(x)$ is constant and $\varphi_t(x)$ varies, then the phase of the illuminating wave is modulated and the hologram is a *phase* hologram. The phase of the illuminating wave can be modulated in one or both of two different ways. If a surface relief is found as the result of exposure to the interference fringe pattern, then the thickness of the material is a function of $x$. If the index of refraction of the material is changed in accordance with the exposure, then the optical path of the transmitted light is a function of $x$. In either case, the phase of the transmitted light will vary with $x$ because the optical path $nd$ will be a function of $x$. The phase $\varphi_t$ of the transmitted light is related to the optical path according to

$$\varphi_t=\frac{2\pi}{\lambda}nd. \quad (1.7)$$

A phase hologram results when $\varphi_t$ is a function of the exposure—such as when either the index of refraction $n$ or the thickness $d$, or both, is exposure dependent. From (1.7) it is seen that a change $\Delta\varphi_t$ in $\varphi_t$ is given by

$$\Delta\varphi_t=\frac{2\pi}{\lambda}[d\Delta n+(n-1)\Delta d], \quad (1.8)$$

where the $n-1$ factor appears assuming the hologram to be in air. If the hologram is thin, $d$ is very small and the contribution to $\Delta\varphi_t$ from the term $d\Delta n$ is negligible and

$$\Delta\varphi_t = \frac{2\pi}{\lambda}(n-1)\Delta d. \tag{1.9}$$

This would be a surface relief hologram, achieved mainly by embossing, etching a photo resist material, or with a thermoplastic film (Chap. 3). For a thick hologram, on the other hand, having a negligible surface relief but a varying index, we have

$$\Delta\varphi_t = \frac{2\pi}{\lambda} d\Delta n. \tag{1.10}$$

Typical materials for this type of hologram are bleached photographic emulsions (Chap. 2), dichromated gelatin (Chap. 3), and ferroelectric crystals (Chap. 4). All of these materials (and others) are discussed in detail in the following chapters.

## 1.4 Diffraction Efficiency

Diffraction efficiency is defined as the ratio of the total, useful, image-forming light flux diffracted by the hologram to the total light flux used to illuminate the hologram. If some of the illuminating light is diffracted into an unwanted conjugate (or primary) image, it is still only the light in the primary (or conjugate) image that is pertinent to the stated diffraction efficiency. The importance of producing a hologram with a reasonable diffraction efficiency is obvious, whether the application is simply the viewing of a three-dimensional scene, holographic interferometry, or data storage; efficient use of the illuminating light allows for smaller and more economical systems. Being able to use a smaller and cheaper laser for illumination or readout, for example, could be the deciding factor in determining the feasibility of a holographic data storage system.

However, one must approach the question of diffraction efficiency with some careful thought. In many cases it is the output signal-to-noise ratio (s/n) that is the more important consideration. The noise in a holographic image arises principally from three factors: intrinsic nonlinearity, such as in phase holograms, that causes unwanted light to be diffracted into the image region; the nonlinearity of the amplitude transmittance-exposure characteristic of the recording medium; and the noise caused by the buildup of the granular structure of the recording material. (A discussion of the sources of noise in holographic images follows in Sect. 5.) It is, however, possible to increase the diffraction efficiency without seriously impairing the s/n characteristics of the recording by a judicious choice of recording material and processing technique.

The equations describing the diffraction efficiency of the various hologram types are divided into two groups, those for plane holograms and those for volume holograms. This is because the description of the diffraction process for the two cases is fundamentally different, even though it has been shown that in the common case of amplitude holograms, for most practical recording situations the resulting diffraction efficiences are equivalent [1.15]. This is definitely not the case for phase holograms, however. The distinction between thick and thin holograms is made with the aid of the $Q$-parameter defined earlier [(1.3)]. The hologram is considered to be thick when $Q \gtrsim 10$ and thin otherwise.

### 1.4.1 Plane Holograms

#### Amplitude Holograms

We begin by assuming an irradiance distribution at the hologram (for a diffuse object) given by

$$\begin{aligned} H(x) &= |O_0(x)\mathrm{e}^{\mathrm{i}\varphi_0(x)} + R_0\mathrm{e}^{\mathrm{i}\omega x}|^2 \\ &= O_0^2(x) + R_0^2 + 2O_0(x)\,R_0\cos(\omega x - \varphi_0)\,, \end{aligned} \quad (1.11)$$

where $O_0(x)$ and $R_0$ are the (real) amplitudes and $\varphi_0(x)$ and $\omega x$ are the phase of the object and reference waves, respectively. The object is assumed to be diffuse so that both the phase and amplitude of the object wave are functions of the space coordinate $x$. The average spatial frequency of this distribution is given by

$$\bar{v} = \frac{1}{2\pi}\left\langle \frac{d}{dx}\,[\omega x - \varphi_0(x)] \right\rangle, \quad (1.12)$$

where the angle brackets indicate a spatial average. If we take into account the MTF of the recording material, the effective exposure is

$$E = tH = t[\langle O_0^2(x)\rangle + R_0^2][1 + mM(\bar{v})\cos(\omega x - \varphi_0)]\,, \quad (1.13)$$

where $t$ is the exposure time and $m$ is the exposure modulation given by

$$m = \frac{2[\langle O_0^2(x)\rangle R_0^2]^{1/2}}{\langle O_0^2(x)\rangle + R_0^2}\,, \quad (1.14)$$

and $M(\bar{v})$ is the MTF. In (1.14) we have used $\langle O_0^2(x)\rangle$ rather than the more rigorous $O_0^2(x)$, and so we are neglecting the self-interference or speckle caused by the diffuse object. However, the effects of the speckle are not excessive as long as the modulation $m$ is not too large. The modulation of the fringes represented by the cosine term of (1.13) is reduced by the factor $M(\bar{v})$, the MTF of the recording material at the average frequency $\bar{v}$. This is a good approximation as long as the object size is such that it subtends a very small angle at the hologram. If this is the case, $\varphi_0(x)$ is nearly a constant, or a constant plus a term proportional to $x$, and the bandwidth of the recorded signal is small and centered around $\bar{\nu}$. If we assume the amplitude transmittance-log exposure curve ($T_a - \log E$) to be essentially linear with slope $\alpha$ in the vicinity of the average exposure $E_0 = t[\langle O_0^2(x)\rangle + R_0^2]$, we can write for the amplitude transmittance

$$T_a(E) = T_a(E_0) + \alpha \log(E/E_0), \tag{1.15}$$

so that

$$T_a(x) = T_a(E_0) + \alpha \log[1 + mM(\bar{\nu})\cos(\omega x - \varphi_0)]. \tag{1.16}$$

Expanding the logarithmic function, we have

$$\begin{aligned} T_a(x) = T_a(E_0) + 0.43\alpha[&mM(\bar{\nu})\cos(\omega x - \varphi_0) \\ &- \frac{m^2M^2(\bar{\nu})}{2}\cos^2(\omega x - \varphi_0) \\ &+ \frac{m^3M^3(\bar{\nu})}{3}\cos^3(\omega x - \varphi_0) - \ldots]. \end{aligned} \tag{1.17}$$

The $T_a - \log E$ curve is linear only in the vicinity of $E_0$, which implies a moderately small exposure modulation $m$. Therefore, we must assume the $m$ is small enough that we can neglect all except the first-order term in (1.17) and arrive at

$$T_a(x) = T_a(E_0) + 0.43\alpha mM(\bar{\nu})\cos(\omega x - \varphi_0). \tag{1.18}$$

The object wave is reconstructed by illuminating the hologram with the reference wave $R_0 \exp(\mathrm{i}\omega x)$, yielding a conjugate image term

$$\psi_c(x) = \frac{0.434}{2} R_0 \alpha mM(\bar{\nu}) \mathrm{e}^{\mathrm{i}\varphi_0}. \tag{1.19}$$

The irradiance in the reconstructed field is

$$H_c = R_0^2 \left[\frac{0.434}{2} \alpha mM(\bar{\nu})\right] \tag{1.20}$$

and the diffraction efficiency is

$$\eta = H_c/R_0^2 = \left[\frac{0.434}{2} \alpha mM(\bar{\nu})\right]^2. \tag{1.21}$$

Now

$$m=2[\langle O_0^2(x)\rangle R_0^2]^{1/2}/[\langle O_0^2(x)\rangle+R_0^2]=\frac{2\sqrt{K}}{1+K}, \tag{1.22}$$

where $K$ is the beam ratio $R_0^2/\langle O_0^2(x)\rangle$. Since we are already restricted to moderately small values of $m$, $K$ is somewhat greater than unity, and so $m\approx 2/\sqrt{K}$ giving for the diffraction efficiency

$$\eta=0.188\frac{\alpha^2}{K}M^2(\bar{v})\,. \tag{1.23}$$

This equation shows that the diffraction efficiency will be a maximum at an average exposure for which the $T_a-\log E$ curve of the recording material has maximum slope.

Because we have assumed that the exposure modulation $m$ is not too large, (1.23) does not indicate the maximum diffraction efficiency obtainable with thin amplitude holograms. To get an idea of this upper limit, we write for the amplitude transmittance of the exposed and processed hologram

$$T_a(x)=T_a(E_0)+T_1\cos(\omega x-\varphi_0)\,, \tag{1.24}$$

where $T_a(E_0)$ is the average transmittance due to the average exposure $E_0=t[\langle O_0^2(x)\rangle+R_0^2]$ and $T_1$ is the amplitude of the spatially varying portion of the exposure,

$$t[\langle O_0^2(x)\rangle+R_0^2]mM(\bar{v})\cos(\omega x-\varphi_0)\,.$$

To achieve the maximum possible diffraction efficiency, $T_a(x)$ should vary between 0 and 1. Thus

$$\begin{aligned}T_a(x)&=\tfrac{1}{2}+\tfrac{1}{2}\cos(\omega x-\varphi_0)\\&=\tfrac{1}{2}+\tfrac{1}{4}\mathrm{e}^{\mathrm{i}(\omega x-\varphi_0)}+\tfrac{1}{4}\mathrm{e}^{-\mathrm{i}(\omega x-\varphi_0)}\,.\end{aligned} \tag{1.25}$$

From this equation it is evident that the maximum diffracted amplitude in either the primary or conjugate image is only one-fourth the amplitude of the illuminating wave. The useful light flux in either image is therefore only one-sixteenth of the illuminating flux, yielding a maximum diffraction efficiency of 0.0625. This is somewhat artificial, however, since there are really no practical materials that are linear over an exposure range so large that $T_a(x)$ will vary from 0 to 1. Some nonlinearity or clipping will always occur when the exposure modulation gets large. Therefore, the maximum efficiency of 0.0625 cannot be achieved when it is necessary that the reconstructed wave amplitude be proportional to the object wave amplitude.

### Phase Holograms

If the phase of the transmitted wave is proportional to the exposure

$$\varphi_t(x)=\gamma E(x) \tag{1.26}$$

then a phase hologram results. If

$$E(x)\propto|O+R|^2 \tag{1.27}$$

and

$$\begin{aligned} O(x)&=O_0\mathrm{e}^{\mathrm{i}\varphi_0(x)}\\ R(x)&=R_0\mathrm{e}^{\mathrm{i}\beta x}\end{aligned} \tag{1.28}$$

then

$$\begin{aligned} T_\mathrm{a}(x)&=b_t\mathrm{e}^{\mathrm{i}\varphi_t(x)}\\ &=b_t\exp(\mathrm{i}\gamma O_0^2)\exp(\mathrm{i}\gamma R_0^2)\\ &\quad\cdot\exp\{2\mathrm{i}\gamma O_0R_0[\cos(\varphi_0-\beta x)]\}\,.\end{aligned} \tag{1.29}$$

Simplifying the notation we set

$$\begin{aligned} K&\equiv b_t\exp[\mathrm{i}\gamma(O_0^2+R_0^2)]\\ a&\equiv 2\gamma O_0R_0\\ \Theta&\equiv\varphi_0-\beta x\end{aligned} \tag{1.30}$$

so that

$$T_\mathrm{a}(x)=K\mathrm{e}^{\mathrm{i}a\cos\Theta}\,. \tag{1.31}$$

Now by using the Bessel function expansions

$$\begin{aligned} \cos(a\cos\Theta)&=J_0(a)+2\textstyle\sum_{n=1}^{\infty}(-1)^nJ_{2n}(a)\cos 2n\Theta\\ \sin(a\cos\Theta)&=2\textstyle\sum_{n=0}^{\infty}(-1)^{n+2}J_{2n+1}(a)\cos[(2n+1)\Theta]\end{aligned} \tag{1.32}$$

and the relation

$$\mathrm{e}^{\mathrm{i}x}=\cos x+\mathrm{i}\sin x \tag{1.33}$$

we can write

$$\begin{aligned} T_\mathrm{a}(x)=K\{&J_0(a)+2\textstyle\sum_{n=1}^{\infty}(-1)^nJ_{2n}(a)\cos 2n\Theta\\ &+2\mathrm{i}\textstyle\sum_{n=0}^{\infty}(-1)^{n+2}J_{2n+1}(a)\cos[(2n+1)\Theta]\}\,.\end{aligned} \tag{1.34}$$

The $J_n(a)$ are Bessel functions of the first kind. Each is the amplitude of the $n$-th diffracted order. The term leading to the images of interest is $K[2iJ_1(a)\cos\Theta]$ which can be written as

$$2iKJ_1(a)\left(\frac{e^{i\Theta}+e^{-i\Theta}}{2}\right)=iKJ_1(a)[e^{i(\varphi_0-\beta x)}+e^{-i(\varphi_0-\beta x)}]. \quad (1.35)$$

These two terms represent the primary and conjugate image waves. The transmission term leading to the primary image is

$$T_a(x)=b_t\exp[i(\gamma O_0^2+\gamma R_0^2+\pi/2)] \cdot J_1(2\gamma O_0R_0)\exp[i(\varphi_0-\beta x)]. \quad (1.36)$$

When the hologram is illuminated with a wave of unit amplitude, the first-order diffracted wave has an amplitude $T_a$ given by (1.36). Since we are assuming for simplicity that we have a pure phase hologram, that is, no absorption, $b_t=1.0$ and the diffracted wave amplitude is simply $(J_1(2\gamma O_0R_0)$. The diffraction efficiency is equal to the square of this quantity. The maximum value is 0.339, which is considerably higher than the maximum efficiency for amplitude holograms. It will be seen from the following sections that the efficiency of thick phase grating is even higher yet, which explains the very great interest in producing low noise phase holograms.

### 1.4.2 Volume Holograms

#### Coupled Wave Theory

The preceding theories for thin holograms cannot apply when the diffraction efficiency becomes high, because for high diffraction efficiencies the illuminating wave will be strongly depleted as it passes through the grating. In some way one must take into account the fact that at some point within the grating there will be two mutually coherent waves of comparable magnitude traveling together. Such an account is the basis of the coupled wave theory, so aptly applied to the problem of diffraction from thick holograms by *Kogelnik* [1.16]. This elegant analysis gives closed form results for the angular and wavelength sensitivities for all of the possible hologram types—transmission and reflection, amplitude or phase, with and without loss, and with slanted or unslanted fringe planes. The equations also give, of course, the maximum diffraction efficiency achievable when the grating is illuminated at the Bragg angle and the fringe planes are unslanted. The equations are the result of a theory that assumes that the gratings are relatively thick so that there are only two waves in the medium to be considered, that is, that the Bragg effects are rather strong. Nevertheless, the equations are surprisingly accurate over a very large range of $Q$-values, including values considerably less than 10.

For a complete treatment of the theory one should refer to [1.16]. Here we will only outline briefly the underlying ideas of the theory and give the principal results.

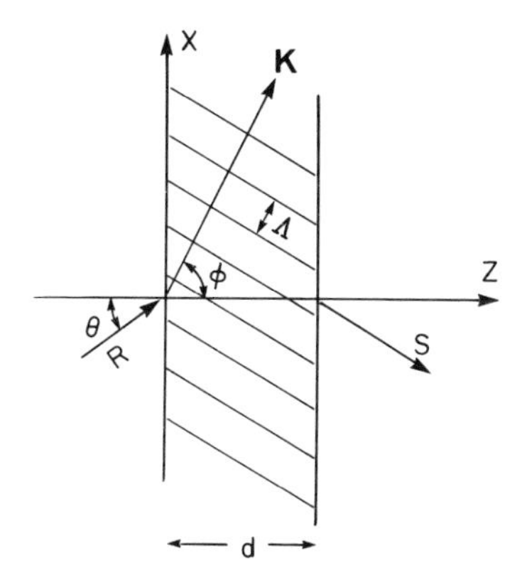


Fig. 1.1. Notation used to define a thick grating

The coupled wave theory assumes that there are only two waves present in the grating: the illuminating wave $C$ and the diffracted signal wave $S$. It is assumed that the Bragg condition is approximately satisfied by these two waves and that all other oders strongly violate the Bragg condition and hence are not present. The equations that are derived express the coherent interaction between the waves $C$ and $S$.

Figure 1.1 defines the grating assumed by *Kogelnik* for his analysis. The $z$-axis is perpendicular to the surfaces of the medium, the $x$-axis is in the plane of incidence and parallel to the medium boundaries, and the $y$-axis is perpendicular to the page. The fringe planes are oriented perpendicularly to the plane of incidence and slanted respect to the medium boundaries at an angle $\varphi$. The grating vector $\boldsymbol{K}$ is oriented perpendicularly to the fringe planes and is of length $|\boldsymbol{K}| = 2\pi/\Lambda$, where $\Lambda$ is the period of the grating. The angle of incidence for the illuminating wave $C$ in the medium is $\Theta$. The Bragg angle is $\Theta_0$. It is assumed that the fringes are sinusoidal variations of the index of refraction or of the absorption constant, or both for the case of mixed gratings. The assumed equations are therefore

$$n_0 = n + n_1 \cos(\boldsymbol{K} \cdot \boldsymbol{X}) \tag{1.37}$$

and

$$a_0 = a + a_1 \cos(\boldsymbol{K} \cdot \boldsymbol{X}), \tag{1.38}$$

where $\boldsymbol{K} \cdot \boldsymbol{X} = \omega x$ and $\omega$ is $2\pi$ divided by the fringe spacing in the $x$-direction. The quantities $n$ and $a$ are the average values of the index and absorption constant, respectively. The slant of the fringe planes is described by the constants $c_R$ and $c_S$, which are given by

$$c_R = \cos\Theta, \tag{1.39}$$

$$c_S = \cos\Theta - \frac{K}{\beta} \cos\Theta, \tag{1.40}$$

where $\beta = 2\pi n/\lambda$, and $\lambda$ is the free space wavelength.

When the Bragg condition is satisfied we have

$$\cos(\varphi-\Theta_0)=K/2\beta \tag{1.41}$$

and

$$c_R=\cos\Theta_0 \tag{1.42}$$

$$c_S=-\cos(2\varphi-\Theta_0)\,. \tag{1.43}$$

In the case where the Bragg condition is not satisfied, either by a deviation $\Delta\Theta$ from the Bragg angle $\Theta_0$ or by a deviation $\Delta\lambda$ from the correct wavelength $\lambda_0$, where both $\Delta\Theta$ and $\Delta\lambda$ are assumed to be small, *Kogelnik* introduces a dephasing measure $\mathcal{O}$. This parameter is a measure of the rate at which the illuminating wave and the diffracted wave get out of phase, resulting in a destructive interaction between them. The dephasing measure is given by

$$\mathcal{O}=\Delta\Theta\cdot K\cdot\sin(\varphi-\Theta_0)-\Delta\lambda K^2/4\pi n\,, \tag{1.44}$$

where

$$\Delta\Theta=\Theta-\Theta_0 \tag{1.45}$$

and

$$\Delta\lambda=\lambda-\lambda_0\,. \tag{1.46}$$

The coupled wave equations that result from this theory are

$$c_R R'+\alpha R=-\mathrm{i}kS \tag{1.47}$$

$$c_S S'+(a+\mathrm{i}\mathcal{O})S=-\mathrm{i}kR\,, \tag{1.48}$$

where $k$ is the constant defined by

$$k=\frac{\pi n_1}{\lambda}-\frac{\mathrm{i}a_1}{2}\,, \tag{1.49}$$

and the primes denote differentiation with respect to $z$, and $\mathrm{i}=\sqrt{-1}$. The physical interpretation of these equations as described by *Kogelnik* is as follows:

As the illuminating and diffracted waves travel through the grating in the $z$-direction their amplitudes are changing. These changes are caused by the absorption of the medium as described by the terms $a_R$ and $a_S$, or by the coupling of the waves through the terms $kR$ and $kS$. When the Bragg condition is not satisfied, the dephasing measure $\mathcal{O}$ is nonzero and the two waves become out of phase and interact destructively.

The solutions to these equations take on various forms depending on the type of hologram (grating) considered. When transmission holograms are considered, we say the fringe planes are unslanted when they are perpendicular

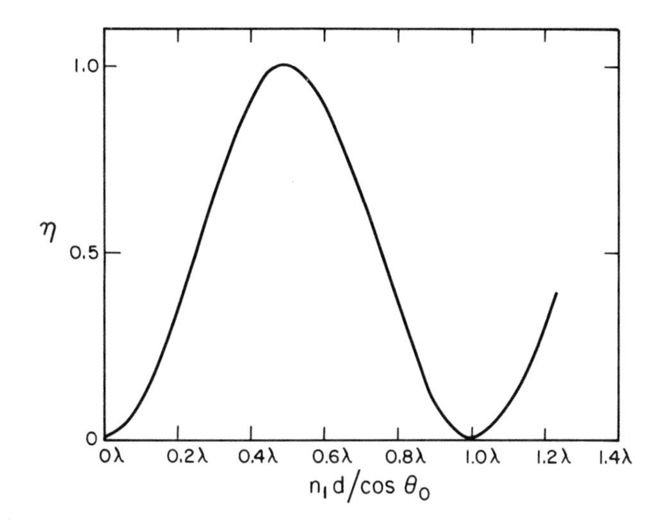


Fig. 1.2. Diffraction efficiency as a function of the optical path variation $n_1 d/\cos\theta_0$ in units of the wavelength for a thick phase hologram viewed in transmission

to the surface. This condition is described by $c_R = c_S = \cos\Theta$ since $\varphi = \pi/2$ in this case. If the Bragg condition is also satisfied, then of course $\Theta = \Theta_0$. For reflection holograms, on the other hand, no slant means that the fringe planes are parallel to the surface, so $\varphi = 0$. If the Bragg condition also holds, then $c_R = -c_S = \cos\Theta_0$.

## Transmission Holograms

*Pure Phase Holograms.* For a pure phase, transmission hologram the absorption constant $a_0 = 0$ and the solution of the coupled wave equations leads to a diffraction efficiency, for the general case of slanted fringes and Bragg condition not satisfied, given by

$$\begin{aligned} \eta &= \sin^2(\nu^2 + \zeta^2)^{1/2}/(1 + \zeta^2/\nu^2) \\ \nu &= \pi n_1 d/\lambda (c_R c_S)^{1/2} \\ \zeta &= \vartheta d/2c_S = \Delta\Theta K d \sin(\varphi - \Theta_0)/2c_S \\ &= -\Delta\lambda K^2 d/8\pi n c_S . \end{aligned} \tag{1.50}$$

In the case for which there is no slant and the Bragg condition is satisfied, the formula reduces to the well-known equation

$$\eta = \sin^2(\pi n_1 d/\lambda \cos\Theta_0) . \tag{1.51}$$

A graph of this equation as a function of the optical path in units of the free space wavelength $\lambda$ is shown in Fig 1.2. It is seen that diffraction efficiencies of 1.0 are possible with this type of hologram.

*Pure Phase Holograms with Loss.* For this case we consider the effect of some loss on a phase hologram. The loss takes the form of some residual absorption, such as might be caused by a chemical stain that has not been washed out or perhaps

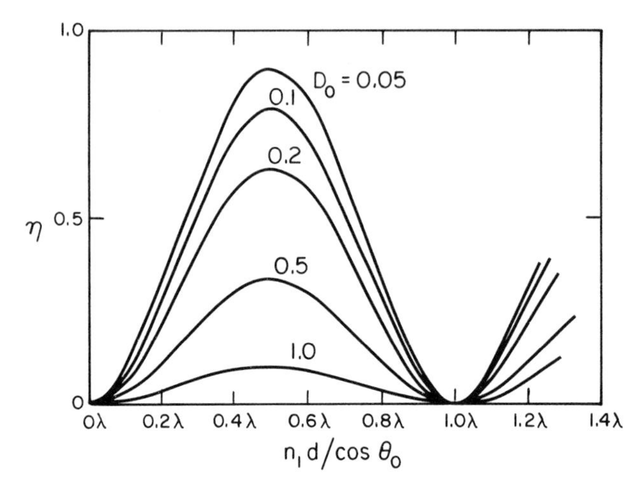


Fig. 1.3. Diffraction efficiency as a function of the optical path variation $n_1 d/\cos\theta_0$ in units of the wavelength for a thick phase hologram with loss, for several values of optical density $D_0$

some intrinsic absorption at the readout wavelength. For this sort of grating, $a_1$ is zero but $a$ is not. The equation for the diffraction efficiency takes the form

$$\begin{aligned} \eta &= \mathrm{e}^{-2ad/\cos\Theta}\sin^2(v^2+\zeta^2)^{1/2}/(1+\zeta^2/v^2)^{1/2} \\ v &= \pi n_1 d/\lambda\cos\Theta \\ \zeta &= \vartheta d/2\cos\Theta = \Delta\Theta\beta d\sin\Theta_0\,, \end{aligned} \tag{1.52}$$

where we assume that the grating is unslanted and that the Bragg condition is not satisfied. The absorption evidenced by the exponential term in front and leading to an overall decrease in the efficiency to Bragg mismatch is not much different from that of the lossless hologram. In the case where the Bragg condition is satisfied, the equation simplifies to

$$\eta = \mathrm{e}^{-2ad/\cos\Theta_0}\sin^2(\pi n_1 d/\lambda\cos\Theta_0)\,. \tag{1.53}$$

The effect of the absorption is simply to decrease the overall efficiency by the exponential factor in front. A curve of this function is shown in Fig. 1.3 for several values of the residual density $D_0 = 0.868\,ad/\cos\Theta_0$, which is just the optical density measured in the direction of the illuminating wave.

*Amplitude Holograms.* For a pure amplitude hologram there is no modulation of the refractive index, so $n_1 = 0$. The solution to the coupled wave equations in this case is

$$\begin{aligned} \eta &= \frac{c_R}{c_S}\exp\cdot\left[-ad\left(\frac{1}{c_R}+\frac{1}{c_S}\right)\right]\sinh^2(v^2+\zeta^2)^{1/2}/(1+\zeta^2/v^2) \\ v &= a_1 d/2(c_R c_S)^{1/2} \\ \zeta &= \frac{1}{2}ad\left(\frac{1}{c_R}-\frac{1}{c_S}\right) \end{aligned} \tag{1.54}$$

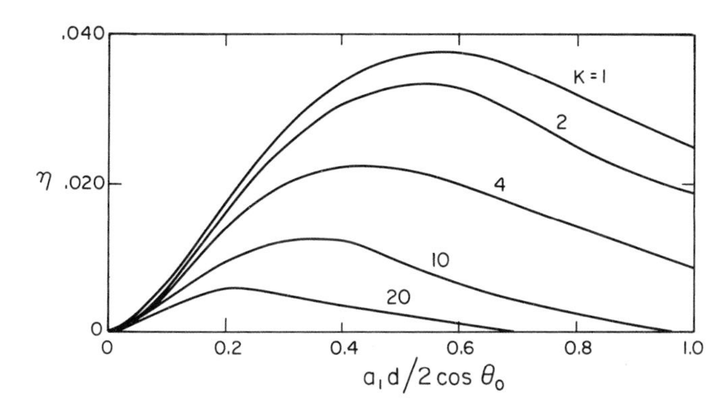


Fig. 1.4. Diffraction efficiency as a function of the parameter $a_1 d/2\cos\theta_0$ for thick amplitude holograms viewed in transmission. The curves are plotted for several values of the reference-to-object beam ratio $K$, which is related to the absorption constants $a$ and $a_1$ through $a/a_1 = 2\sqrt{K}/(1+K)$. The average specular transmittance of the hologram is $\exp(-2ad/\cos\theta_0)$

for the case where the Bragg condition is fulfilled but the fringe planes are slanted. For the case where the fringes are unslanted but the Bragg condition is not satisfied, the equation for the diffraction efficiency becomes

$$\begin{aligned} \eta &= e^{-2ad/c_R} \sinh^2(v^2-\zeta^2)/(1-\zeta^2/v^2) \\ v &= a_1 d/2\cos\Theta \\ \zeta &= \vartheta d/2\cos\Theta \cong \Delta\Theta\beta d \sin\Theta_0 \\ &= -\frac{1}{2}\frac{\Delta\lambda}{\lambda} K d \tan\Theta_0 \end{aligned} \tag{1.55}$$

and in case the Bragg condition is satisfied this simplifies to

$$\eta = e^{-2ad/\cos\Theta_0} \sinh^2(a_1 d/2\cos\Theta_0) \tag{1.56}$$

which is plotted in Fig. 1.4. The maximum diffraction efficiency is achieved when $a_1 = a$ and $ad/\cos\Theta_0 = \ln 3 = 1.1$, and is 0.037. This is considerably lower than the theoretical maximum for a thin amplitude hologram (0.0652). The maximum is reached when $ad/\cos\Theta_0 = 1.1$, which corresponds to an average density of 0.955 (measured in the direction $\Theta_0$) or an amplitude transmittance of 0.34.

*Mixed Holograms.* A mixed hologram is one for which there is a phase modulation in addition to the amplitude modulation. In this case both $a_1$ and $n_1$ are nonzero. This situation is of some practical importance because any real hologram will be, more or less, a mixed hologram. For example, for a grating produced on a photographic emulsion there is quite likely to be an exposure-dependent variation in the refractive index or a surface relief image in addition to the transmittance variation. This then would result in a mixed hologram. The relative contribution of the phase and amplitude portions can be predicted from coupled wave theory. Interestingly enough, it turns out that the contributions of the phase and amplitude parts are simply additive, at least for the case of an unslanted grating and Bragg incidence. The equation for the diffraction efficiency is

$$\eta = [\sin^2(\pi n_1 d/\lambda \cos\Theta_0) + \sinh^2(a_1/2\cos\Theta_0)]\mathrm{e}^{-2ad/\cos\Theta_0}, \tag{1.57}$$

where all of the symbols have been previously defined.

## Reflection Holograms

*Pure Phase Holograms.* The general equation for the diffraction efficiency of a pure phase reflection hologram with slanted fringe planes and non-Bragg illumination is

$$\begin{aligned}\eta &= [1 + (1 - \zeta^2/v^2)/\sinh^2(v^2 - \zeta^2)^{1/2}]^{-1} \\ v &= \mathrm{i}\pi n_1 d/\lambda(c_R c_S)^{1/2} \\ \zeta &= -\vartheta d/2c_S \\ &= \Delta\Theta K d \sin(\Theta_0 - \varphi)/2c_S \\ &= \Delta\lambda K^2 d/8\pi n c_S .\end{aligned} \tag{1.58}$$

Note that $v$ is real since $c_S$ is negative. For an unslanted grating and Bragg incidence (1.57) becomes

$$\eta = \tanh^2(\pi n_1 d/\lambda \cos\Theta_0). \tag{1.59}$$

Figure 1.5 shows a plot of this equation. It is seen that it is possible to have a diffraction efficiency of 1.0.

*Amplitude Holograms.* For a reflection hologram of the amplitude type, where $n_1 = 0$ and $a_1$ is nonzero, the diffraction efficiency for unslanted fringes ($\varphi = 0$) and Bragg incidence is given by

$$\begin{aligned}\eta &= \frac{c_R}{c_S}\left[\frac{\zeta}{v} + \left(\frac{\zeta^2}{v^2} - 1\right)^{1/2} \coth(\zeta^2 - v^2)^{1/2}\right]^{-2} \\ v &= \mathrm{i}a_1 d/2(c_R c_S)^{1/2} \\ \zeta &= D_0 - \mathrm{i}\zeta_0 \\ D_0 &= ad/\cos\Theta_0 \\ \zeta_0 &= \Delta\Theta\beta d \sin\Theta_0 = \frac{1}{2}\frac{\Delta\lambda}{\lambda} K d .\end{aligned} \tag{1.60}$$

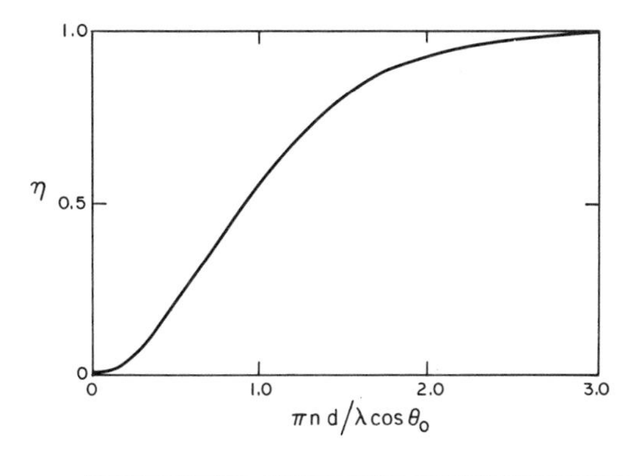


Fig. 1.5. Diffraction efficiency as a function of the parameter $\pi n_1 d/\lambda \cos\theta_0$ for a pure phase reflection hologram

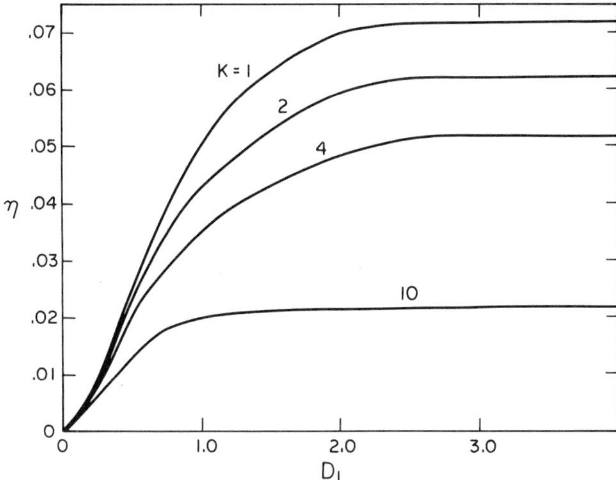


Fig. 1.6. Diffraction efficiency as a function of $D_1 = a_1 d/\cos\theta_0$ (the specular density) for several values of the beam ratio $K$, which is related to the densities $D_0$ and $D_1$ through $D_1/D_0 = 2\sqrt{K}/(1+K)$

If the Bragg condition is satisfied, then this equation simplifies to

$$\eta = D_1^2\{2[D_0^2 + (D_0^2 - D_{1/4}^2)^{1/2}\coth(D_0^2 - D_{1/4}^2)^{1/2}]\}^{-2}, \tag{1.61}$$

where $D_1 = a_1 d/\cos\Theta_0$. This equation is plotted in Fig. 1.6. The largest allowable modulation is for $D_1 = D_0$ or equivalently, $a_1 = a$. In this case the maximum diffraction efficiency is 0.072, which is achieved in the limiting case $D_1 = D_0 = \infty$.

## Summary

Figure 1.7 summarizes the principal results. The numbers shown are the maximum theoretical diffraction efficiency obtainable for each type of hologram. With the exception of the thick, phase, reflection hologram, each of the theoretical maxima has been very nearly achieved experimentally. For a reason that is a yet unexplained, no holographic material and process have yet come close to the theoretical 1.0 diffraction efficiency for this type of hologram.

THEORETICAL MAXIMUM DIFFRACTION EFFICIENCY

| Hologram Type: | Thin Transmission | | Thick Transmission | | Thick Reflection | |
|---|---|---|---|---|---|---|
| Modulation : | Amplitude | Phase | Amplitude | Phase | Amplitude | Phase |
| Efficiency : | 0.0625 | 0.339 | 0.037 | 1.000 | 0.072 | 1.000 |

Fig. 1.7. Table of maximum diffraction efficiency for various hologram types

Clearly the highest efficiencies are achieved for phase-type holograms, and these can be formed quite readily on many types of holographic recording materials. Also, for each of the hologram types considered, it was assumed that some sort of linear relationship existed between input and output. For the case of the thick holograms it was between the amplitude transmittance and the exposure. Unfortunately, it is impossible in any real system to maintain this linearity over the exposure range required to produce maximum efficiency. Thus in most real situations one has to balance an increase in nonlinearity noise against an increase in diffraction efficiency. According to *Kogelnik* [1.16], the efficiencies that are possible while maintaining the assumed linear response are smaller by about 50% than the corresponding maximum efficiencies, depending on how strictly linearity is defined.

## 1.5 Noise

The noise associated with a holographic image is any unwanted light that is scattered or diffracted into the direction of the desired image during the reconstruction process. This noise light can take the form of a veiling glare that reduces the contrast of the image, or it can be in the form of an unwanted, spurious image. All of the forms of noise present in holography are signal dependent in that they depend either on the amount or form of the object light; because of the coherent nature of the reconstruction process, the noise light adds coherently to the signal or image light.

The principal sources of noise light are as follows:

*1) Granularity.* When a photographic emulsion is used as the recording material, the information recorded on the hologram is built up from millions of tiny clumps of silver called grains. These photographic grains appear microscopically as fine black specks on a clear background. When light is transmitted through this pattern it is scattered over a wide range of directions, and this is the light that forms a veiling glare around the desired holographic image. There are some recording materials that are either grainless or essentially grainless, and hence do not scatter light. A thermoplastic material, for example, utilizes the surface relief to diffract the light and the material itself is grainless, and so one would expect no noise of this type. The photochromic materials have grains that are molecular in size. These are so small that the material is essentially grainless and scatters no light.

*2) Scattering from the Support and/or Binder Material.* Scattering can arise even in the grainless materials because of the optical imperfections of the host material. The photochromics are crystalline materials that are subject to imperfections that scatter light. In the case of the thermoplastics it would be the plastic itself plus the photoconductor layer plus the support material that would scatter the light. In the case of the photographic types of materials one often has to consider the scattering from the support material and from the gelatin binder itself.

*3) Phase Noise.* Phase noise is intrinsic to phase holograms, but occurs only in the case of diffuse objects. The self-interference of the object light is recorded as a low frequency pattern on the hologram. This low-frequency phase pattern diffracts unwanted light into and around the holographic image.

*4) Nonlinearity Noise.* Nonlinearity noise arises in cases where the final amplitude transmittance of the hologram is not strictly proportional to the recording exposure. The presence of recorded information on the hologram that is proportional to the second and higher orders of the exposure gives rise to unwanted light diffracted in the image direction. This type of noise occurs mainly when the exposure modulation is high so that the range of exposures exceeds the dynamic range of the recording material.

*5) Speckle Noise.* This is a type of noise associated with holographic images, but it does not strictly fit the definition given in the first paragraph above. In a strict sense speckle noise is not really noise because it arises precisely because of an accurate reconstruction of the object wave. However, it is noise in the sense that the image is definitely degraded by its presence. It arises whenever a diffuse wavefront of coherent light is limited in extent by the apertures of the system. When this happens the phase undulations of the diffuse wavefront become amplitude variations, which have a granular or salt-and-pepper appearance. However, since this noise is not a function of the recording material, it will be ignored in this volume.

A more detailed discussion of the various types of noise is presented with the discussions of the different recording materials in the following chapters and so we will not elaborate more here.

## References


1.1 D. Gabor: Nature **161**, 777 (1948)
1.2 E. N. Leith, J. Upatnieks: J. Opt. Soc. Am. **52**, 1123 (1962)
1.3 E. N. Leith, J. Upatnieks: J. Opt. Soc. Am. **54**, 1295 (1964)
1.4 M. J. Beurger: J. Appl. Phys. **21**, 909 (1950)
1.5 W. L. Bragg, G. L. Rogers: Nature **167**, 190 (1951)
1.6 H. M. A. El Sum: *Reconstructed Wavefront Microscopy*, Ph. D. Thesis (Stanford Univ. 1952)

1.7 M. E. Haine, J. Dyson: Nature **166**, 315 (1950)
1.8 M. E. Haine, T. Mulvey: J. Opt. Soc. Am. **42**, 763 (1952)
1.9 A. V. Baez: J. Opt. Soc. Am. **42**, 756 (1952)
1.10 G. L. Rogers: Nature **166**, 237 (1950)
1.11 P. Kirkpatrick, H. M. A. El Sum: J. Opt. Soc. Am. **46**, 825 (1956)
1.12 Y. N. Denisyuk: Soviet Phys.-Doklady **7**, 543 (1963)
1.13 P. J. Van Heerden: Appl. Opt. **7**, 393 (1963)
1.14 F. G. Kaspar: J. Opt. Soc. Am. **63**, 37 (1973)
1.15 H. M. Smith: J. Opt. Soc. Am. **62**, 802 (1973)
1.16 H. Kogelnik: Bell Syst. Tech. J. **48**, 2909 (1969)

# 2. Silver Halide Photographic Materials

K. Biedermann

With 25 Figures

For one hundred years, silver halide gelatin emulsion have been used for recording visible and invisible radiation in general photography and cinemato-graphy, astronomy and spectroscopy, x-ray photography and electron micro-scopy, and in many other fields where information is to be stored and retrieved in pictorial, graphical or digitally encoded form. This chapter deals with the rôle of photographic silver halide emulsions in holographic wavefront recording and reconstruction, a new technique that poses quite different requirements on the material and a challenge to meet them.

When the coherent light of the He–Ne laser became accessible for holography about fifteen years ago, the photographic emulsion was the only material readily available that provided the necessary high resolution and sufficient sensitivity to the red light of this laser. Since then, lasers generating radiation of shorter wavelengths have been developed, making other light-sensitive processes useful for hologram recording. A great number of nonsilver photographic materials have been improved or adapted for holography, as to be reported on in the following chapters of this book.

Silver halide materials have a long tradition, they are versatile, commercially available with a choice of data and sizes, can be handled and processed with a minimum of equipment, and are suitable for making amplitude and phase, color and Lippmann holograms. Hence, there exists more practical experience with their use and more investigations of their properties have been reported than for other materials. Their sensitivity is unequalled by other materials, achieved, however, by means of the time-consuming wet developing process, which makes it also unlikely that silver halide emulsions ever might provide the possibility of local erasure and rewriting, as desired for optical memories.

This chapter is intended to give a state-of-the-art review of silver halide photographic materials in holography with preference to facts and problems of interest to the user. The presentation includes assessment of materials and methods for evaluating their characteristic parameters, and discussion of effects that might lead to properties of the holograms or spatial filters not intended by their designer. Research into basic questions with remote connection to practical applications is given shorter treatment with commented references to the literature.

## 2.1 The Holographic Process

Both amplitude and phase of a wavefield can be stored in coded form only according to Gabor's interferometric two-step process. The requirements on the photographic layer in holography, hence, are as follows.

In the first step, the layer has to be sensitive to the wavelength from the laser used and has to be capable of recording the microscopic interference structures without appreciable deterioration. The carrier frequency $\nu$ around which the image information is encoded is given by

$$\nu = \frac{\sin\theta_o - \sin\theta_r}{\lambda} \ [\mathrm{c/mm}], \tag{2.1}$$

where $\lambda$ is the wavelength in air and $\theta_o$ and $\theta_r$ are the angles between the propagation directions of the object and the reference field and the recording plane normal. For a very simple and usual hologram setup with a He–Ne laser ($\lambda = 633$ nm), like for nondestructive testing, the angles may be $\theta_o = 0^0$, $\theta_r = 45^0$, which gives $\nu \approx 1120$ [(cyc./mm)]. For Lippmann holograms, the interference fringes are oriented in the depth of the emulsion, where the wavelength is shorter by the index of refraction, $n = 1.64$, and $\theta_o \approx 90^0$, $\theta_r \approx -90^0$; for a blue Ar-laser wavelength of $\lambda = 457.6$ nm, the recording of $\nu \approx 7170$ [cyc./mm] may be required. These two figures give a rough first idea of the resolution requirements. They may be compared with conventional photography or microfilm, which put similar demands on detail rendition at about 40 and 200 [(cyc./mm)], respectively.

By this exposure and subsequent processing, changes of the optical properties of the layer have to be introduced, which make the result, a hologram, capable of executing the second step: to spatially modulate an incident wavefront so as to reconstruct or generate another wavefront of certain amplitude and phase distribution. For this process, *Gabor* [2.1] derived a demand for linear relation between exposure and amplitude transmittance of the developed plate. Fortunately, practice should prove that fulfillment of the rigorous Gabor condition, $\gamma = -2$, can be replaced by approximately linear transfer even with usual negative reproduction.

It is seen that these requirements are quite different from those prevailing in ordinary photography. A large part of this chapter will deal with the behavior of the silver halide emulsion in, and its adaptation to, holography.

## 2.2 The Silver Halide Emulsion

A photographic emulsion [2.2–4] consists of gelatin as a protective colloid in which microscopic or submicroscopic crystals of silver halide, predominantly silver bromide, are precipitated.

High resolution or high values of the modulation transfer function (MTF) at high spatial frequencies, as required in holography, is a question not only of grain size but still more of light scattering in the emulsion. The index of refraction of gelatin is about 1.5, that of AgBr is 2.25. In conventional emulsions, the size of the embedded crystals is of the order of 1 μm, somewhat larger than the wavelength in the medium, hence Mie theory applies and scattering and light diffusion are strong. In order to decrease scattering, the particles must be made much smaller than the wavelength and Rayleigh theory starts to become applicable. Typical values of the grain size of emulsion for holography are 0.08–0.03 μm [2.5]. However, the sensitivity of fine grain emulsions is inherently very low. The probability of absorbing a certain number of photons is proportional to the projected area of a crystal and its depth, hence to the volume or the third power of its "mean diameter". Some margin is in the intrinsic sensitivity of the single grains.

The theories of latent image formation [2.2–4] may be summarized to a very coarse explanation: Absorption of a photon supplies energy to free an electron

$$\mathrm{Br}^- + h\nu \rightarrow \mathrm{Br} + e^-$$

which can move through the crystal lattice. At one of the crystal imperfections (which have to be suitably distributed through the crystal) it is trapped and attracts an interstitial silver ion

$$e^- + \mathrm{Ag}^+ \rightarrow \mathrm{Ag}\,.$$

This single silver atom will trap another liberated electron and the cycle repeats itself several times during the exposure. One single silver atom has a lifetime of about one second, a subspeck of two atoms is stable, a latent image speck of at least three or four silver atoms is necessary for catalyzing development [2.6].

This mechanism of latent image formation explains also that there must be an optimum rate of photon absorption per time interval. In holography with $Q$-switched pulse lasers, the exposure may last for a few nanoseconds only, and attention has to be paid to possible reciprocity failure [2.7].

The intrinsic sensitivity of silver halides is restricted to the blue and ultraviolet regions of the spectrum. Fortunately, certain dyes absorbing in regions of longer wavelengths, are, when adsorbed to silver halide crystals, capable of transferring the absorbed energy to the crystals for latent image formation. The silver halide emulsion is the only recording material, which, thanks to spectral sensitization, can be used for making holograms with the radiation from the ruby-laser ($\lambda = 694$ nm). Figure 2.1 shows examples of spectral sensitivity curves of some emulsions intended for recording holograms at different laser wavelengths.

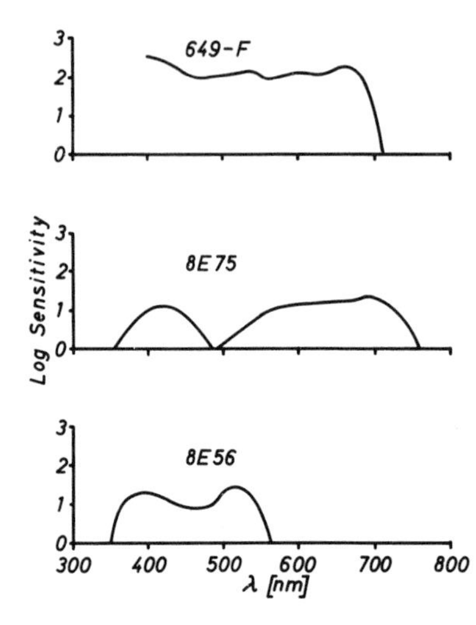


Fig. 2.1. Spectral sensitivity curves of three types of emulsions used in holography: Panchromatic spectroscopic plate Kodak 649-F; Agfa-Gevaert Scientia ("Holotest") 8E75, sensitized for the wavelengths 694 nm of the ruby laser and 633 nm of the He–Ne-laser; Scientia 8E56 sensitized for the wavelengths 454 to 514 nm of the Ar laser

The latent image is made visible by development. In the commonly used chemical process, the liquid developer soaks the gelatin and gets the agent in contact with the silver halide crystals. By catalytic action, those crystals containing latent image specks of at least critical size are completely reduced to metallic silver, while the others are not. They are dissolved and removed in the subsequent fixing and rinsing baths.

Development of silver halides constitutes an amplification process which is unparalleled by other light-sensitive systems. A silver bromide grain with a volume of $1\,\mu m^3$ contains $2\cdot 10^{10}$ silver ions which can be reduced after the absorption of a few photons. For this grain, the amplification is about $10^{10}$; for holographic emulsions it can hence be estimated to be of the order of $10^6$.

This great amplification has to be paid for by the inconvenience and delay of the wet development process and drying. However, there is an important advantage in the latent image, especially in storing many holograms by multiple exposure: The recording medium does not start to modulate incident fields as soon as it is being exposed.

Processing can also affect the gelatin which may become tanned. The resulting surface relief modulates the phase of an incident wavefield, an unintended effect which can be disturbing, especially in spatial filters.

The result of exposure and development is traditionally presented as the well known $D$-log$E$ curve, density ($=-\log$ transmittance) as a function of the logarithm of exposure. The form and especially the steepness of this function is the result of a number of statistical processes [2.8, 9]—grain size distribution, distribution of number of quanta required for developability, etc. Since, in holography, the $D$-log$E$ curve should be very steep in the toe (Sect. 2.4.2), emulsions had to be cultivated with as narrow distributions of these parameters as possible.

## 2.3 The Practical Procedure of Making Holograms and Spatial Filters

In this section, we will deal with some aspects of the practical procedure of making holograms with photographic emulsions. We do not treat general questions of holography, like stability of the setup, coherence requirements, etc.

### 2.3.1 Photographic Materials Available

Table 2.1 gives some idea of the choice of photographic silver halide emulsions designed or suitable for hologram recording as compiled at the date of writing. It may be interesting to compare the data for silver halide emulsions with those of the other hologram recording materials as presented in tabular form in [2.10]. Of the plates and films listed in Table 2.1, certain emulsions and sizes are available on special order only. On the other hand, experimental coatings have been made by both Eastman Kodak and Agfa-Gevaert to customer's request, e.g., with thicknesses exceeding 30 μm (see Sect. 2.8).

Table 2.1. Commercial silver halide emulsions for holography (sources: manufacturers' brochures, [2.5], and laboratory measurements)

| Type | Substrate | Emulsion thickness [μm] | Suitable λ [nm] | Exposure [μJ/cm²] | Suitable for spatial frequencies [cyc./mm] |
|---|---|---|---|---|---|
| **Agfa-Gevaert** | | | | | |
| Pan 300 | plate | 15 | panchrom. | 200–400 | ≫2000 |
| Scientia 8E75 | plate/film | 7/ 5 | 633, 694 | 10 | >2000 |
| Scientia 8E75B | plate | 15 | | | |
| Scientia 8E56 | plate/film | 7/ 5 | 400–550 | 30 | >2000 |
| Scientia 10E75 | plate/film | 7/ 5 | 633, 694 | 3 | <2000 |
| Scientia 10E56 | plate/film | 7/ 5 | 400–550 | 2 | <2000 |
| Copex Pan | film | | 514 | 0.07 | < 500 |
| **Eastman Kodak** | | | | | |
| Spectroscopic Plate/Film 649F | plate/film | 17/5 | panchrom. | 50 | >2000 |
| Holographic Plate 120–02 | plate | 6 | 633, 694 | 20 | >2000 |
| High-Speed Holographic Film SO-253 | film | 9 | 633 | 0.4 | <2000 |
| Microfile AHU | film | 5 | 514 | 0.05 | < 500 |

Comments: It will be clear from p. 36 that it is the MTF that characterizes the performance of an emulsion at a certain spatial frequency. Hence, here only a very coarse classification is indicated. Agfa-Gevaert materials are now being marketed under the name "Holotest". In this book we continue to use the names under which the materials are identified in the scientific literature.

### 2.3.2 Recording a Hologram

#### Aspects on Material, Size, and Geometry

The plate size most commonly in stock is 9 cm × 12 cm or 10 cm × 12.5 cm (4″ × 5″). However, the choice of the material depends very much on the purpose. When, as one example, holograms of three-dimensional scenes are made, viewing becomes a real delight first when the "window" is at least 24 cm × 30 cm wide. For this purpose, plates of the class of least grain size suggest themselves since they show high diffraction efficiency and low scattered flux and ensure brilliant, crisp images. Even some preconditioning (Sect. 2.4.5) may reward the effort with a slight further improvement of image quality.

#### A Deviation on Antireflection Treatment

For good visual image quality (and also for uniform diffraction efficiency, e.g., when holographic interferograms are to be evaluated by point scanning of the hologram), exposure variations from internal reflection within the emulsion support have to be avoided. The high coherence of the laser light causes light reflected at the rear support-air interface to interfere with the incoming light. The exposure variation may be about ±50% resulting in a disturbing density variation in the hologram, which then looks like a piece of grainy wood and shows a corresponding variation in diffraction efficiency. Most of the plates can be obtained with and without antihalation backing. However, not all of these antihalation backings appear to be sufficient for coherent systems (as well as the usual antireflection coatings on lenses and beam splitters), where reflections of even as small as 0.25% still create interference fringes of 10% modulation. Many suggestions for preventing reflections have been published. A good compromise between expense and efficiency seems to be the application of a black lacquer (even on top of the factory AH-coating), which can be stripped off after processing and rinsing. The graphics industry uses stripping lacquers of this type, but also many other black lacquers can easily be peeled off from glass after having gone through the processing cycle.

Good index-matching between glass and backing appears to be achieved by a coating made from 100 g polyvinylalcohol in 1 l of water (dissolves slowly, boil for about one hour). The antihalo technique had been developed for making diffraction gratings for spectroscopy in photoresist, where freedom from spurious fields and stray light is crucial. The absorbing dye in the coating must not scatter, hence dyes soluble both in water and polyvinylalcohol have to be added to the liquid, like metanil yellow for Ar-laser wavelengths or methylene blue for He–Ne laser light. The strong film is easily peeled off before processing [2.11].

#### More on Choosing Material, Size, and Geometry

As an example of another application of holography, hologram interferometry has quite varying demands. If vibrations of an object under investigation are to be mapped with high accuracy, many interferogram fringes must be identified in spite of their irradiance decreasing with the square of a zero-order Bessel function $J_0^2(\Omega)$. In this case, emulsions of very fine grain, like the ones recommended for the display application, are preferable for good signal-to-background ratio [2.12].

For routine measurements in holographic nondestructive testing (HNDT), on the other hand, where only the presence or basic structure of an interferogram is determined from one fixed observation direction, lower signal-to-background ratio is sufficient. In addition, the hologram area often does not need to be larger than the eye or (TV-) camera lens viewing the interferogram. Medium-speed, perforated, holographic 35-mm film will be an economical choice. A body of a miniature camera 24 mm × 36 mm or less will serve well for holding and transporting the film. It is appropriate to adjust the camera carefully in such a manner that the normal to the film plane bisects the angle between object and reference beam. Then, slight motion of the film out of its plane will be essentially along the orientation of the interference fringes in space and will not blur the hologram. (The adjustment is easily done by looking from the reference source via the reflection from a blank film in the camera to the object.)

If high dimensional fidelity is needed in the holographic image, distortion from slight changes in film buckling between hologram recording and reconstruction can be minimized by using a good quality camera lens that projects an image of the object near the hologram plane [2.13]. For reconstruction, a glassless film carrier from an enlarger can be used to hold the film flat. Film can be held more stably and reproducibly for recording and reconstruction, if, instead of using an existing camera body, a holder is made that bends the film around film guide rails that form a vertical cylinder of about 10 cm radius [2.13].

While a hologram 24 mm × 36 mm on holographic film is more than a factor of ten less expensive than a plate 9 cm × 12 cm, for routine recording, e.g., HNDT at a production line, another appreciable factor can be saved by using commercial microfilm. Since the MTF of these films decreases rapidly already at low spatial frequencies (see Fig. 2.10), the angle of view would be too limited. Instead, image plane holograms with minimum bandwidth requirements can be made in an arrangement where the reference source is located in the center of the exit pupil of the imaging system. Overlap of intermodulation and image terms can be prevented by means of a double-slit aperture. Reconstruction and projection of the holographic interferograms is possible with nonlaser light [2.14].

### Intensity Ratio *K* and Exposure

As shown in Chapter 1, exposure of a recording medium to an interference pattern of modulation close to one, especially with an extended object, results in strong stray light from intermodulation and nonlinearity terms. Hence, the ratio $K$ of the intensity in the reference beam to the intensity in the object beam at the recording plane, $K = I_r/I_o$, can be made greater than one. For values of $K$ up to about 5, the decrease in brightness of the holographic image, viz., diffraction efficiency, is less than reciprocal to $K$ [2.15]. At the same time, much more laser light becomes available in the recording plane and allows one to use shorter exposure times, since losses in the reference beam path are very small compared to those in the path from the beam splitter via the object to the recording material [2.16]. Hence, for recording pictorial holograms, beam intensity ratios of $K \approx 5$ to 15 can be regarded as a good compromise. For holograms to be bleached for phase holograms, $K$ should be chosen larger than 10 [2.17, 18].

It should be borne in mind that only the electric vector component of the object field parallel to that of the reference field contributes to hologram formation (the orthogonal one, however, creates intermodulation and should be eliminated where possible). The exact way is to determine $K$ with a polarizer in front of the detector and with normal incidence for each beam, while for determining exposure, irradiance in the recording plane has to be measured, and that without regard to polarization and origin.

The exposure should lead to a mean density of 0.7–0.8 for amplitude holograms, since at that density, diffraction efficiency is maximum [2.19], while for holograms to be bleached the density before bleaching should be greater than 1 or 1.5 [2.16, 19]. For amplitude holograms, exposure time should be adjusted carefully. A deviation from that exposure that leads to maximum diffraction efficiency, by 40% (photographers would say, half an $f$-stop), drops diffraction efficiency to less than one-half. It is possible to monitor mean density to its optimum value by adjusting development time. In this way, exposure latitude is increased to about 1:4 (Sect. 2.3.3).

### 2.3.3 Processing

#### Standard Procedures

For normal demands, processing is straightforward and uncritical. However, quite complicated procedures have been published starting with treatments of conditioning and gelatin hardening before exposure and aiming at phase holograms via a multistage conversion process. Ways of making phase holograms in photographic emulsions will be discussed in Section 2.7.

It is recommended to develop holograms immediately after exposure, since the extremely fine grain of these emulsions may be subject to latent image fading corresponding to a density difference of up to 0.2 density unit for the first hour of delay [2.20].

It seems that most of the strong metol (elon)-hydroquinone developers used for graphic arts films (and some formulas for paper) yield satisfying results. The important properties for a developer are that it gives very low fog and a steep $D$-$\log E$-curve which is sharply bent in the toe (cf. p. 35). Eastman Kodak

Table 2.2. Developers suitable for processing holograms

| | Agfa 80 | | Kodak D-19 |
|---|---|---|---|
| Water | | 750 ml | |
| Metol (Elon, 4-Methyl-aminophenolsulfate) | 2.5 g | | 2 g |
| Sodium sulfite (cryst.) | 100 g | | 90 g |
| Hydroquinone | 10 g | | 8 g |
| Potassium carbonate | 60 g | | |
| Sodium carbonate | | | 50 g |
| Potassium bromide | 4 g | | 5 g |
| Add water to make | | 1000 ml | |

recommends developer D-19, development time 5–8 min at 20 °C. The author's group prefers the formula Agfa 80, which has a slight advantage over the common other developers with respect to fog and steepness of the *D*-log *E*-curve. For convenience, the formulae for the two most recommended developers are reproduced in Table 2.2.

In work where high reproducibility of the sensitometric data is required, as for certain spatial filters or for measurement of MTF and other material data, all development parameters should be kept under control. Prepare developer from the formula, not from factory-packed mixed powders, use distilled water and degas it by boiling, avoid bubbles when stirring, agitate during development, and discard developer after use. In Section 2.5.2 we shall report on adjacency effects in hologram development; we experienced occurrence of appreciable and quite varying amounts of adjacency effects if any one of the recommendations just given was not observed.

For general holography, the exact duration of development and the temperature are not critical, but they should be kept within some 10 s and 1/2 °C once optimum exposure has been established. Prolonged development and increased temperature compensate for underexposure (as mentioned in the preceding subsection); they increase diffraction efficiency and granularity and hence scattered flux (see Sect. 2.4). An example for $T_a - \log E$ and $\eta$-$\log E$ curves with development time as a parameter is given in Fig. 2.2.

When density is monitored for optimization by adjusting development time, the emulsion has to be transilluminated in some way. The wide choice of emulsions available today, most of them more or less panchromatic, makes the recommendation of a safelight filter for visual inspection impossible. A dark blue-green safelight with maximum emission around 507 mm, the sensitivity peak of the dark-adapted eye, should be acceptable for universal use. A CdS exposure meter, like a Gossen Luna-Pro, is well suited to measure density in safelight illumination, when the distant bulb is imaged onto the photoresistor by means of a condenser lens close to the hologram [2.19]. A special densitometer would use infrared radiation from a photodiode and a phototransistor as a detector.

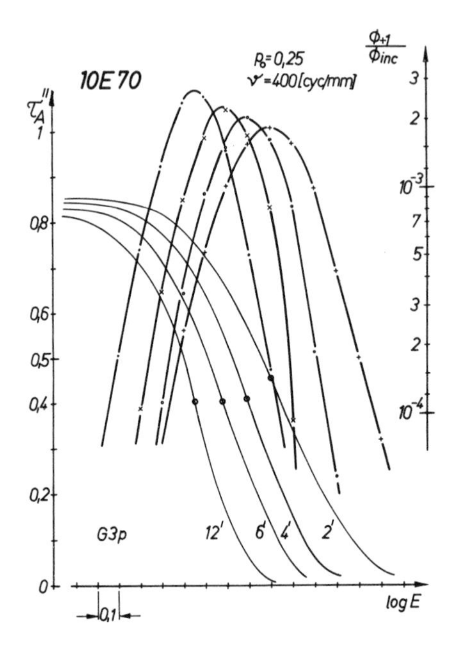


Fig. 2.2. $T_a - E$ and diffraction efficiency curves for Scientia 10E70 plates developed in Agfa-Gevaert G3p for 2, 4, 6, and 12 min [2.19]

After development, an acid stop-bath (20 ml acetic acid in 1 l of water) should be applied for about 30 s, followed by a short rinse in water. For fixing, the suppliers recommend Kodak Rapid Fixer for 5 min, or Agfa-Gevaert G 334c for 4 min, respectively. Other film fix formulae can be used as well. The fixing times recommended seem to have quite large safety margin. Fixing for a longer time is useless, but etches the tiny grains of these emulsions and lowers density. The final washing time is 15–25 min; it can be shortened to 5 min after use of an alkaline hypo clearing bath.

Residual sensitizing dye in the emulsion can be removed by bathing the plate in alcohol for some minutes. An optimum has been found to be a mixture of 3 parts of ethyl alcohol with 1 part of water [2.21].

When good visual image quality is desired, care should be spent on drying. Tap water sediments or dust dried into the emulsion scatter light and can hardly be removed by repeated washing. Single drops left on the layer or rolling across it give rise to dark spots looking like miniature footprints of small gnawers. These marks are due to a distortion of the gelatin; they light up at a different reference beam or observation angle. Inattentiveness can, in this way, spoil a beautiful large display hologram, as there is no way to iron out these marks again. For that reason, it was found worth the effort to give valuable holograms a final bath in distilled water, spray them off with distilled water from a wash bottle, remove all drops at edges, corners, and clamps, and dry the plates in a cabinet not much above room temperature. If a wetting agent is used with the distilled water, its concentration should be one third of that recommended for film drying (e.g., 1 ml Agfa-Gevaert Agepon/1 l $H_2O$ dest.) in order to avoid scattering from a residual film.

### Special Developers

Types of developers other than the rapid developers have not proven useful. For spatial filters and other purposes, extended exposure latitude or dynamic range is desirable. Use of low-$\gamma$ developers, however, results in loss of sensitivity, extreme decrease of diffraction efficiency, and adjacency effects (see Sect. 2.5.2).

Monobath developers, i.e., processing liquids that combine development and fixation [2.22], to our experience, can be formulated for a certain emulsion (or perhaps a batch of that emulsion) and rigorously controlled conditions. The slightest deviation, however, is liable to result in failure due among many other reasons to dichroic fog.

The very strong nonlinear behavior of such a monobath developer has been used to increase the dynamic range of a complex spatial filter for image deblurring [2.23].

Similarly inconsistent results have been allotted to studies for reversal development procedures. Even if a fair macroscopic $D$-log $E$ curve had been achieved, holograms could fail to show the diffraction efficiency expected from that curve (cf. Sect. 2.4.1). It seems that the very narrow size distribution of the extremely small grains in the hologram emulsions is very responsive to the silver halide solvents present in monobath and reversal developers. Also use of reducer (e.g., Farmers reducer) on overexposed holograms or in attempts to change the $D$-log $E$ curve did not give improvements in diffraction efficiency. Improvement of diffraction efficiency from addition of silver halide solvents to normal chemical developers has been reported by *Smith* et al. [2.24].

## 2.3.4 Reconstruction

There are two points deserving a short discussion under this heading: avoiding effects from gelatin surface relief and choosing material for photographing holographic images.

Amplitude holograms in photographic emulsions do in reality show a complex transmittance distribution due to a phase modulation at index of refraction variations within the gelatin layer and at a surface relief (cf. Sect. 2.5.3). In certain applications, holograms are designed to generate well-defined wavefronts, e.g., matched spatial filters or computer-generated holograms for spatial filtering or testing aspherical optical elements. Little can be done about the internal phase image, which, fortunately, has much less effect than the gelatin surface relief. The phase changes from the surface relief and also from the thickness variations of the glass plate or film substrate are eliminated by means of a liquid gate. The outer surfaces of the windows are optically flat and parallel to one another, and as immersion liquids matching the index of refraction of the developed layer ($n \approx 1.52$), among others, ethyl benzoil acetate ($n = 1.523$) [2.18] has been used. In [2.25] a number of immersion liquids, for both dry and wet film, have been investigated. It has also been recommended that intermodulation noise can be decreased by coating the hologram with a lacquer primarily intended as a UV-absorbing layer for color prints [2.26] (cf. Sect. 2.5.3).

Photographing images from wavefront reconstructions differs from ordinary photography in two aspects (aside from the fact that the camera can be used without a lens when a real image is photographed): the monochromatic light and laser speckles have to be taken into account when choosing the film.

Film speed indication in ASA or DIN is related to exposure with a standardized continuous spectral distribution ("mean daylight"). Very few conventional panchromatic films show appreciable sensitivity at the krypton line 647 nm. The neon line 633 nm is at about the edge of usual sensitization curves and the speed of films with equal ASA number may come out quite different. It may also happen that the dip between the intrinsic sensitivity of the silver halide and the green sensitizer may fall somewhere close to the 514 nm argon line. Exposure meters work well, once they have been calibrated to a certain film at a certain wavelength.

In monochromatic light, images of diffusely reflecting objects are inevitably disturbed by speckles [2.27]. When viewing a hologram, stray light from the plate will be one more source of speckles of high modulation (cf. Sect. 2.4.3). The size of the speckles is of the order of *f*-stop times wavelength. Hence there is a trade-off: no increase in depth of field without increased "graininess" in the picture. When recording holographic interferograms with a camera $24 \times 36\,mm^2$, f:5.6 gives just perceptible speckling at eight times enlargement. This means also that there is no point in using slow fine grain film; films of 100 to 400 ASA, like Agfapan 100 and 400 and Kodak Tri-X, proved both adequate and convenient for $\lambda = 633$ nm. In about the same sensitivity class is Kodak Photomicrography Monochrome Film SO-410, which, due to increased $\gamma$, is particularly suitable for enhancing fringe contrast when photographing holographic interferograms of objects vibrating at high mechanical amplitudes.

Photographing holographic images with color reversal film makes it quite apparent that holography is able to reproduce much higher object luminance ratios than common photography. In order to enhance color saturation, color reversal films have increased $\gamma$-values (up to $-1.6$). This type of tone reproduction does improve fringe visibility in (correctly exposed) holographic interferograms, but spoils highlights and shadows in the color slides of holographically presented technical subjects.

### 2.3.5 Real Time Interferometer with in Situ Development

Users of hologram interferometry would like to do their experiments "in real time", i.e., to observe the interferograms of an object under deformation or vibration at the same time as these changes are executed. Photographic

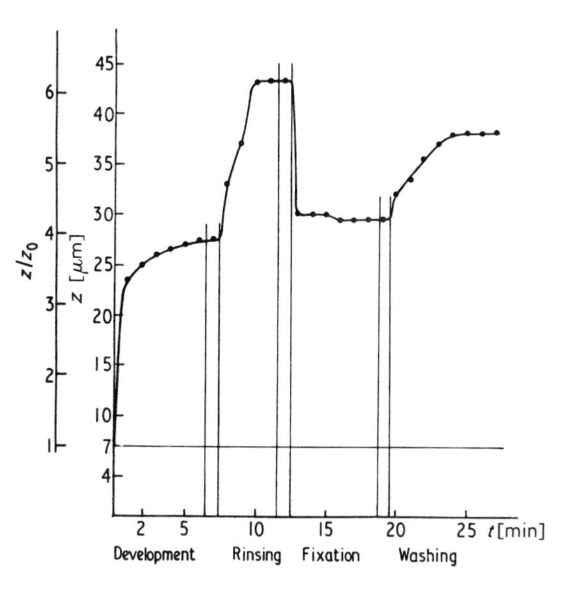


Fig. 2.3. The thickness $Z$ of a Scientia 8E70 layer during processing as a function of time and solutions. Double vertical lines indicate change of solution. The left ordinate relates actual thickness $Z$ to thickness $Z_0$ of the dry emulsion layer [2.28]

emulsions have the disadvantage of a delay of at least 20 min after exposure for processing, drying and relocating the plate in the hologram setup so as to bring the reconstructed and the actual object fields into register. This time gap can be shortened, when the plate is developed *in situ*, eliminating relocation, and still more when drying is disposed of by having the plate in a swollen condition in a liquid gate during both exposure and reconstruction.

Several liquid gate plate holders have appeared on the market. The process, however, is not a trivial one and considerable gain in time and diffraction efficiency can be made by using an optimized procedure [2.28].

The procedure consists of filling the liquid gate with developer, which swells the emulsion to four to seven times its initial thickness, exposure, rinsing, filling the gate with fixer, and reconstruction with the plate in the fixer.

For optimum results, developer and fixer have to be matched in two respects. For fulfillment of the Bragg condition, the gelatin layer has to have the same thickness during exposure in the developer and during reconstruction in the fixer. (Figure 2.3 gives an idea of the thickness variations in an ordinary processing cycle.) For equal optical path length through the cell at both occasions, the index of refraction of both solutions should be equal. Reference [2.28] gives formulae for a rapid developer and a matched fixer, and describes in detail the measures for achieving fulfillment of the Bragg condition with emulsions of different swelling properties. Apart from this inconvenience, these swollen holograms have advantages, e.g., suppression of the intermodulation term. With the formulae elaborated in [2.28], a gain in speed by a factor of 2–3 due to hypersensitization and a time interval of 100–180 s between exposure and start of interferometric evaluation have been achieved.

## 2.4 Characteristic Parameters of Thin Amplitude Holograms and Their Evaluation

The mathematical formulation of the principle of holography (Chap. 1) shows in a clear and convincing way that, upon reconstruction, one of the fields emerging from the (fictitious) hologram is proportional to the original object field. In this section, we shall go just one step from the abstract principle to practical realization and analyze what determines the proportionality factor in the case of a thin, pure amplitude hologram of low modulation in a photographic layer.

### 2.4.1 The Two-Step Transfer Process

Figure 2.4 shows the transfer of a signal from exposure to density of transmittance for a negative material according to Frieser's two-step model [2.9]. When the emulsion is exposed to the interference pattern containing the holographically encoded information, light spread within the emulsion lowers the modulation of the signal. As in conventional photography, this change in modulation is suitably described as a function of spatial frequency $\nu$ [cyc./mm] by means of the modulation transfer function (MTF), $M(\nu)$. In the case of contact copying holograms with incoherent light or demagnifying computer-generated holograms to final size, the MTF of this process (e.g., also the MTF of the reduction lens) has already affected the incident exposure; the cascaded MTF's are multiplied. This first step is a purely optical process linear in intensity.

In the second step, effective exposure is transferred to density via the $D$-$\log E$ curve (a relation that holds strictly only for large uniformly exposed areas, cf. Sect. 2.5.2). Since, in amplitude holograms, the amplitude of the incident reconstructing field is modulated, instead of density $D$ amplitude transmittance $T_a$ is appropriate

$$|T_a| = \sqrt{T} = 10^{-D/2}.$$

The diffraction efficiency of a thin amplitude hologram is given by (1.21)

$$\eta = [\tfrac{1}{2} \cdot 0.434 \cdot \alpha(\bar{E}) \cdot m(\nu) \cdot M(\nu)]^2. \quad (2.2)$$

Diffraction efficiency of thin amplitude holograms is proportional to the squares of the macroscopic characteristic $\alpha(\bar{E})$, the input modulation, and the MTF. These functions will be discussed in a little more detail in the following sections.

### 2.4.2 The Macroscopic Characteristic

In the beginning of holography, the demand for linear transfer from exposure to amplitude transmittance [2.1] suggested the use of the plot $T_a$ vs. $E$ for characterizing the macroscopic transfer by photographic emulsions. However, it

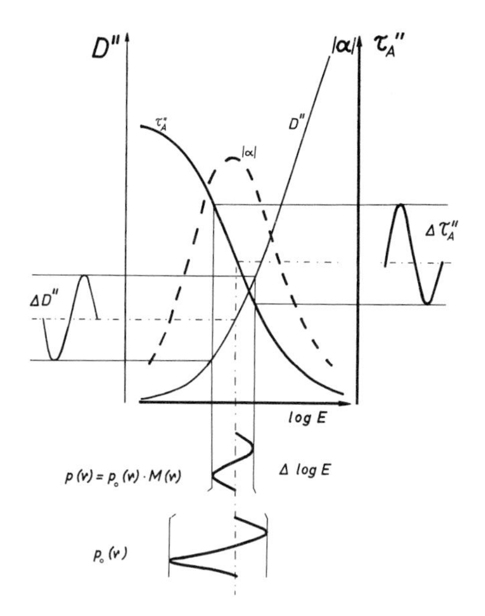


Fig. 2.4. The relation between exposure $E$ and density $D$ or amplitude transmittance $T_a$ in the developed hologram according to the two-step model. $|\alpha|$ is the derivative of the $T_a$-log $E$ function [2.29]

has been shown [2.15, 29] that, in fact, the plot $T_a$ vs. $\log E$ is more appropriate with respect to diffraction efficiency. This presentation, fortunately, also avoids some problems of the $T_a$-$E$ characteristic. In the $T_a$-$\log E$ plot, emulsions of different sensitivity or different developers can easily be compared in the same diagram; the signal amplitude does not change in proportion to mean exposure, rather modulation or $\Delta \log E$ can move along the $\log E$ abscissa to explore diffraction efficiency as a function of mean exposure.

The important characteristic for diffraction efficiency is the function $\alpha(E)$, the derivative of $T_a$ with respect to $\log E$ [2.15, 29], since $\eta \sim \alpha^2(E)$

$$\alpha(E) = \left(\frac{dT_a}{d\log E}\right)_{\bar{E}} = -\frac{\ln 10}{2} \cdot T_a(\bar{E}) \cdot \gamma(\bar{E}). \tag{2.3}$$

The second relation indicates the interpretation that diffracted flux is proportional to the square of the local gradient $\gamma(E)$ of the $D$-log $E$-curve at any working point attenuated by the mean intensity transmittance of the hologram. The dependence on these two variables that move into opposite directions with increasing exposure makes clear that the maximum of diffraction efficiency will be obtained in the toe of the conventional $D$-log $E$-curve, that maximum $\gamma$ in the straight-line portion of the $D$-log $E$-curve is of less interest than a steep bend in the toe, and that fog (base density) should be as low as possible.

Figure 2.5 shows especially for this chapter measured $T_a - \log E$ and $\alpha^2 - \log E$ curves for a number of materials available at the time of writing. It can be seen that the maximum of $\alpha^2$ of the different materials is roughly between 1 and 3 with a corresponding effect on diffraction efficiency. Interestingly, the transmittance value at which the maximum occurs is about 0.4 with very little variation between different emulsions and also development procedures, as

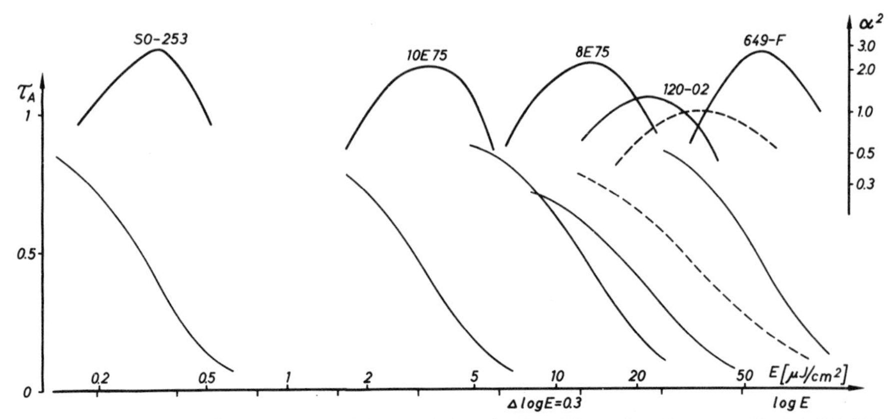


Fig. 2.5. $T_a$-log $E$ and $\alpha^2$-log $E$ curves of five emulsions for $\lambda = 633$ nm. Development: Kodak SO-253 and Agfa-Gevaert Scientia 8E75 in Kodak D-19 for 5 min; Agfa-Gevaert Scientia 10E75 and Kodak 649-F in Agfa 80 for 5 min; Kodak 120-02 in D-19 for 6 min (dashed curve) and 8 min (full curve). Temperature 20° C

confirmed, with little larger variations, also by other authors [2.30]. Hence it is recommended in Section 2.3.2 to expose for density 0.7–0.8, and in Section 2.3.3 to compensate for over- and underexposure by adjusting developing time, a procedure which has been checked over the range of 2–12 min (Fig. 2.2).

It is suitable to choose the exposure at which $\alpha^2$ is maximum as a criterion for sensitivity of photographic emulsions to be used for making amplitude holograms. If we derive a numerical value for sensitivity from this criterion, we may call it sensitometric sensitivity or sensitometric speed, since we will see in Section 2.4.5 that the total laser output required depends also on $\alpha^2_{max}$ and the MTF of the emulsion. With this $\alpha^2_{max}$-criterion, the exposure requirements of the emulsions represented in Figure 2.5 span such a wide range as 0.3–60 $\mu J/cm^2$ of He–Ne laser radiation ($\lambda = 633$ nm).

### 2.4.3 The Microscopic Characteristic

#### The Modulation Transfer Function (MTF)

As seen in Section 2.1, the carrier frequency of holograms is usually around one thousand cyc./mm or even much more. Equation (1.21) indicates that the MTF of the material in the spatial frequency range used lowers diffraction efficiency by its square. This means that, ideally, the MTF should be as close to one as possible.

The MTF describes the consequence of light scattering in the emulsion during exposure. Emulsions for conventional photography show an approximately exponential line spread function decreasing to 1/10 at a width $k$ of about 20 μm [2.9]. It is unrealistic to think of decreasing the mean diffusion length to less than one micron in order to achieve appreciable MTF at 1000 cyc./mm (the

MTF is the Fourier transform of the line-spread-function). Rather very transparent emulsions of extremely fine grain are made (mean grain diameter, e.g., about 80 nm for the 10 *E*-materials to 30 nm for the Pan 300 emulsion of Agfa-Gevaert [2.5]). The mean diffusion length of scattered light is increased by an order of magnitude [2.5, 31], which means that most of the photons scattered have a greater probability to leave the emulsion than to become absorbed. The part of the MTF due to light scattering decreases rapidly within some ten cyc./mm. What constitutes the effective exposure are, in fact, the photons absorbed at their first interaction with a silver halide grain. Consequently, their part in the spread function is represented by a $\delta$-function, whose Fourier transform is a constant, extending the MTF to very high spatial frequencies as a plateau. This two-term version for an MTF suggested by *Frieser* [2.9] is

$$M(v)=\varrho+\frac{1-\varrho}{1+(\pi k v/2.3)^2}, \tag{2.4}$$

where $\varrho$ is the fraction of the exposing photons absorbed at their first interaction, the fraction $(1-\varrho)$ is absorbed after scattering. The parameter for the exponential line-spread-function is $k$, as mentioned, the width at 1/10 of the maximum, or $k/2.3$ at $1/e$. Of the light incident, only one third or less is absorbed by these fine grain emulsions [2.31]. The plateau height is approximately constant. Starting at some thousand cyc./mm, the influence of the finite grain diameter makes a decrease of the MTF noticeable [2.5].

### The MTF in Coherent Light

The height $\varrho$ of the MTF plateau is important not merely for diffraction efficiency. Since $\varrho$ is a function of the ratio of absorption cross section to scattering cross section of the emulsion [2.32], it depends heavily on the wavelength of the exposing light, because spectral sensitivity cannot be of uniformly high efficiency over the entire spectrum. The dependence of $\varrho$ on wavelength is demonstrated in [2.5, 33] (the absolute values of $\varrho$ of these measurements came out somewhat low, cf. p. 44).

One might use an emulsion at a less suitable laser wavelength and put up with the loss in diffraction efficiency. However, though the fraction $(1-\varrho)$ is certainly lost for the formation of the hologram, it has not lost its ability to interfere. Hence, these fields scattered within the emulsion interfere with the reference field and form a transmittance structure ("diffusion mottle" [2.31]), from which they are reconstructed as a kind of scattered flux (Sect. 2.4.4). Seen in the realm of transfer theory, the process of exposure of an emulsion with light scattering is linear in intensity in incoherent light, but linear in complex amplitude in coherent light. Due to the statistical variations in relative phase between the scattered light and the other incident light, the spread function is not unique but shows statistical fluctuations around the mean distribution determined in incoherent light. When photographic emulsions are exposed to coherent light, spread function and MTF are used as a statistical description.

### Methods for Measuring MTF at High Spatial Frequencies

For assessing the performance of photographic emulsions, ideally, one should be able to compare just the effective exposure distribution to the incident exposure distribution. However, so far, no other photometer has become known for measuring exposure distributions in a photographic emulsion than the emulsion itself. This means that the characteristics of the "instrument" have to be determined accurately for eliminating them: processing, the second step in the two-step model of photographic image formation, is described by the $D$-log $E$ or $\alpha^2(E)$-curve, but it introduces also less evident changes like adjacency effect (Sect. 2.5.2) and surface relief (Sect. 2.5.3).

The principle of MTF-measurements at high spatial frequencies is to expose sinusoidal irradiance distributions of known modulation $m_0$ generated preferably by two-beam interference, at several spatial frequencies $\nu$. The modulation $m(\nu)$ of the effective exposure has to be determined in some way. In all methods but one, first the layer has to be developed.

It has been suggested to circumvent the problems arising from processing by measuring diffraction efficiency of the photolytically formed latent image [2.34]. However, excessive exposure is necessary before a very faint reconstructed field is detected submerged in stray light scattered from the silver halide emulsion.

Hence, the problem of measuring the MTF is a two-field one, in accordance with the two-step model of photographic transfer: to measure the transmittance distribution of the sample, and to determine the modulation of effective exposure from it.

We shall discuss three methods: the classical microdensitometer, the diffractometer, and the multiple-sine-slit microdensitometer (Fig. 2.6).

### The Microdensitometer

The microdensitometer is the most common instrument for converting graphical information from photographic records to electrical signals [2.9]; its principle is outlined in Figure 2.6a [2.35]. In the case of one-dimensional test patterns, a long narrow slit is the usual scanning aperture. There are a number of sources for spread in the photometry of fine structures, such as: size of the scanning area, diffraction and aberration limitations of the microscope objective used, inaccuracy of focus and sample flatness, and misalignment of the slit to the pattern. These effects can, to some extent, be taken into account by an MTF of the scanning procedure. For a slit width of 1–2 μm and a sufficient degree of incoherence, a typical value of the MTF of a microdensitometer is about 0.6 at 200 cyc./mm. Already this figure shows that the spatial frequencies typical for holography are beyond the capabilities of the classical microdensitometer. To decrease slit width gives little effect in comparison to the limitations set by the lens optics; on the contrary, inaccuracy from granularity will increase, since the standard deviation of granularity is proportional to the square root of the

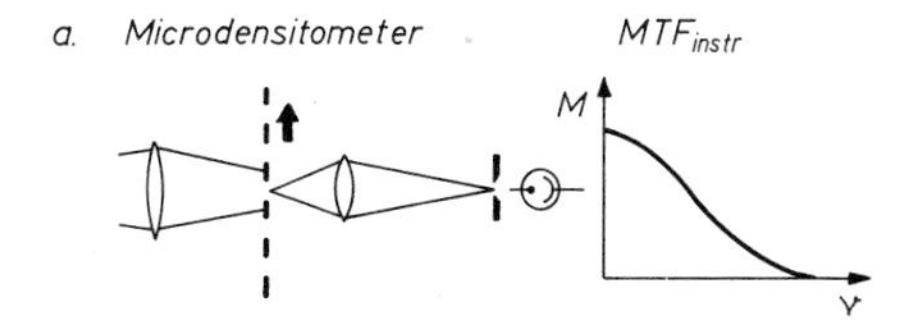


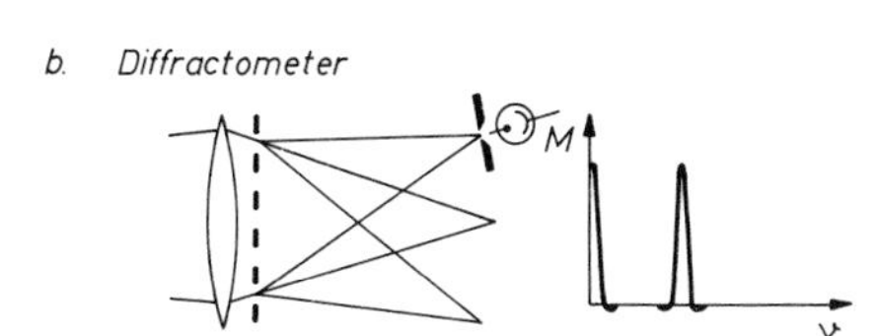


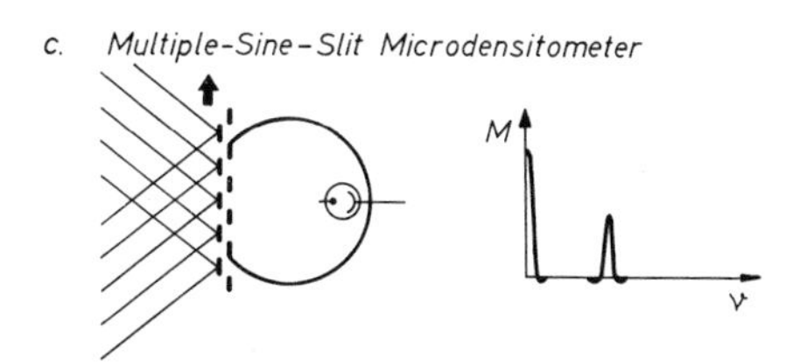


Fig. 2.6. The principles of three methods for measuring transmittance modulation. The instrument MTF's to the right indicate the transfer of the spatial frequencies of the sample [2.35]

scanning area. The density distribution plotted by the microdensitometer is then traced back over the $D$-log$E$ curve, determined from the same sample, and coverted into exposure, in principle, according to Figure 2.4. However, in the case of very high spatial frequencies, the MTF of the scanning procedure will be quite low and, hence, the density distribution plotted by the microdensitometer will differ from the distribution in the sample by strong spatial filtering. Compensation for the losses can only approximately restore the original modulation of the density distribution [2.36]. Hence, we may conclude, that a classical microdensitometer is not a sufficient tool for satisfactory MTF determination in the spatial frequency range of interest for holography.

### The Diffraction Method

Of all characteristics of materials for holography, the MTF is of extraordinary interest. This is due to the fact that the square of the MTF enters into diffraction efficiency (1.21). Hence, it is possible just to measure diffraction efficiency as a function of spatial frequency [or angle between two coherent sources as seen from the hologram according to (2.1)]. This was done by a number of early workers in holography. Also, before lasers were available, two-beam interference was performed as a test exposure by means of a mercury arc lamp and a Lloyd mirror. MTF also was determined from the flux into higher orders in Fraunhofer diffraction at a contact copy of a low-frequency Ronchi ruling (e.g., [2.9], pp. 240 and 254). In the beginning of holography it was suggested that MTF's be derived from a series of diffraction efficiency measurements, where the relative function obtained was normalized to 1 at low spatial frequencies [2.37,

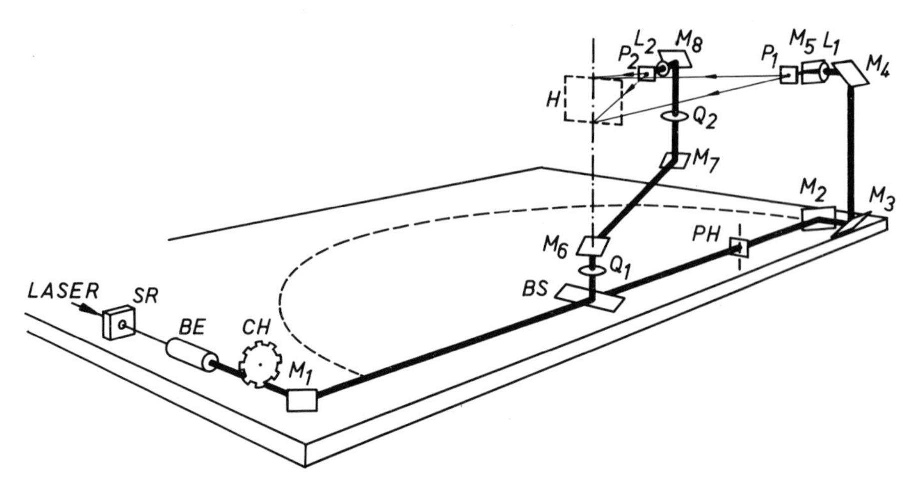


Fig. 2.7. Arrangement for the generation of interference patterns with continuously adjustable spatial frequency [2.39]

38]. This method, however, has to presume that the influence from all the other variables on diffraction efficiency could be kept sufficiently constant throughout the spatial frequency series.

The subsequent application of Frieser's two-step model to holography showed via (1.21) that a carefully measured value for the macroscopic characteristic $\alpha(\bar{E})$ at the working point (mean exposure) $\bar{E}$ is needed in order to evaluate $M(v)$ from the diffraction efficiency measured. In practice, not just one exposure per spatial frequency will be made but rather a series of equally increasing exposure steps with an increase of, suitably, 0.1 log unit. In order to justify the assumption of linear transfer, low modulation has to be chosen. A beam intensity ratio $K=65$, e.g., modulation $m_0=0.25$, proved very convenient [2.16, 19, 29]. In order to avoid volume effects from possible thickness changes of the layer during processing, the sample plane should be kept normal to the bisector of the angle between the directions of the two interfering fields. Figure 2.7 shows the principle of a system for exposing recording materials to two-beam interference patterns of spatial frequencies that are continuously adjustable from a few cyc./mm to $v=2/\lambda$ [2.39].

The evaluation is schematically shown in Fig. 2.6b. The technical solution in connection with the apparatus of Fig. 2.7 is outlined in Fig. 2.8: The measurement of diffraction efficiency in each exposure series yields a series of values $\eta(E)$, of which an example was given in Fig. 2.2. Such a plot of a series of values $\log \eta$ vs. $\log E$ also provides redundancy and some check of possible stray shots caused by environmental disturbances during exposure.

Under ideal and very careful working conditions, the shape of the $\log \eta - \log E$ curve is identical to that of the $\log \alpha^2 - \log E$ curve taken from the same sample. The two curves are brought to overlap; the shift along the ordinate is read off; and the MTF value is easily calculated from (1.21). Since diffraction

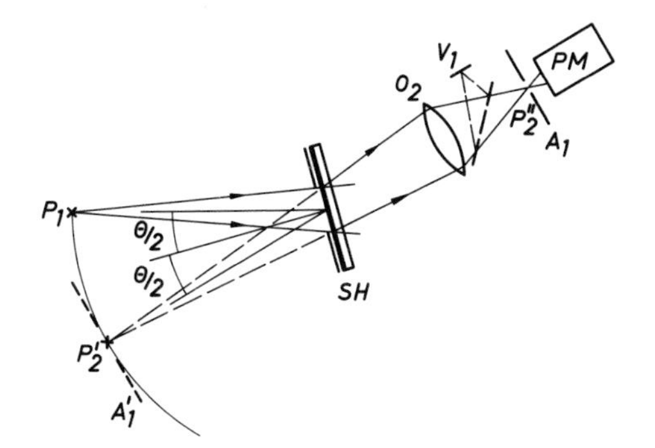


Fig. 2.8. Optical arrangement for the evaluation of diffraction efficiency of holograms of point objects [2.39]

from the low-modulation gratings is usually small, the $\alpha^2(E)$ values can be derived with sufficient accuracy from $T_a(E)$ values measured in the zero order of the samples.

In comparison to microdensitometry the diffractometric method has the advantages of a large measuring area that efficiently averages variations from granularity (especially appreciable with conventional, coarse-grain materials), and the improved signal-to-noise ratio allows reduction of the modulation of the incident exposure to very low values for practically linear transfer. There is no influence from an instrument MTF, and the spatial frequency range covers all frequencies occurring in holography.

Careful measurements down to spatial frequencies of a few cyc./mm for checking the experimental fulfillment of $M(0)=1$, however, reveal that the assumption of pure amplitude modulation in the photographic layer is not fulfilled. Rather, a spatial frequency dependent phase function contributes to diffraction efficiency, and (1.21) has to be modified [2.16, 40]

$$\eta=\eta_r+\eta_i=[\tfrac{1}{2}\cdot m_0\cdot M(\nu)\cdot\log e\cdot\alpha(\bar{E})]^2\cdot\left[1+\left(2\cdot\log e\,\frac{d\varphi(\nu,\lambda)}{dD}\right)^2\right], \tag{2.5}$$

where the subscripts r and i stand for the real and the imaginary part of the complex amplitude transmittance of the hologram and $d\varphi/dD$ is the dependence of the phase shift on density modulation. This phenomenon will be discussed in some more detail in Section 2.5.3. At this point, we have to state that the MTF evaluated by means of diffraction efficiency measurements and (1.21) should be called "as-if-MTF" [2.19], an MTF calculated as if only amplitude modulation caused diffraction.

## The Multiple-Sine-Slit Microdensitometer

The influence of the imaginary part of the transmittance distribution can be determined in several ways (Sect. 2.5.3). We will give here a short discussion of an instrument that eliminates phase effects by performing the measurements in the sample plane rather than in a transform plane.

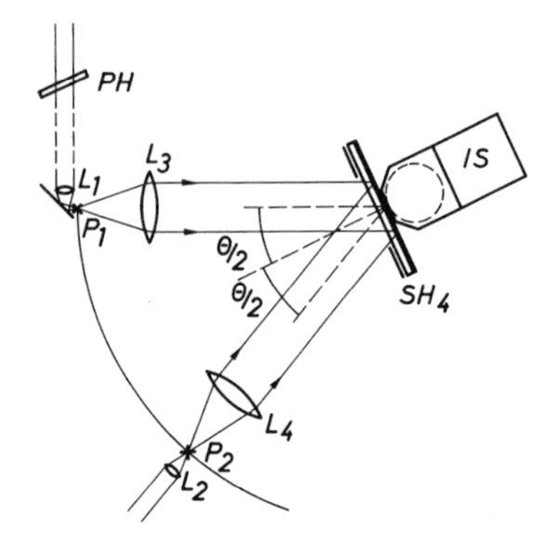


Fig. 2.9. Principle of the MSSM for scanning a sample in plane $SH_4$ with two-beam interference fringes. PH indicates a phase shifting device, IS is a detector with an intergrating sphere in contact with the sample [2.39]

In principle, the sample, having been exposed to two-beam interference and developed, is irradiated by a two-beam interference distribution of the same spatial frequency and orientation again [2.41]. All the light transmitted by the sample is collected by an integrating sphere, as outlined in Fig. 2.6c and shown in more detail in Fig. 2.9. When the phase of one of the interferring fields is changed, the interference pattern moves across the sample and the light flux detected in the integrating sphere varies periodically.

The working principle of this method can be treated in several ways. Scanning the sample with an sinusoidal irradiance distribution many periods wide can be compared to the common microdensitometer principle. This interpretation gave the name "Multiple-Sine-Slit Microdensitometer (MSSM)" to the instrument [2.35]. We may also note that the transmittance function of the sample is multiplied by the cosine function of the two-beam interference, and then the flux resulting from the product is integrated over the entire sample by the integrating sphere. Hence, the instrument performs a Fourier analysis of the transmittance function of the sample [2.42]. We can regard this operation as a cross-correlation as well, or as scaling down of high spatial frequencies to a moiré fringe of sufficiently low spatial frequency to be conveniently scanned by the aperture of the integrating sphere. Finally, the principle can also be treated as diffraction of two coherent fields at a complex grating and subsequent addition of the diffracted fields while the relative phase of the incident fields is changing [2.43]. Also, from this model, conditions can be derived for measurements suitable to separate the contributions of the amplitude and the phase structures to the total diffracted flux [2.44] (Sect. 2.5.3).

Any of the derivations yields as the final result for the total flux detected

$$\phi_{\text{tot}}(\nu, \Delta\varphi) = 2 \cdot \phi \cdot \{q(0) + [q(\nu)/2] \cdot \cos \Delta\varphi\}, \tag{2.6}$$

where $\phi$ is the flux from each one of the irradiating fields measured without sample, $\Delta\varphi$ is the momentary phase difference between the interferring fields, $q(0)$ is the Fourier coefficient of frequency zero, i.e., the mean intensity transmittance of the sample, and $q(\nu)$ is the Fourier coefficient of the intensity transmittance distribution for spatial frequency $\nu$ of the scanning interference

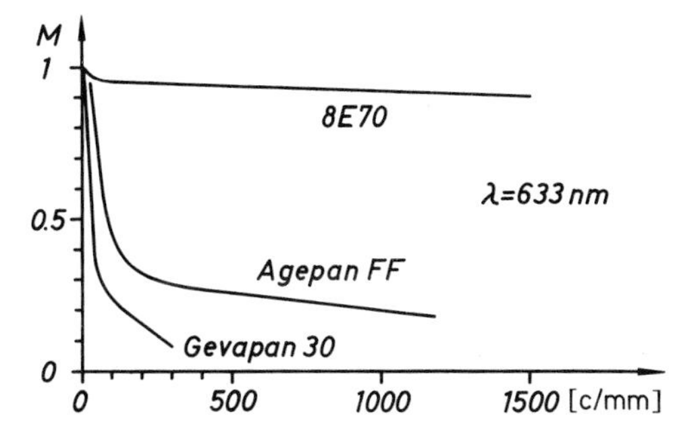


Fig. 2.10. Typical MTF's of three different kinds of emulsions: Gevapan 30, a plate of 23 DIN ≙ 160 ASA for pictorial photography; Agepan FF, a microfilm; Scientia 8E70, a Lippmann-type plate optimized for hologram recording at $\lambda = 633$ nm [2.10]

field. We see that only intensity transmittance is represented, not a possible imaginary part of the transmittance fields distribution. In the measurement, only the maxima and minima of $\phi(\nu, \Delta\varphi)$ have to be recorded and the transmittance modulation $m(\nu)$ of the frequency $\nu$ in the sample is simply

$$m(\nu) = \frac{q(\nu)}{q(0)} = 2 \cdot \frac{\phi_{\max} - \phi_{\min}}{\phi_{\max} + \phi_{\min}}. \tag{2.7}$$

It is seen that instrument data do not enter the relation and that the output signal of the MSSM is independent of spatial frequency. Since, as indicated in Fig. 2.6, the method has a narrow-band characteristic similar to the diffraction method, granularity noise is suppressed and signals of very low modulation can be measured with good accuracy, such as the higher harmonics generated by nonlinear transfer. Due to nonlinear transfer, only for transmittance modulations $m(\nu) < 0.4$ can the way back over the $D$-log $E$ curve be approximated by division by $\gamma(\bar{E})$. For higher modulation, a correction term $k$ takes account of the departures resulting from using a constant slope $\gamma(\bar{E})$ and disregarding higher harmonics. Values for $k$ can be derived from a chart in [2.43]. The MTF is obtained from

$$M(\nu) = [m(\nu)]/[(1+k) \cdot \gamma(\bar{E}) \cdot m_0]. \tag{2.8}$$

The design of the measuring system is described in detail in [2.39].

## Modulation Transfer Data of Silver Halide Materials

With the measuring methods established, we may now give some data on the MTF of silver halide materials. Figure 2.10 compares MTF's that are typical for a conventional emulsion of 23 DIN = 160 ASA, a microfilm, and a plate designed for holography (exposure to He–Ne-laser light, $\lambda = 633$ nm). These samples have been developed with utmost care in order to obtain MTF's as free from external effect as possible (developer Agfa 80, vigorous agitation, 5 min, cf. p. 29 and Sect. 2.5.2). It is seen that the microfilm may be used for making holograms with angular extent between object and reference source of less than 20°; the diffraction efficiency, however, will be about 13 times lower than for the special hologram plate due to the difference in MTF according to (1.21). In a special

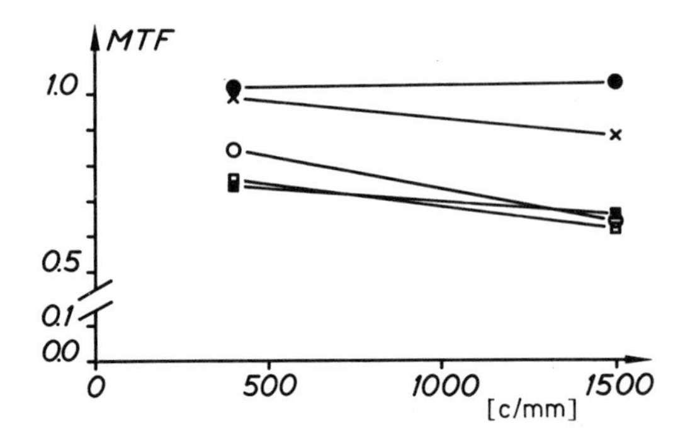


Fig. 2.11. MTF values for $\lambda = 633\,\mu m$ evaluated at 400 $c/mm$ and 1500 $c/mm$ by means of the MSSM: ● Kodak Holographic Plate 120–02; × Agfa-Gevaert Scientia 8E75; ○ Kodak SO-253; □ Agfa-Gevaert Scientia 10E75; ■ Kodak 649-F

geometry for making image plane holograms with a carrier frequency of 100 cyc./mm, this film proved suitable for hologram interferometry [2.14]. Since MTF evaluations showed that the special fine-grain emulsions for holography, in fact, are characterized by a marked plateau in their MTF (2.4), the values at two frequencies only, $\nu = 400$ and 1500 cyc./mm, are given in Fig. 2.11. The data are taken from [2.43] and recent measurements. Values of the MTF a few percent above one for the Kodak plate Type 120-02 and recent verification of this with a different emulsion batch and developer Kodak D-19 indicate a liability of this emulsion to adjacency effect. Hence, this MTF should be regarded as an "apparent MTF" (cf. Sect. 2.5.2).

### 2.4.4 Scattered Flux Spectrum

When the reconstructing field strikes a hologram, the well-known four fields are generated, among them the desired replica of the original object field. However, some light more or less randomly scattered from the hologram is also superimposed on the reconstructed field. Since this scattered flux is coherent to the reconstructed field, interference and speckles will occur.

In the presence of an average scattered irradiance $\langle I_N \rangle$ from a multitude of randomly phased amplitudes, the signal-to-noise ratio is given by the ratio of the deterministic image irradiance $I_i$ to the standard deviation $\sigma$ of the total irradiance [2.45]

$$\frac{I_i}{\sigma} = (I_i/\langle I_N \rangle)/(1 + 2I_i/\langle I_N \rangle)^{1/2} . \tag{2.9}$$

If the irradiance of the signal field is $10^3$ times as strong as the average scattered irradiance, the contrast in large areas is certainly about 1000:1, but in small details the signal-to-noise ratio will be as low as 22.7:1. It is obvious that the scattered flux is stronger the greater the granularity of the hologram. The strange consequence is that the speckles make the images projected by holograms made on grainy film look "grainy", although the real grains are several orders of magnitude below resolution of the eye.

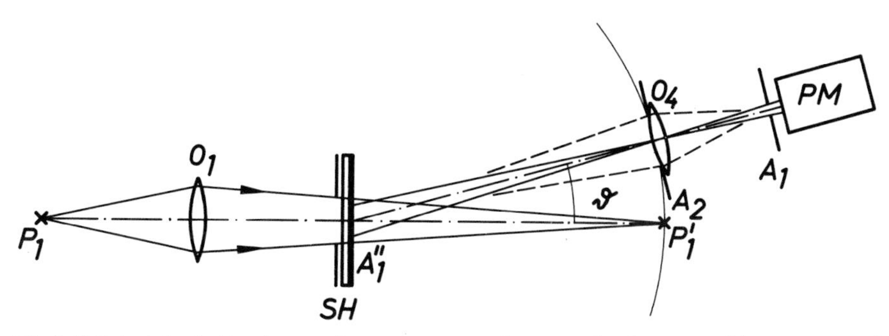


Fig. 2.12. Principle of a Fourier transform arrangement for measuring the scattered flux spectrum of the sample SH [2.39]

The characterization of granularity by means of the Wiener spectrum is less suitable in coherent optical systems [2.31, 46]. There are also other sources of stray light in addition to the diffraction spectrum of the photographic grains. As shown above, in coherent systems, "noise" is not additive. For these reasons, we prefer to measure the angular distribution of light scattered into a solid angle of $1 \times 1$ (cyc./mm)$^2$ two-dimensional spatial bandwidth [2.31, 46, 47] and to call this experimental function "scattered flux spectrum".

The principle of the measurement is seen in Fig. 2.12 [2.31, 39]. The holograms or uniformly exposed areas are brought into a Fourier transform arrangement where the probing field is focused to a point at distance $R = 750$ mm further downstream. The scattered flux $\phi_N(\theta)$ is defined for normal incidence and normalized on $\phi_{\mathrm{inc}}$ in order to eliminate the sample area. $\phi_N(\theta)$ is measured as a function of angle $\theta$ to the normal on the sample. The collecting, quadratic aperture $A_2$ of size $\Delta\xi \cdot \Delta\xi$ has a bandwidth in the plane of the scanning circle with radius $R$

$$\Delta v_1 = \frac{\Delta\xi}{R \cdot \lambda} \cdot \cos\theta \tag{2.10}$$

and in a direction vertical to it

$$\Delta v_2 = \frac{\Delta\xi}{R \cdot \lambda}.$$

In order to keep diffraction at the physical borders of the sample from contributing to the flux measured, lens $O_4$ images the center of the sample onto aperture $A_1$. The projection of the image $A_1''$ of $A_1$ onto the sample is the effective measuring area. Since it is proportional to $1/\cos\theta$, the decrease with $\cos\theta$ of the bandwidth of the collecting aperture [(2.10)] is compensated for, and the normalized scattered flux spectrum is given by

$$\frac{\phi_N(\theta)}{\phi_{\mathrm{inc}}} = \frac{\phi(\theta)}{\phi_{\mathrm{inc}}} \cdot \left(\frac{R \cdot \lambda}{\Delta\xi}\right)^2 [\mathrm{cyc./mm}]^{-2}, \tag{2.11}$$

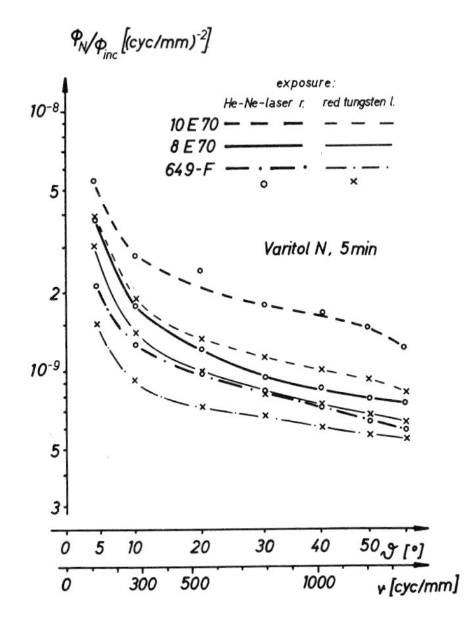


Fig. 2.13a–c. Scattered flux spectra at $\lambda = 633$ nm of three plates for holography exposed by He–Ne-laser radiation (○, heavy lines) and red tungsten light (×, light lines). Mean transmittance is $\tau_i = 0.2$ [2.31]

where $\phi(\theta)$ is measured with the aperture $A_2$ of size $(\Delta\xi)^2$, and $\phi_{\text{inc}}$ is measured with the same aperture at $\theta = 0$ without sample.

Figure 2.13 shows complete normalized scattered flux spectra for three materials. (These curves were measured in 1969; they are, however, still typical for these classes of emulsions, as can be seen from the recently measured values given in Table 2.3.) The scattered flux is, of course, dependent on mean transmittance (cf. Fig. 2.15). The measurements reported in Fig. 2.13 and Table 2.3 were obtained with samples of mean amplitude transmittance $T_a \approx 0.45$, i.e., $D \approx 0.7$, where diffraction efficiency is about maximum. From the data, we may make a number of observations.

It is obvious that scattered flux is higher for plates of higher sensitivity (cf. Table 2.1 and Fig. 2.5).

Table 2.3. Normalized scattered flux of He-Ne laser radiation measured at 30° from plate normal ($\nu \approx 800$ cyc./mm) for holographic plates uniformly exposed to He-Ne laser radiation. Amplitude transmittance $T_a \approx 0.45$ ($D \approx 0.7$)

| | Plate | Developer [20 °C] | Time [min] | $(\phi N/\phi_{\text{inc}})_{T_a=0.45}$ [cyc./mm]$^2$ |
|---|---|---|---|---|
| Agfa-Gevaert | 10E75 | Agfa 80 | 5 | $1.75 \times 10^{-9}$ |
| Agfa-Gevaert | 8E75 | Agfa 80 | 5 | $1.0 \times 10^{-9}$ |
| Kodak | 649-F | Agfa 80 | 5 | $0.95 \times 10^{-9}$ |
| Kodak | 120-02 | Agfa 80 | 5 | $0.60 \times 10^{-9}$ |
| Kodak | 120-02 | D-19 | 5 | $0.60 \times 10^{-9}$ |
| Kodak | 120-02 | D-19 | 8 | $0.85 \times 10^{-9}$ |
| Kodak | SO-253 | D-19 | 6 | $1.70 \times 10^{-9}$ |

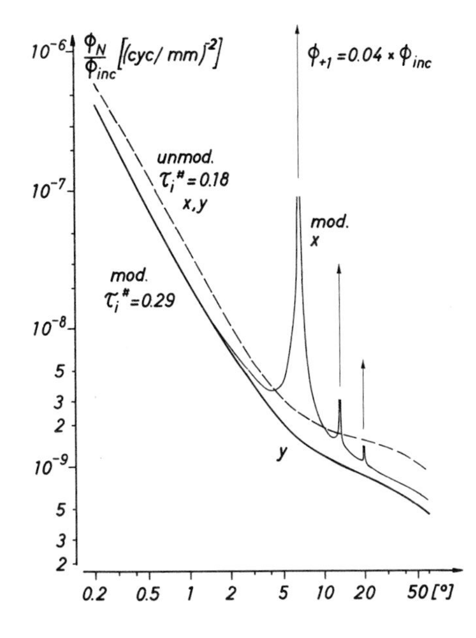


Fig. 2.14. Scattered flux spectrum around images of the reference source and an image point and its higher order reconstructions shows low-frequency spread from gelatin scattering. Light line: Scanning through the plane of the sources. Heavy line: Scanning through perpendicular plane. Dashed line: Unmodulated plate with same mean exposure [2.31]

Samples exposed to laser radiation show increased scattered flux compared to samples exposed to incoherent thermal light of the same wavelength. This phenomenon has been studied in detail [2.31] and has been explained to be scattering from a speckle structure exposed into the emulsion by coherent radiation scattered during exposure. This transmittance structure ("diffusion mottle") is a hologram of scattered light and cannot be separated from the holograms of signal fields or filter functions recorded. The relation of this phenomenon to film MTF has been discussed on p. 37.

The scattered flux spectrum is essentially flat, as is to be expected as far as the scattering at the photographic grains is concerned. Scattering from glass substrates is minor [2.31, 46], while flexible film supports of acetate or estar themselves may scatter as much light as emulsions do [2.46]. The steep increase of the scattered flux spectra towards low angle could be attributed to the gelatin [2.31]. This small-angle scattering from the gelatin results in some sort of spread function in the reconstructed image. Experimental evidence is given in Fig. 2.14, where, from a hologram of a point source, the spectrum of the diffracted light has been recorded including the reconstruction of the point source and of two spurious images from higher harmonics. In the double logarithmic scale we note the exponential spread of scattered light around the reconstruction source at the origin of the angular scale, as well as an equal spread around the reconstructed point source.

The relation between scattered flux and development time seen in one example of Table 2.3 is in agreement with earlier observations where for every one minute increase of development time the scattered flux increased very roughly by about 12%.

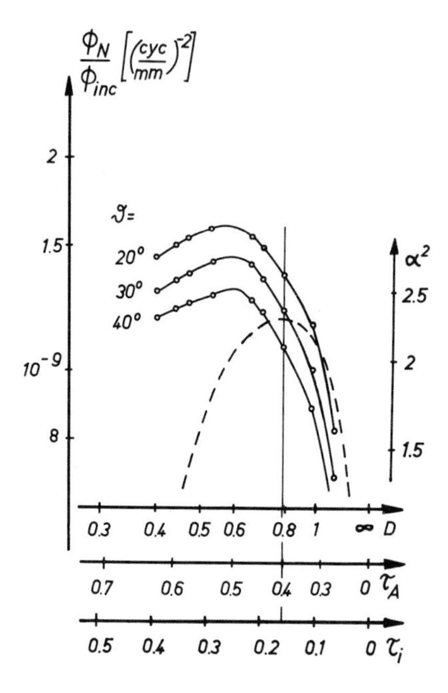


Fig. 2.15. Scattered flux and film characteristic $\alpha^2$ as a function of mean density. Scattered flux decreases faster with increasing density than does diffraction efficiency [2.31]

The dependence of scattered flux on density has been studied using the checkerboard and the circular-grain models for the photographic grain [2.45–47]. These models predict maximum scattered flux around $T_a \approx 0.5$ and $T_a \approx 0.6$, respectively. Figure 2.15 shows an experimental result. Fortunately, maximum diffraction efficiency occurs at higher density than maximum scattered flux. The effect of density-dependent scattered flux can also be observed in Fig. 2.14. The holographic grating with variations between very low and very high density values shows, upon scanning normal to the plane of the sources, a lower scattered flux spectrum than does a plate of the same mean exposure without any modulation.

### 2.4.5 Figure of Merit and Attempts to Improve Photographic Materials

#### Holographic Exposure Index

When materials for holography are to be compared, a criterion should be used that measures the expense of laser light necessary for achieving a given diffraction efficiency [2.16] or a given ratio of diffraction efficiency to scattered flux [2.48]. While the laser light transmitted through the reference beam path reaches the plane of the recording medium with little loss, only a few percent of the light scattered from an object is used for exposure. Hence, the higher $\alpha^2_{max}$ and $M^2(\nu)$ and the lower $\phi_N(\nu)$, the higher the beam intensity ratio $K$ can be chosen, the greater can be the share of laser light channelled through the reference beam path, and the shorter will be the exposure time with a given laser power, when a certain luminance and contrast of the hologram are desired. Figures of merit based on these relations are discussed in [2.16, 48].

## Ways of Increasing the Holographic Exposure Index

There are essentially four ways in which one could think of improving the holographic exposure index by improving one or several of the characteristics involved: the sensitometric speed, MTF, or $\alpha^2_{max}$ [2.16].

Three ways hold little promise: an increase of the MTF can hardly be achieved by the user; extended development increases speed and $\alpha^2_{max}$ but also scattered flux (there is no gain compared to using an emulsion of higher speed from the beginning), and pre- and postexposure lowers $\alpha$ by a factor equal to the ratio of coherent exposure to total exposure.

The fourth way is hypersensitization by bathing. The effect of hypersensitization is mainly due to removal of excess bromide ions and/or an increase in the relative concentration of free silver ions in the emulsion. One consequence is reduced storage time and increased fog. For application in holography, an extension of the toe of the $D$-log $E$ curve to lower exposures would decrease $\alpha^2_{max}$ as it would increase fog. Hence, hypersensitization procedures have to be studied and improved using holographic exposure index criteria. As a result of a study [2.16], a relatively simple procedure can be recommended:

1) bathing in a solution containing 10 ml ammonia/liter of distilled water (ammonia 25%, spec. density 0.91) at 20 °C for 3 min,

2) washing in distilled water containing 1 ml/l inert wetting agent (e.g., Agfa-Gevaert Agepon) for 5 min,

3) careful, uniform drying (avoid drops) in a drying cabinet with a fan at room temperature or slightly higher (30°).

The ammonia-treated plates should be kept for a few hours only before exposure and development, water-treated plates [step 1) omitted] not more than a few days at low temperature. There is a slight gain in $\alpha^2_{max}$ some decrease in scattered flux, and an increase in sensitometric speed by, e.g., a factor of 2.6 and 2.2 for 8E70 plates treated in ammonia or only water, respectively.

## Extending Sensitivity to Infrared Radiation

Although special photographic emulsions with spectral sensitization for the infrared region up to about 1.3 μm are available, their MTF and granularity characteristics make them unsuitable for recording holograms with the radiation of 1.06 μm from Nd: YAG lasers.

It has been shown, however, that some of the special emulsions for holography show susceptibility for Herschel effect, i.e., destruction of a latent image by exposure to infrared radiation [2.49]. Kodak plates SO 120-01 obtained a uniform pre-exposure sufficient to yield a density of 4 upon development. This uniform latent image was then exposed to two-beam interference of $\nu = 1300$ cyc./mm generated by the radiation of $\lambda = 1.06$ µm from a Nd:YAG laser. After the coherent IR exposure and development, samples were bleached, and diffraction efficiency values up to 25% were measured with He–Ne laser radiation ($\lambda = 633$ nm). From the result, it is assumed that the MTF for IR exposure should be comparable to that holding for visible light. The sensitivity to IR radiation, however, was, in this experiment, at least six orders of magnitude lower than that to visible light.

### 2.4.6 Holography vs. Photography

*Gabor* [2.50, 51] compared holography and photography and suggested that the amplification provided by a coherent background should make it possible to record a hologram with 1/10 or even 1/100 of the light flux from the object which otherwise would be required to take a conventional photograph, where the total exposure has to be provided by the object. The only analysis known to that date [2.52] was made at a time when still very little experience and data on recording materials in coherent light were available. From the discussions of this chapter, we have seen that holography and photography are not comparable merely on the basis of sensitometric sensitivity. For its carrier frequency technique, holography requires a high MTF at very high spatial frequencies. The value of the MTF enters into diffraction efficiency (and hence signal strength) with its square. (For comparison: hologram plate: exposure about 10 $\mu J/cm^2$, MTF about 0.9 at $\nu = 1500$ cyc./mm. Conventional film of 21 DIN $\triangleq$ 100 ASA: exposure about $2 \cdot 10^{-3}$ $\mu J/cm^2$, MTF about 0.5 at $\nu = 50$ cyc./mm.) The high granularity of conventional films has a correspondence in the scattered flux which interferes with the signal field and generates speckles (Sect. 2.3.4). In other words, the amplification of the signal in the hologram recording step is followed by an amplification of the noise in the reconstruction step. Hence, we can ask the interesting question: When we compare holographic and conventional photographic imaging at the same resolution and signal-to-noise ratio in the final image, is there a real gain in the holographic method? Or could there be some kind of invariance between both methods and different sensitivity classes and sizes if we take all the implications into account? So far, a more detailed investigation does not seem to exist.

## 2.5 Effects

In Section 2.4, the hologram forming process was treated in an idealized manner. Transfer from exposure to amplitude transmittance was approximated as linear by regarding small modulations only and disregarding development effects. We dealt with purely absorptive holograms; any phase effects were neglected. Under the heading "Effects", this section will report on deviations from the idealized image which have to be taken into account in practice. The four subsections deal with macroscopic and microscopic nonlinear transfer, formation of complex gratings, and effects ascribed to stress in the emulsion.

### 2.5.1 Macroscopic Nonlinear Transfer

The basic equations of holography show (Chap. 1) that the irradiance distribution to which a hologram recording material is exposed consists of a sum of four terms.

$$I \sim RR^* + RO^* + OR^* + OO^* . \tag{2.12}$$

Only one of them contains the encoded information on the wavefield of interest, in (2.12) it is the third one, if we denote the signal field by $O$ and the reference field by $R$. If we provide or assume linear transfer from exposure to amplitude transmittance, amplitude transmittance can be described by the sum of the same four terms (plus a constant and a factor), which modulate the reconstructing field and make the hologram generate four fields, one of which is the desired reconstructed wavefield, $ORR^*$.

If the transfer from exposure to amplitude transmittance is nonlinear, higher-order terms appear, among them multiple autoconvolutions and autocorrelations of the desired reconstructed field, with the result that false images and convolved images are also projected onto the first-order image [2.53, 54]. An illustration is given in Fig. 2.16.

These problems of nonlinear transfer will occur with any recording material, when the modulation exceeds a certain value. In particular, the transfer by materials leading to complex or phase holograms is inherently nonlinear (cf. Chap. 1 and Sect. 2.7.1).

For amplitude holograms in photographic emulsions, a linear $T_a - E$ relationship would fulfill the condition for linear transfer. It is less obvious that a straight-line $T_a - \log E$ curve would also eliminate first-order intermodulation terms, provided that, in addition, the MTF is constant and equal to 1 [2.15, 55, 137]. Due to the statistical nature of the image-forming process in silver halide emulsions [2.2, 3, 8, 9], there is no way to realize these hypothetical exposure-transmittance relations.

A number of papers have dealt with the nonlinear transfer by photographic emulsions. It seems, however, that these investigations have not yet resulted in advice on how to control emulsion, exposure and processing parameters in such a way that higher diffraction efficiency is achieved by lowering the beam intensity ratio $K$ without increasing the disturbing background on the reconstruction. The studies may assume the object to be described as a number of point sources (e.g., [2.56–58]), or as diffusely transmitting or reflecting (e.g., [2.53, 59–62]). There are a great many ways to describe the transmittance-exposure relationship: as a $\nu$-th law device (e.g., [2.53, 56]); by polynomial

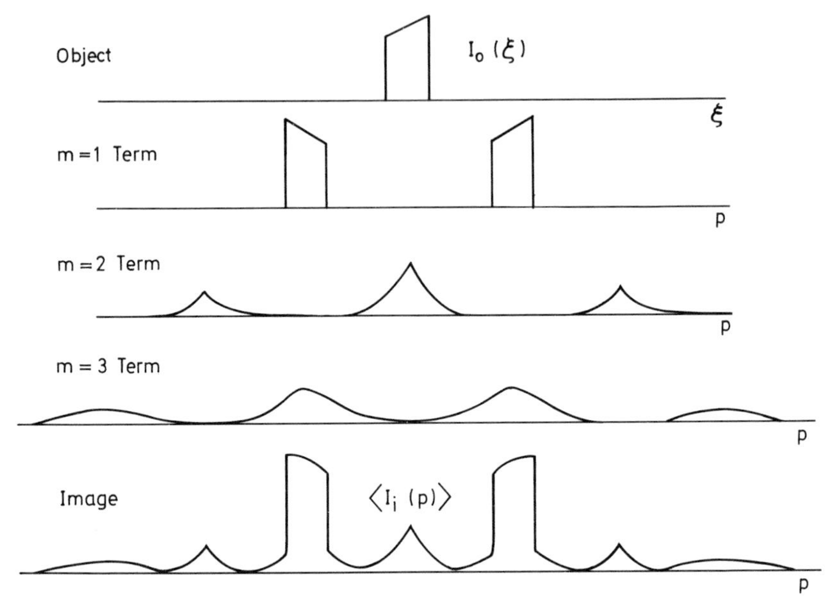


Fig. 2.16. Schematic pictorial representation of the terms generated by film nonlinearities and their contribution to the final holographic image [2.53]

approximations (e.g., [2.53, 57, 59, 60, 62]); by the transform method (e.g., [2.62, 64, 65]); by five points [2.58]; or by a suitable matched analytical function [2.66].

Other approaches to describe linearity behavior of amplitude recording materials have been reported, which are suited for direct experimental measurements and do not have to assume a certain form of the macroscopic characteristic or a certain working point on that curve. The presentation relates the amplitude of the reconstructed wave to the amplitude of the recorded wave. According to (1.21), both $\sqrt{\eta}$ vs. modulation of incident exposure, $m_0$, at constant exposure [2.67] and $\sqrt{\eta}$ vs. relative object field amplitude, i.e., $\sqrt{1/K}$, with reference field exposure as a parameter [2.68] are suitable plots to estimate linearity from the range of proportionality. These plots are derived from holograms obtained by interference of two plane wavefields; for diffuse objects, an estimate must be made.

An analysis of nonlinear transfer is, obviously, quite complex, and it is uncertain to what accuracy it can be made with regard to the quite large variations in emulsion, exposure, and processing parameters. Two properties make the analysis still more difficult. One is the thickness of the emulsion which weights the diffraction efficiency of the diffracted field as a function of the angular deviation of the reconstructing beam from the Bragg condition, but may, in addition, introduce coupling between fields (cf. Sect. 2.8.3). The second influence, which cannot be neglected, comes from the MTF of the material. The intermodulation term $O_sO_s^*$ [(2.12)] is centered around frequency zero, where the

MTF is always one (and surface relief generation is maximum, cf. Sect. 2.5.3), while the signal term is modulated on a high carrier frequency with lower MTF, $M(\nu) \approx \varrho$ (2.4). Hence, the intermodulation term introduces a low-frequency modulation and an unattenuated variation of the exposure mean value of the attenuated signal term. We can conclude that the noise in holograms in emulsions of higher sensitivity must be larger due both to increased granularity and to increased nonlinearity effects, when the loss in diffraction efficiency due to lower MTF is precompensated by lower beam intensity ratio $K$ (cf. p. 48).

### 2.5.2 Microscopic Nonlinear Transfer

Whereas light scattering in the emulsion during exposure is a linear process, which can be described by an MTF, processing often introduces not only a nonlinear transfer to transmittance but also a change of the modulation that depends on spatial frequency and cannot be explained by means of the $T_a - E$ relationship.

This development effect, called "adjacency effect" or "neighborhood effect", was observed very early in stellar photography and spectrography [2.2].

Diffusion of developer and reaction products in the swelled gelatin causes isolated, high-exposure details and edges to develop to a higher density than large areas that had obtained the same exposure, or, isolated, low-exposure details end up with too low a density due to general developer exhaustion.

Figure 2.17 [2.69] shows a very illustrative experiment designed to demonstrate the effect of adjacency effect on the rendition of sinusoidal exposure variations. An Agfa-Gevaert Scientia 8 E75 plate was exposed to a series of two-beam interference exposures of $\nu = 35$ cyc./mm matched in such a way that, in each frame throughout the series, the minima of the sinusoidal exposure distributions were exactly equal to the maxima in the preceding frame. The plate was cut in half; one piece was developed in the strong developer formula Agfa 80, the other one in the universal developer Agfa Rodinal diluted with 100 parts of water. Figure 2.17 shows the macroscopic $D - \log E$ curves determined at large, uniformly exposed areas, and the density traces obtained from a microdensitometer. While the traces for the Agfa 80 development are about as expected from the two-step model, the density distributions obtained with the diluted Rodinal developer overlap in a way which cannot be described by a $D - \log E$ curve. The experiment provides evidence for the nonlinearity caused by adjacency effect; namely, that elements of equal exposure come out with different densities.

Adjacency effects had previously been observed mainly with conventional emulsions of high sensitivity. The spatial frequency response obtained with adjacency effect, which is called "apparent MTF", usually showed increased values between a few and about 25 cyc./mm. The investigation of [2.69] also revealed strong adjacency effects for the extremely fine grain emulsions designed for holography. Figure 2.18 shows that the apparent MTF increases rapidly at

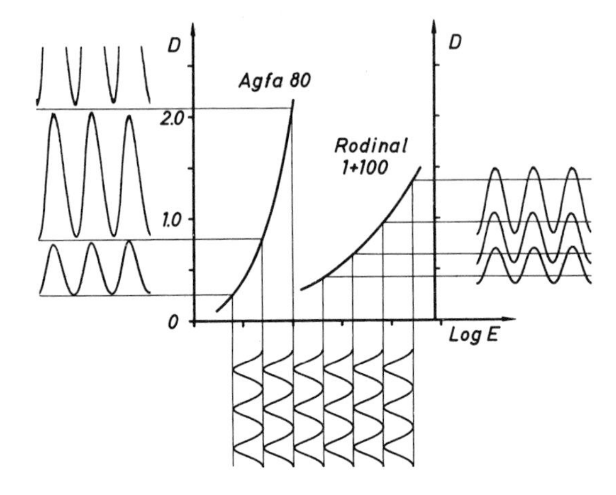


Fig. 2.17. Experiment demonstrating development without (Agfa 80) and with adjacency effect (Rodinal 1 + 100). Further explanations are given in the text [2.69]

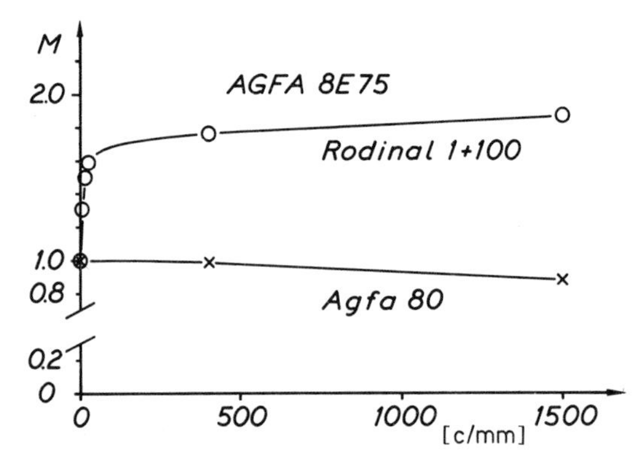


Fig. 2.18. Apparent MTF of Agfa 8E75 plate developed in strong developer Agfa 80 (×) and weak developer Agfa Rodinal diluted 1 + 100 (○). (Both 5 min at 20° C) [2.69]

low spatial frequencies, and than keeps a very elevated value up to the highest spatial frequency measured (1500 cyc./mm).

The dependence of the diffraction efficiency on the square of the MTF (1.21) might suggest a gain from the increase in MTF due to adjacency effect. However, the simultaneous decrease in $\alpha^2_{max}$, as seen in Fig. 2.19, gives, as a net result, a diffraction efficiency that is only one third of that obtained with the strong developer.

There are often circumstances in holography and spatial filtering where large exposure latitude or extended dynamic range is desired. Efforts to lower photographic $\gamma$ include use of weak, diluted or cooled, developers or short developing times. Also a very-low-gamma developer named POTA has been recommended. The investigations of [2.69] show that low $\gamma$ attained by any means implicates incomplete development, and hence a base for adjacency effects. The POTA developer, in particular, did keep the $\gamma$ of 8E75 plates as low as 0.5, but exhibited a very strong dependence of the apparent MTF on mean density, namely $M^{(400)}_{app} \approx 3$ at mean density 0.4 to $M^{(400)}_{app} \approx 1$ at $\bar{D} \approx 0.8$ (Ref. [2.69], Fig. 5).

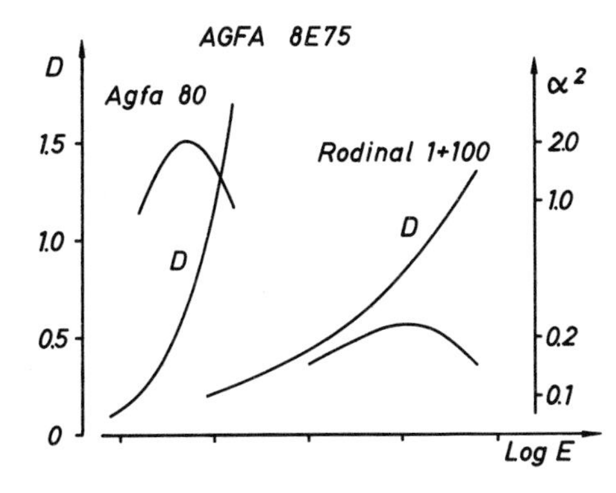


Fig. 2.19. Comparison of $D$-log $E$ and $\alpha^2$-log $E$ curves of the two samples leading to the MTF's of Fig. 2.18 [2.69]

Some observations and conclusions from [2.69] are: Development effects introduce nonlinearities, especially spatial filters may seriously deviate from their design. Strong developers are least inclined to cause development effects, but oxidation, repeated use and extended storage result in a marked increase (cf. pp. 29 and 44). In spite of an apparent MTF that is higher than the true MTF, diffraction efficiency is lower due to lower $\alpha^2$. There may be some promise in development effects obtained in ways other than with weak developers, e.g., with monobath developers [2.23] or developers with silver halide solvents added [2.24].

### 2.5.3 Complex Grating Formation

As discussed in some detail in Sect. 2.3.4, there is a phase modulation in developed photographic layers in addition to the desired amplitude modulation. In the region of low spatial frequencies, used in conventional photography, these phase effects have been studied using interference microscope techniques [2.70–74]. The predominant effects is a surface relief in the gelatin caused by the tanning action of the reaction products of certain developers (and also bleaching agents); it shows a marked dependence on spatial frequency. This effect has been used for deliberately producing relief images with arbitrary profile [2.75], and as one method for making phase holograms (p. 62). Since this unwanted effect in amplitude holograms is a problem with spatial filters, or when measuring the MTF of emulsions, it is desirable to determine the phase effect quantitatively also in the spatial frequency range not accessible to microdensitometry and interference microscopy.

The contributions of the amplitude structure and of the phase structure to diffraction efficiency can be evaluated with certain assumptions from diffraction efficiency measurements at two different wavelengths [2.76]. Since the Multiple-Sine-Slit Microdensitometer measures the modulation of the real part of the amplitude transmittance (p.43), these results may be compared with those from diffraction efficiency measurements (p.41) Figure 2.20 compares the MTF, $M(\nu)$, obtained by means of the MSSM with the "as-if MTF", "$M(\nu)$", resulting from

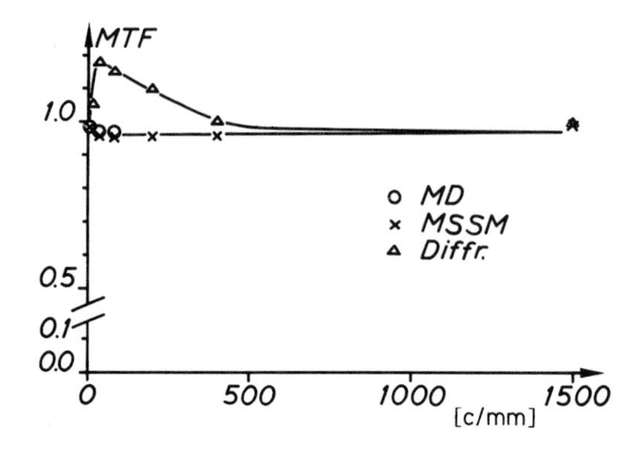


Fig. 2.20. Comparison of "as-if MTF" of Scientia 8E70 plate evaluated from diffraction efficiency measurements (△) with MSSM evaluation (×). Data points (○) have been obtained from conventional microdensitometer measurements [2.40]

diffraction efficiency evaluation. Reference [2.40] derives a simple relationship for calculating the phase retardation in a complex grating in rad per unit density difference from the two MTF's:

$$\frac{d\varphi}{dD}(\lambda, \nu) = 1.15 \cdot \sqrt{\left[\frac{\text{“}M(\nu)\text{”}}{M(\nu)}\right]^2 - 1}\,. \tag{2.13}$$

Here, as in the formula for $\eta$ of a complex grating [(2.5)], the well verified assumption is used that phase retardation $\Delta\varphi$ is proportional to the variation of mass of silver per unit area and, hence, approximately proportional to density variation $\Delta D$ (e.g., [2.71, 72]). Figure 2.21 shows the phase retardation function for $\lambda = 633$ nm for a Scientia 8E70 plate developed in Agfa-Gevaert G3p developer. At spatial frequencies beyond 600 cyc./mm, the phase effects become negligible. For a grating of exposure modulation 0.2, the phase effect contributed 30% to the total diffraction efficiency of $\eta_{tot} = 2.2 \cdot 10^{-3}$ at $\nu = 35$ cyc./mm, and 7% of $\eta_{tot} = 1.62 \cdot 10^{-3}$ at $\nu = 400$ cyc./mm.

Also a measurement of the phase effect by means of a modified MSSM has been suggested [2.40, 77].

In holograms processed in the usual way, phase effects do not appreciably contribute to diffraction efficiency. However, a strong, irregular gelatin relief is generated by the intermodulation term centered around frequency zero, which, in this way, affects also the contrast in the desired reconstructed fields, (e.g., [2.60]). Means for eliminating the effect of the surface relief are described in Sect. 2.3.4.

### 2.5.4 "Emulsion Stress"

Inconsistent results with hologram plates have led to the view that, in manufacturing and drying photographic emulsions, an anisotropic stress may be introduced. This stress is released when the emulsion swells in the processing solutions and it causes nonlinear displacements of parts of the layer containing

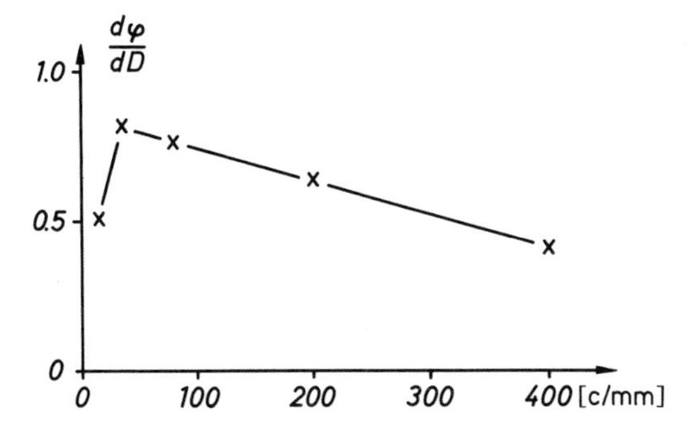


Fig. 2.21. Phase retardation per unit density difference $d\varphi/dD$ for $\lambda=633$nm in the Scientia 8E70 plate as a function of spatial frequency $\nu$, evaluated from the MTF data of Fig. 2.20 [2.40]

the hologram structure. Recommendations have been made to improve hologram quality by preconditioning the emulsions. Stress relief may be obtained by soaking the plates in water [2.78], or suspending them in a high humidity atmosphere overnight [2.79]. At least part of the improvement observed is due to hypersensitization and the accompanying increase of $\alpha^2$ and decrease of scattered flux (p. 49).

## 2.6 Experimental Data on Diffraction Efficiency and Signal-to-Background Ratio of Thin Amplitude Holograms

A perfect grating with sinusoidal amplitude transmittance of modulation $l$ diffracts 6.25% of the incident radiation into both the $+1$ and $-1$ diffraction orders (Chap. 1). With photographic plates for holography exposed to two plane waves of equal strength ($K=1$), the maximum of diffraction efficiency in an exposure series is usually around 4% (e.g., [2.18]). This figure, which, in practice, is of little interest, is the only figure that can be measured rather unambiguously.

As soon as wavefields from extended objects are recorded, a strong intermodulation term $O_s O_s^*$ (2.12) limits modulation for the image term and, upon reconstruction, competes with the image terms for diffracted flux. Hence, in an experiment with a groundglass as the object, the maximum of diffraction efficiency was observed with a beam intensity ratio $K \approx 2$ [2.15] rather than $K=1$; but with $K \approx 7.5$, $\eta$ became lower by only one-fourth. A balance between diffraction efficiency and background flux has to be found; there are a great number of parameters involved, and the optimum depends on the requirements of the application.

Mainly two applications have a demand for high signal-to-background ratio. (The term "signal-to-noise ratio" would, probably, be more appropriate for applications involving resolution of small luminance differences in fine details comparable in size to the speckles from interference between the signal field and the coherent scattered background.)

One application is hologram interferometry of vibrating surfaces. In an interferogram of a sinusoidally vibrating surface, the luminance of the fringes

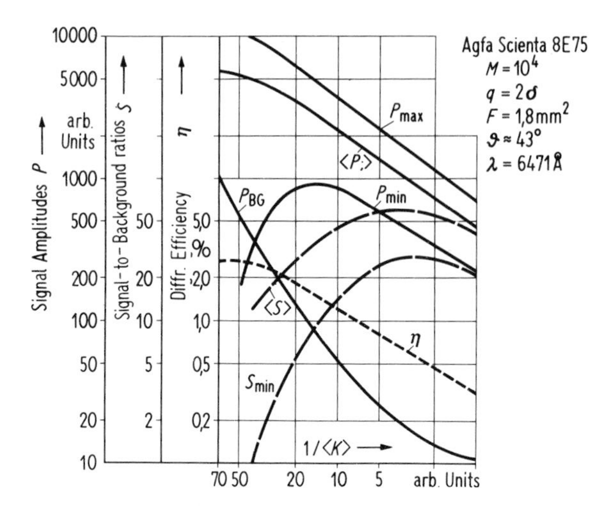


Fig. 2.22. Signal ($P_{max}$, $P_{min}$, $\langle P_i \rangle$), background ($P_{BG}$) and signal-to-background ratio ($\langle S \rangle$, $S_{min}$) of a data mask as a function of $1/\langle K \rangle$ as described in the text [2.83]

decreases following the square of a zero-order Bessel function $J_0^2$. The number of fringes discernible is directly related to the accuracy attainable in determining vibration amplitudes. Consequently, a test object consisting of a piece of vibrating, silver-painted aluminum angle stock has been proposed [2.12]. Large-amplitude vibrations smooth out the intermodulation term; hence $K=1$ is appropriate, and maximum interferogram fringe visibility was found at average densities of 1.5 to 2 (cf. Fig. 2.15) for Scientia 10E70 perforated film. Interestingly, comparisons made with this test object resulted in higher visually resolvable fringe numbers for bleached holograms than for amplitude holograms.

The other application requiring quantitative data on signal and scattered background is holographic data storage. Since the data are coded in arrays of light spots, the periodicity in the object causes the characteristic peaked patterns of multiple interference in the recording plane (e.g., [2.80–83]). Among the remedial measures is the use of defocused Fourier transform holograms [2.83]. Since the information consists in the presence or absence of light spots in the reconstructed array, the simplified signal-to-background scheme is adequate. Figure 2.22 gives, as one typical example, the curves obtained in an experimental investigation of the parameters governing the signal-to-background ratio [2.83]. In an Agfa-Gevaert Scientia 8E75 plate, units of $M=10^4$ data points are stored in defocused Fourier transform holograms of area $F=1.8\ \mathrm{mm}^2$. Due to the difficulty of defining and measuring beam intensity ratio $K$ in this application, mean radiance in the object plane was controlled and a value $1/\langle K \rangle$ was plotted in arbitrary units. The magnitudes given in Fig. 2.22 are: $\eta$, diffraction efficiency; $P_{max}$, $P_{min}$, $\langle P_i \rangle$, the maximum, minimum and average signal measured at data points "on"; $P_{BG}$, the highest background signal measured at the positions of data points "zero", it is composed of scattered flux, higher diffraction orders of adjacent spots, and intermodulation flux; $\langle S \rangle = \langle P_i \rangle / P_{BG}$ is the average signal-to-background ratio; $S_{min} = P_{min}/P_{BG}$ is the lowest signal-to-background ratio observed; it is relevant for the reliability of the data readout. The data of this diagram are typical for amplitude holograms, and we may summarize: Maximum $\eta$ is about 2.5%; however, maximum $\langle S \rangle$, about 60, and $S_{min}$, about 30, occur at $\eta \approx 0.6-0.7\%$. At higher $K$ (to the right), $\eta$ decreases and scattered flux is the predominant source of background. At lower $K$, background flux from nonlinearity effects increases faster than $\eta$.

As a second example, data reported on a similar data storage system include [2.84]: Maximum signal-to-background ratio of 23 dB and $\eta = 0.9\%$ for Kodak High Resolution Plates exposed with $K$ between 16 and 32, similarly $\eta = 0.9\%$ and $K = 16$ for Agfa-Gevaert 10E56 film.

Comparable relations between $\eta$, signal-to-background ratio, and $K$ have been obtained with test objects consisting of a dark square in a bright field [2.60, 85], a checkerboard of $4 \cdot 10^6$ points [2.85], and a resolution test chart [2.30].

## 2.7 Phase Holograms

According to elementary diffraction theory (Chap. 1), thin amplitude gratings diffract up to 6.25% of the incident light into one diffraction order, thick amplitude transmission gratings 3.7%, while the corresponding maximum values for phase gratings are 33.9 and 100%, respectively. Hence, there has been seen considerable promise in materials and procedures producing phase holograms. Fortunately, photographic emulsions are suitable for making phase holograms. The procedure consists of additional steps subsequent to development in which the absorbing silver distribution in the layer is removed with simultaneous generation of a surface relief or internal index of refraction variations.

### 2.7.1 Principles of Phase Grating Generation

The generation of a surface relief due to tanning of the emulsion gelatin (increased cross-linking between gelatin molecules) was discussed in Sect. 2.5.3. While this is an unwanted effect in connection with amplitude holograms, it may be cultivated for bleached phase holograms. The relevant references on the principles are given in Sect. 2.5.3, and the practical procedures will be discussed on p. 61. We may recall here that the phase shift $\Delta\varphi$ created is proportional to the mass of silver developed or present before bleaching and hence approximately proportional to density difference $\Delta D$. For low phase modulations $\Delta\varphi$, diffraction efficiency $\eta$ is proportional to $(\Delta\varphi)^2$ (Chap. 1) and also $(\Delta D)^2$, hence, for these phase holograms, the square of

$$\gamma(\bar{E}) = \left(\frac{dD}{d\log E}\right)_{\bar{E}} \tag{2.14}$$

will control diffraction efficiency in much the same way as $\alpha^2(\bar{E})$ does for amplitude holograms [(1.21)]. In fact, this relation is embodied in (2.5) [cf.

(2.17)]. Since the relief generation is affected by surface tension and very much dependent on spatial frequency, its practical importance has diminished in favor of methods that convert the photographic silver to dielectric compounds.

For this conversion, or "bleaching" process, it is essential that the silver distribution generated by exposure and development be converted to a large refractive index modulation, since (Chap. 1)

$$\begin{aligned} &\eta = J_1^2(\Delta\varphi) && \text{for thin sinusoidal phase gratings,} \\ &\text{and} && \\ &\eta = \sin^2(\Delta\varphi/2) && \text{for thick sinusoidal phase gratings,} \\ &\eta \approx \tfrac{1}{4}(\Delta\varphi)^2 && \text{for } \Delta\varphi \lesssim 0.6 \approx (2\pi/\lambda)\cdot 0.1\lambda \text{ in both cases.} \end{aligned} \tag{2.15}$$

When we regard the photographic layer after bleaching as an approximately homogeneous mixture of gelatin and dielectric silver salt, we can calculate the local index of refraction from the concentrations of the contributing substances by means of the Lorentz-Lorenz equation (e.g., [2.86–88])

$$n^2 = \frac{1 + 8\pi \cdot \sum_i \alpha_i N_i/3}{1 - 4\pi \cdot \sum_i \alpha_i N_i/3}, \tag{2.16}$$

where $N_i$ is the partial molecular concentration and $\alpha_i$ the electrical polarizability of substance $i$.

In [2.88] it is shown that the application of the Lorentz-Lorenz formula to bleached photographic layers ultimately leads to a very simple relation for the phase shift in the emulsion (at higher spatial frequencies, where surface relief generation can be disregarded)

$$\Delta\varphi/\Delta D = \frac{1}{a} \cdot 8.7 \cdot K \cdot q \cdot \alpha_b\,. \tag{2.17}$$

In this equation, $D$ is the photometric density before bleaching, $a$ relates density to silver mass, $K = 2\pi/\lambda$, $q$ is the number of molecules of silver compound formed from one atom of silver, and $\alpha_b$ is the polarizability of the compound embedded in the gelatin. From (2.15) and (2.17) it is seen that diffraction efficiency is proportional to the square of the polarizability of the compound produced by the bleach process. Table 2.4 [2.88] compares some of the most commonly used silver compounds with respect to their data relevant for (2.17) and predicted relative diffraction efficiencies for low, equal density modulation. It is worth noting that the refractive index of the much used $Ag_4Fe(CN)_6$ differs very little from that of gelatin.

With the dependence of $\Delta\varphi$ and hence $\eta$ on $\Delta D$ established in (2.17), $\gamma(\bar{E})$ is also the characteristic responsible for diffraction efficiency for these holograms.

Table 2.4. Some silver compounds produced by bleaching and the expected relative diffraction efficiency [2.88]

| | $n$ | $q$ | $\alpha_b(10^{-30}\ \mathrm{m}^3)$ | $\eta(\Delta D_0)$ relative |
|---|---|---|---|---|
| AgCl | 2.07 | 1 | 5.3 | 1 |
| AgBr | 2.25 | 1 | 6.6 | 1.6 |
| $Ag_4Fe(CN)_6$ | 1.56 | 1/4 | 35.9 | 2.9 |
| AgI | 2.20 | 1 | 9.2 | 3.0 |
| $AgHgCl_2$ | 1.82 | 1 | 12.4 | 5.5 |

In fact, for about ten bleach processes, it was found that for low modulations [2.16]

$$\eta(\bar{E}) \sim \gamma^2(E) \cdot \tau\,, \tag{2.18}$$

where $\gamma(E)$ is the local gradient of the $D$-$\log E$ curve before bleaching (2.14), $\tau$ is the specular intensity transmittance of the sample after bleaching, i.e., a measure for the attenuation in the bleached layer. [Hence, (2.18) should not be confused with (2.3).] Consequently, the working point for maximum diffraction efficiency will be somewhere close to the maximum gradient of the $D$-$\log E$ curve, typically between density 2.5–4 before bleaching. This means that the higher diffraction efficiency of a phase hologram in comparison to an amplitude hologram has to be paid for with higher exposure, typically by a factor of about 3.

### 2.7.2 Bleaching Procedures

A sea of bleaching baths has flooded holographers during the recent years, part of them being existing, sometimes commercial bleaches belonging to intensifying, reducing, toning, color processing and similar procedures, part of them brought forth by pure alchemy, others formulated and improved with competence. It seems, unfortunately, that nobody so far has undertaken the task of a systematic comparison of bleaching procedures. The main obstacle ought to be the difficulty of defining relevant criteria for the assessment, since the variables are so many and the requirements on the results (diffraction efficiency and signal-to-background ratio in the first place) depend so much on the object and purpose of the hologram. In the following, a survey of some bleach processes is given.

#### Gelatine Relief Formation

These bleach procedures are based on the tanning action ammonium dichromate exerts on the gelatin. Often the formula for the Kodak Bleach Bath R-10 has been used (e.g., [2.90]), also with modifications [2.91]. In our own experiments, ammonium dichromate treatments with gelatin tanning by subsequent uniform UV exposure were formulated that yielded clear, transparent gratings of extreme diffraction efficiency, but failed completely when finally used to record holograms of extended objects, where reticulation shreds the gelatin.

The strong spatial frequency dependence makes gelatin relief procedures less suitable for general holography. One useful application can be the production of reflecting diffraction gratings with a He–Ne laser as an inexpensive substitute for gratings in photoresist, which would require coating facilities and high laser power at short wavelengths. Particularly high modulation of surface depth can be achieved by developing the gratings in Kodak Tanning Developer followed by bleaching in Kodak Bleach R – 10 [2.11].

Dichromated gelatin (Chap. 3) excels by extremely low scattered flux. Suggestions for holograms in pure hardened gelatin have been published [2.92, 93] which use the high speed and panchromatic sensitivity of silver halide emulsions in the exposure step. Then, either in bleaching the hologram a dichromate is introduced [2.92], or the unbleached hologram is sensitized by a dichromate [2.93], and a second, uniform exposure to light of short wavelength produces an imagewise tanning of the gelatin. All the silver is removed from the layer by bleaching and clearing. By dehydration in isopropanol, volume phase effects can be achieved.

### Bleaching

The most common procedures soak the emulsion with a salt solution which oxidizes the metallic silver in the emulsion

$$Ag - e^- \rightarrow Ag^+$$

while a metal ion in the oxidizing agent reduces its valency. The silver ion forms, together with negative ions, a silver salt. (On p. 63, procedures will be mentioned in which halogens directly oxidize the silver.)

We may here just give some references to work reporting formulae and instructions for bleaching procedures [2.18, 86–88, 94–104].

The dielectric silver compounds formed are light sensitive; hence extended exposure to light, especially to shorter wavelengths, forms photolytical silver and darkens the hologram. Therefore, attention has been focused on methods that increase the stability of bleached holograms to light, e.g., by means of adding desensitizing dyes to the bleached emulsion [2.99–103].

Our own favorite bleach recipes are far from the sophistication of many of the procedures just referred to. Silver chloride is obtained in the bleach bath of a process intended for decreasing granularity and lowering $\gamma$ in conventional photography (formula Agfa/Orwo 710/I, e.g., [2.105]):

| | |
|---|---|
| Water | 800 ml |
| copper sulfate cryst. | 100 g |
| sodium chloride | 100 g |
| sulfuric acid conc. | 25 ml |
| add water to | 1000 ml |

The silver chloride obtained in this bleach shows the least diffraction efficiency of a great number of procedures studied, but also the least scattered flux, and the best signal-to-background ratio. Exposed to daylight, the layer grows purplish, but bleaching can be repeated. Bathing these holograms for 2 min in a solution of 20 g silver bromide in 1 l of water converts AgCl to AgBr with higher diffraction efficiency and higher scattered flux. Bathing for 2 min in a solution of 20 g silver iodide in 1 l water converts either AgCl or AgBr to AgI with still higher diffraction efficiency and scattered flux, but excellent light stability. Data in between AgBr and AgI are obtained from $Ag_4Fe(CN)_6$, which is formed when the silver layer is bathed for several minutes in a solution of 15–30 g potassium ferricyanide in 1 l of water. The highest diffraction efficiency and scattered flux from gratings with low exposure modulation was observed in this comparison with $AgHgCl_2$, converted from silver in a solution of 20 g mercuric chloride in 1 l of water. $AgHgCl_2$ holograms darken fast in daylight. (HgCl is poisonous!) For everyday laboratory use, the silver chloride and the ferricyanide bleaches combine good results with easy use.

A spatial frequency-dependent degradation occurs in the bleaching process. Not only does $\Delta D$ decrease with increasing spatial frequency due to the MTF of the material, but also $\Delta\varphi/\Delta D$ decreases. This decrease is probably due to mechanical rearrangements in the swollen gelatin layer when every silver atom is replaced by a larger molecule. A number of years ago in preliminary tests on some kind of bleaching transfer function like $[\Delta\varphi/\Delta D(\nu)]/[\Delta\varphi/\Delta D(0)]$, large decrease with spatial frequency was observed for the ferricyanide bleach (leading to $Ag_4Fe(CN)_6$), little decrease for mercuric chloride bleach (leading to $AgHgCl_2$). Systematic studies do not seem to exist yet.

### Gaseous Bleaching

The wet bleaching process and the subsequent drying cause considerable thickness variations and mechanical distortion to the gelatin layer and may contribute to the spatial frequency dependent degradation mentioned at the end of the preceding section.

Therefore, experiments have been made to bleach photographic layers by exposure to bromine vapor [2.106]. Experiments are described in some detail in [2.107], where both bromine vapor and chlorine gas have been employed. The process is straightforward. The developed, dry hologram is placed in a closed vessel containing bromine liquid and is there exposed to the vapor emanating from the liquid for about 15 min. Because of the toxicity of the halogen vapors, the process should be applied with care in a fume hood. The bleaching could be easily verified. Performance data are given in [2.107]. However, they do not permit a direct comparison with conventional wet processes, in particular, because two-beam interference with plane wavefields and $K=1$ were used, which should give a high modulation with several maxima and minima of diffraction efficiency for the pre-bleach density range between 0 and 7 studied [2.89].

**Reversal Processes**

Another consideration, how to obtain phase holograms with the smallest and most compact grains possible, arrives at the conclusion that a hologram ought to consist of a distribution of those silver halide grains that were present in the emulsion from the beginning, while the exposed and developed grains should be removed [2.86]. To this end, a successful reversal process has been devised, which simultaneously provides for suppression of the intermodulation term [2.108]. The intermodulation term is centered around frequency zero, where the tendency to forming a surface relief is greatest. A deliberately generated low-frequency gelatin relief in the first step, the negative hologram, may be able to compensate for the corresponding phase variations from the residual silver halide constituting the final, positive hologram.

We may quote here a later modification of the reversal process (which has been less accessible internationally because of its non-English publication [2.109]). Scientia 8E75 had been used in working out the procedure. In the formulae all quantities refer to solutions of 1 l final volume.

1) Exposure for density 2–4
   Development 5 min at 20° C (Agfa-Gevaert G 3p)
2) Stopbath, 2 min (20 ml acetic acid)
3) Rinse, 5 min
4) Bleach, 2 min (5 g potassium dichromate, 5 ml sulfuric acid, conc.)
5) Rinse, 5 min
6) Clear, 1 min (50 g sodium sulfite, sicc., 1 g sodium hydroxide)
7) Rinse, 5 min
8) Desensitize and dehydrate, 10 min
   (880 ml ethyl alcohol
   100 ml water
   20 ml glycerine
   50 mg potassium bromide
   200 mg phenosafranine).

Another modification of a reversal bleach process is described in [2.110].

### 2.7.3 Experimental Data on Diffraction Efficiency and Signal-to-Background Ratio

The theoretical limits of 33.9 and 100% diffraction efficiency for ideal thin and thick phase gratings have as a counterpart theoretical limits of 18.4 and 64%, respectively, for the reconstruction of diffuse object fields ([2.65, 17], resp.). With simple gratings, diffraction efficiencies close to the theoretical values (with

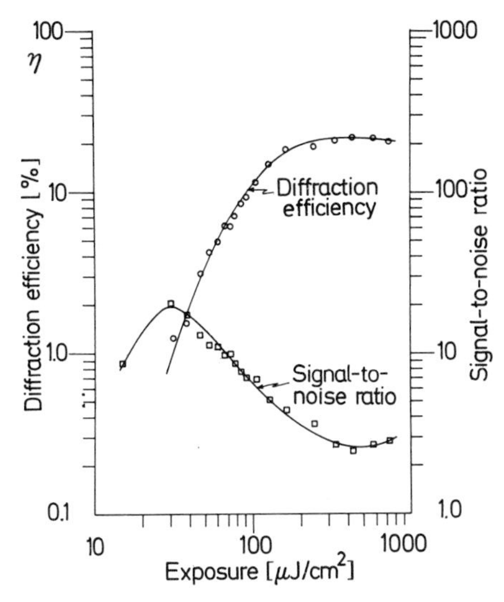


Fig. 2.23. Dependence of diffraction efficiency $\eta$ and signal-to-background ratio on exposure for a diffuse test object recorded with $K=1$ on a bleached 649-F plate. ($\lambda=633$nm) [2.18]

allowance to losses in the medium) have been obtained. For diffuse objects, the 18% limit for thin phase holograms was approached [2.111], while for thicker holograms, the highest efficiencies reported stay below 30% (e.g., [2.17, 18, 97, 99, 111]).

Severe limitations in image contrast arise from the inherent nonlinearity of phase holograms (Chap. 1) and from the low-frequency intermodulation term. The effect of the latter one is so severe because in phase holograms all the terms recorded (2.12) become products instead of the summation valid in absorption holograms. Hence, the intermodulation term distorts the reconstruction of the desired wavefront much like a shower glass [2.17]. Unfortunately, this low-frequency term is favored by film MTF, surface relief generation, and spatial frequency response of the bleaching procedures. Hence, a liquid gate is recommended by many authors for signal-to-background improvement; immersion with a liquid of higher refractive index than that of the gelatin can partly compensate also for the internal intermodulation term [2.17, 18], as attempted with reversal bleaches (Sect. 2.7.3). Also spraying the surface with a lacquer could increase contrast by a factor of up to four for the type of test object used [2.26].

Figure 2.23 is intended to give an idea of the relation between diffraction efficiency and signal-to-background ratio as a function of exposure (or working point on the $D$-$\log E$ curve before bleaching) [2.18]. The test object is a ground glass with an opaque square. At maximum efficiency, $\eta \approx 20\%$, for $K=1$, the signal is only twice as strong as the diffuse background. Maximum signal-to-background ratio, about 120, was found for $K=40$, and $\eta$ was 1.3%. As a good compromise, the authors felt the experimental values $K=10$, $\eta=10\%$, and signal-to-background ratio 57. These results can be regarded as typical for bleached photographic emulsions.

In conclusion, when aiming for maximum signal-to-background ratio, the order of magnitude achieved experimentally is 20 dB, while $\eta$ is of the order of

1 %, both for amplitude holograms (Sect. 2.6) and phase holograms. With some sacrifice of signal-to-background ratio, however, phase holograms can offer much higher diffraction efficiency than amplitude holograms.

## 2.8 Thick Holograms

Gabriel *Lippmann* in 1891 was the first to use the depth of a photographic emulsion for recording optical information. For his physical method of color photography, he had to formulate emulsions that resolved structures with a periodic spacing of $\lambda/2n$ ($n$ is the refractive index of the emulsion) [2.112], as *Denisyuk* did in 1962 [2.113] for his holographic modification of the same principle. The principles of wavefield recording and reconstruction with three-dimensional media (e.g., [2.114–116], and Chap. 1) are the same for all recording materials. Photographic emulsions, however, differ from most of the media in several respects. With a thickness of 5–17 μm, commercially available emulsions show features of thin or thick transmission holograms depending on the spatial frequency range used (Chap. 1); the sensitivity of the Bragg condition to deviation in angle or wavelength is correspondingly lower than for other volume recorders. Photographic emulsions have very high and panchromatic sensitivity, which can facilitate the recording of color holograms (e.g., [2.117–120]); they are available in very large sizes. A unique property of the silver halide emulsion is the formation of a latent image. In multiple image recording in volume holograms, incident wavefields are not modulated by previously recorded holograms, as is the case with media generating absorption or phase structures without development. Also, recorded patterns are stable after development. The wet development and fixing process, on the other hand, entails problems with emulsion swelling and shrinkage.

### 2.8.1 Characteristics of Thick Photographic Emulsions

Although considerable theoretical work has been done on diffraction in thick gratings (Chap. 1), investigations into real photographic materials are far from as extensive as those reported in Sect. 2.4, where the photographic layer is assumed to be thin in the sense of the parameter $Q$ discussed in Chap. 1.

As to the macroscopic characteristic of thick amplitude holograms, a substitution of the $D$-$\log E$ curve into the expression for diffraction efficiency given by the coupled-wave theory (Chap. 1) gave the result that the function $\alpha^2(E)$ also governs the diffraction efficiency of thick amplitude holograms [2.121]. The validity of (1.21) for these holograms requires that the Bragg condition is fulfilled and that only moderate modulations are involved. One conclusion is that, for a hologram illuminated at the Bragg angle, there is, to a first approximation, no dependence on the thickness of the emulsion other than expressed by the $D$-$\log E$ curve. Experiments covered the regions of both thin and thick holograms by using emulsions of coating thicknesses of 4–15 μm and spatial frequencies of 125–2000 cyc./mm. The results verified the independence of the diffraction efficiency curves of the thickness of the layer. Since the maximum possible diffraction efficiency for thin amplitude holograms is 6.25 %, and 3.7 % for thick holograms (Chap. 1), the validity of (1.21) for both types of absorption holograms can only be an approximation for low modulations. Further work [2.122] included testing of two theories of thick gratings for the case of gratings in real photographic emulsions when a complex dielectric constant and nonlinearities have to be taken into account. Not-so-thick holograms, large modulations, and strong intermodulation speckle from diffuse objects continue to impose limitations on the applicability of either theory.

Another problem is how to define an MTF for a recording material of thickness much larger than the reciprocal of the spatial frequencies to be evaluated (e.g., by a factor 25 for 1500 cyc./mm in a layer of 17 μm). That (1.21) can be used for evaluating diffraction efficiency measurements allows a somewhat straightforward practical procedure, although technical problems arise from uneven exposure, input modulation, light scattering ("MTF in elementary layers"), and development [2.114] through the depth of the emulsion.

### 2.8.2 Materials Available; Problems

Although an emulsion thickness of at least 20 μm is desirable [2.86], the thickness around 15 μm of the 649-F plate and the 8E75 B plate seems to be the best compromise with respect to degradations from exposure decreasing with depth (leading to a variation in $\bar{E}$ and $K$ in Lippmann-Bragg holograms!), turbidity, mechanical distortions from swelling and drying. Experiments have been made with special coatings of 42 μm [2.114] and 34 μm thickness [2.123]. Experimental checks of the orientation dependence of diffraction efficiency [2.114, 124] showed that instead of the theoretical sinc$^2$-dependence of $\eta$ on reconstruction angle deviation (Chap. 1), a broader and smoother curve was obtained. This "apodization" of the sinc$^2$-curve indicates that the modulating properties of the layer are not uniform through the thickness of the layer, which, in the theory, described by a rect-function has the sinc$^2$-dependence of $\eta$ as a consequence.

Fixation of a photographic layer removes silver halides and causes shrinkage of the emulsion [2.114] by up to 15 % of the original thickness. A correspondingly shorter wavelength should be used in reconstruction, or is chosen automatically when Lippmann-Bragg holograms recorded with red He-Ne-laser light show green images upon white-light reconstruction. Also for ordinary, thick transmission holograms, the inclination of the interference patterns in the layer changes due to shrinkage, and this changes the Bragg angle and hence optimum reconstruction geometry [2.125]. For restoring the original emulsion thickness, soaking the gelatin in an aqueous solution of about 6 % triethanolamine (normally used as a sensitizer) has become a tradition [2.113, 119].

However, fixing also causes a decrease in bulk refractive index of the emulsion by about 0.1, which is not compensated for by reswelling [2.126].

There are cases where really plane holograms without any Bragg effects would be needed. One of these applications is spatial filtering, where the correlation signal should be independent of object position. However, due to the Bragg effect in usual "thin" emulsions, it is not, and the effect has to be taken into account and kept as low as possible by choosing emulsions as thin as possible and suitable geometry of the spatial filter arrangement [2.127, 128].

### 2.8.3 Holograms of Extended Thickness

By soaking with water, the gelatin of photographic emulsions can increase its thickness several times (Fig. 2.3). Holograms made with the emulsion swollen in a liquid gate show the typical properties of thick holograms, among others: high angular sensitivity, and suppression of the reconstruction from the intermodulation term [2.28]. Also, Lippmann-Bragg holograms with white-light reconstruction have been studied.

Swollen emulsions offer an excellent means for studying the properties of thick holograms [2.123, 129–131]. They have all the advantages of silver halide materials, as they are panchromatic, highly sensitive, and have the possibility of making amplitude and phase holograms. In addition, the thickness can be adjusted at will by the choice of the swelling solutions (e.g., to 400 μm from special Scientia 8E75 emulsions with a dry thickness of 34 μm), and in the same way the refractive index of the matrix can by varied by about 0.12 unit [cf. (2.16)]. The thickness can be stabilized by sealing the layer, or by forming a polymer in the extended layer. As an example, Fig. 2.24 shows the relative diffraction efficiency as a function of angular deviation from Bragg angle for two gratings made from Scientia 8E75 B plates (dry thickness 15 μm). The dashed curve was obtained from a grating stabilized at a thickness of 35 μm by inclusion of a polymer complex within the emulsion, the solid curve refers to a hologram of 110 μm thickness stabilized by overcoating the swollen emulsion with a polyester resin. These curves where measured three months after the stabilization treatment [2.129]. It was also shown that 30 different holograms of a three-dimensional object could be stored sequentially in a hologram of 30 μm thickness and read out individually [2.129]. An investigation of Bragg angle governed scattering by means of these thick holograms will be mentioned in the following section.

### 2.8.4 Experimental Data on Diffraction Efficiency and Signal-to-Background Ratio of Thick Holograms

Thick absorption transmission holograms have a maximum theoretical diffraction efficiency of 3.7%; in reflection, the theoretical maximum is 7.2% (Chap. 1). Since thick transmission and reflection phase holograms both promise 100%

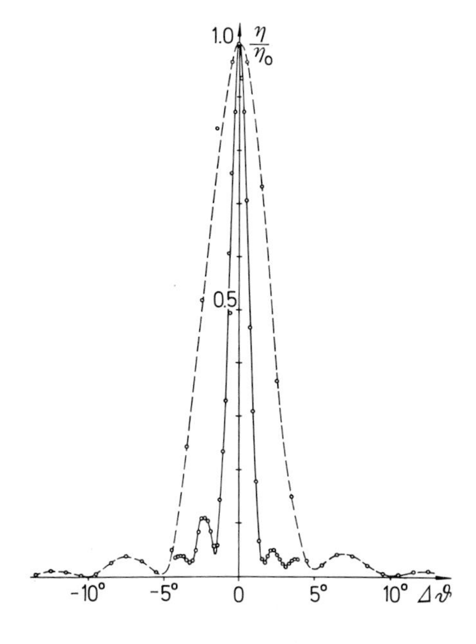


Fig. 2.24. Relative diffraction efficiency versus angular deviation of the reconstructing beam from Bragg angle incidence. Dashed curve: Thickness of the extended emulsion 35 μm. Solid curve: Thickness 110 μm [2.129]

efficiency, large efforts were devoted to improving phase hologram procedures. By bleaching photographic emulsions, a peak-to-peak phase difference $\Delta\varphi$ in sinusoidal exposures well above $\pi$ can be attained. A value of $\pi$ yields 100% diffraction efficiency in thick transmission phase gratings, with absorption and scattering in the bleached emulsions, the measured values for $\eta$ stay below 70%. In practice, the properties of transmission phase holograms in bleached emulsions are closer to those of thin holograms, and results for diffuse objects have been discussed in Sect. 2.7.3. Data on the efficiency of thick reflection phase holograms in bleached photographic emulsions are sparse, values of about 10% [2.132] and 20% [2.133] are reported for nondiffuse wavefronts. With dichromated gelatin, higher values have been obtained both for thick transmission and reflection phase holograms (Chap. 3). For bleached reflection holograms of a diffuse object, the numerical data reported [2.134] are a wavelength shift from 633nm to about 550 nm and a signal-to-background ratio of about 13:1.

Theoretically, thick phase media should be able to store several incoherently superimposed holograms and to reconstruct them with 100% diffraction efficiency [2.135]. However, so far, experimental comparisons between a single exposure and $N$ exposures have not given evidence of diffraction efficiency larger than the drastic $1/N^2$ decrease holding for absorption holograms.

Anticipations of intermodulation suppression and low cross-talk in multiply exposed thick holograms are confronted with the observation of scattering phenomena which are enhanced in those directions that fulfill the Bragg condition for the internally scattered light. Concentration of scattered light into

Fig. 2.25. Scattering phenomena in thick photographic emulsion exposed to two-beam interference [2.136]

cones is observed in very thick media like polymethylmetacrylate and lithium-niobate (Chap. 4). By virtue of the latent image, which excludes mutual modulation of wavefields recorded successively in the same volume, swelled photographic emulsions (Sect. 2.8.3) are particularly suited to a study of scattering effects in volume recordings [2.10, 131, 136]. Figure 2.25 shows an example for diffraction from a volume grating (in swollen photographic emulsion of about 400 μm thickness) that is nonlinear due to recording with $K=1$, monobath development, and bleaching, which also produced scattering silver compounds in the extended matrix. The incident beam does not fulfill the Bragg condition, it rather impinges perpendicularly to the grating structure. The symmetric pattern of scattering rings shows the directions into which the scattered light is concentrated by the Bragg condition. Of particular interest in this example is the appearence of the optical counterpart of Kossel lines together with indications of dark scattering rings, resulting from the presence of a second harmonic in the grating structure.

## 2.9 Summary

Photographic silver halide emulsions are well suited to the requirements of holography and well established as a hologram recording material. In their application to thin amplitude holograms their characteristics are thoroughly investigated and there is good agreement between theory and experiment. If stray light due to nonlinearity effects has to be kept low, the useful diffraction efficiency of both amplitude and phase holograms is as low as about 1% for diffuse wavefields. It is an open question whether there is a difference in the energy requirements of conventional photography and holographic recording in photographic emulsions. Our understanding of the factors determining performance of volume hologram recordings in photographic emulsions is still in its beginning.

## References


2.1 D. Gabor: Proc. Roy. Soc. (London) **A197**, 454 (1949)
2.2 C. E. K. Mees, T. H. James (eds.): *The Theory of the Photographic Process*, 3rd ed. (Macmillan, New York 1966)
2.3 H. Frieser, G. Haase, E. Klein (eds.): *Die Grundlagen der photographischen Prozesse mit Silberhalogeniden* (Akademische Verlagsgesellschaft, Frankfurt am Main 1968)
2.4 P. Kowalisky: *Applied Photographic Theory* (John Wiley and Sons, London, New York etc. 1972)
2.5 H. T. Buschmann: In *Optical and Acoustical Holography*, ed. by E. Camatini (Plenum Press, New York, London 1972) 151
2.6 G. R. Bird, R. C. Jones, A. E. Ames: Appl. Opt. **8**, 2389 (1969)
2.7 H. Nassenstein, H. Dedden, H. J. Metz, H. E. Rieck, D. Schultze: Phot. Sci. Eng. **13**, 194 (1969)
2.8 J. C. Dainty, R. Shaw: *Image Science* (Academic Press London, New York 1974)
2.9 H. Frieser: *Photographic Information Recording – Photographische Informationsaufzeichnung* (Focal Press London, New York – R. Oldenbourg Verlag München, Wien 1975)
2.10 K. Biedermann: Opt. Acta **22**, 103 (1975)
2.11 L.-E. Nilsson: Institute of Optical Research, Stockholm (private communication)
2.12 K. A. Stetson, K. Singh: Opt. Laser Techn. **3**, 104 (1971)
2.13 K. A. Stetson: In *Holographic Nondestructive Testing*, ed. by R. K. Erf (Academic Press New York, London 1974) Chap. 7, pp. 181—220
2.14 K. Biedermann, L. Ek: J. Phys. E: Sci. Instrum. **8**, 571 (1975)
L. Ek, K. Biedermann: J. Phys. E: Sci. Instrum. **8**, 691 (1975)
2.15 F. G. Kaspar, R. L. Lamberts: J. Opt. Soc. Am. **58**, 970 (1968)
F. G. Kaspar, R. L. Lamberts, C. D. Edgett: J. Opt. Soc. Am. **58**, 1289 (1968)
2.16 K. Biedermann: Appl. Opt. **10**, 584 (1971)
2.17 J. Upatnieks, C. Leonard: J. Opt. Soc. Am. **60**, 297 (1970)
2.18 W. S. Colburn, R. G. Zech, L. M. Ralston: "Holographic Optical Elements"; Technical Report AFAL-TR-72-409, Harris Electro-Optics Center of Radiation, Ann. Arbor, Mich. (1973)
2.19 K. Biedermann, K. A. Stetson: Phot. Sci. Eng. **13**, 361 (1969)
2.20 KODAK High Speed Holographic Film S0-253, Data Sheet; Eastman Kodak, Rochester. N.Y. (1975?)
2.21 D. B. Coblitz, J. A. Carney: Appl. Opt. **13**, 1994 (1974)
2.22 G. Haist: *Monobath Manual* (Morgan and Morgan, Inc., Hastings-on-Hudson and New York 1966)
2.23 S.-I. Ragnarsson: Phys. Scripta **2**, 145 (1970)
2.24 H. M. Smith, C. A. Callari, Jr.: Phot. Sci. Eng. **19**, 130 (1975)
2.25 G. P. Pal'tsev, K. A. Stozharova: Sov. J. Opt. Technol. **38**, 43 (1971)
2.26 H. T. Buschmann: Opt. Commun. **6**, 290 (1972)
2.27 J. C. Dainty, ed.: *Laser Speckles and Related Phenomena*, Topics in Applied Physics, Vol. 9 (Springer Berlin, Heidelberg, New York 1975)
2.28 K. Biedermann, N.-E. Molin: J. Phys. E: Sci. Instrum. **3**, 669 (1970)

2.29 K. Biedermann: Optik **28**, 160 (1968)
2.30 M. J. Landry, G. S. Phipps: Appl. Opt. **14**, 2260 (1975)
2.31 K. Biedermann: Opt. **31**, 367 (1970)
2.32 H. J Metz: Phot. Korr. **106**, 37, 56, 70 (1970)
2.33 H. T. Buschmann, H. J. Metz: Opt. Commun. **2**, 373 (1971)
2.34 M. G. Girina, G. A. Sobolev: Opt. Spectr. **32**, 111 (1972)
2.35 K. Biedermann, S. Johansson: Optik **35**, 391 (1972)
2.36 S. Johansson, K. Biedermann: Phot. Sci. Eng. **18**, 151 (1974)
2.37 A. A. Friesem, A. Kozma, G. F. Adams: Appl. Opt. **6**, 851 (1967)
2.38 A van der Lugt, R. H. Mitchel: J. Opt. Soc. Am. **57**, 372 (1967)
2.39 K. Biedermann, S. Johansson: J. Phys. E: Sci. Instrum. **8**, 751 (1975)
2.40 K. Biedermann, S. Johansson: Japan. J. Appl. Phys. **14**, (1975), Suppl. 14—1, p. 241
2.41 J. Ch. Viénot: Proc. Phys. Soc. (London) **72**, 661 (1958)
2.42 S. Johansson, K. Biedermann: Opt. Eng. **13**, 553 (1974)
2.43 S. Johansson, K. Biedermann: Appl. Opt. **13**, 2280 (1974); Appl. Opt. **13**, 2288 (1974)
2.44 R. L. Lamberts, F. C. Eisen, F. G. Kaspar: J. Opt. Soc. Am. **63**, 480A (1973)
2.45 J. W. Goodmann: J. Opt. Soc. Am. **57**, 493 (1967)
2.46 H. M. Smith: Appl. Opt. **11**, 26, 1869 (1972)
2.47 D. H. R. Vilkomerson: Appl. Opt. **9**, 2080 (1970)
2.48 C. D. Leonard, A. L. Smirl: Appl. Opt. **10**, 625 (1971)
2.49 A. Graube: Appl. Phys. Lett. **27**, 136 (1975)
2.50 D. Gabor, W. P. Goss: J. Opt. Soc. Am. **56**, 849 (1966)
2.51 D. Gabor: J. Phys. E: Sci. Instrum. **8**, 161 (1975)
2.52 J. W. Goodman, R. B. Miles, R. B. Kimball: J. Opt. Soc. Am. **58**, 609 (1968)
2.53 J. W. Goodman, G. R. Knight: J. Opt. Soc. Am. **58**, 1276 (1968);
G. R. Knight: Thesis, Stanford University 1967
2.54 O. Bryngdahl, A. Lohmann: J. Opt. Soc. Am. **58**, 1325 (1968)
2.55 J. M. J. Tokarski: Appl. Opt. **7**, 989 (1968); Appl. Opt. **8**, 1510 (1969)
2.56 A. A. Friesem, J. S. Zelenka: Appl. Opt. **6**, 1755 (1967)
2.57 A. Kozma: Opt. Acta **15**, 527 (1968)
2.58 F. T. S. Yu: J. Opt. Soc. Am. **59**, 360 (1969)
2.59 A. Kozma, G. W. Jull, K. O. Hill: Appl. Opt. **9**, 721 (1970)
2.60 K. O. Hill, G. W. Jull: Opt. Acta **18**, 729 (1971)
2.61 C. H. F. Velzel: Opt. Commun. **3**, 133 (1971)
2.62 C. H. F. Velzel: Opt. Acta **20**, 585 (1973)
2.63 F. J. Tischer: Opt. **30**, 488 (1970); Appl. Opt. **9**, 1369 (1970)
2.64 A. Kozma: J. Opt. Soc. Am. **56**, 428 (1966)
2.65 H. Dammann: J. Opt. Soc. Am. **60**, 1635 (1970)
2.66 A. Vander Lugt, F. B. Rotz: Appl. Opt. **9**, 215 (1970)
2.67 L. H. Lin: J. Opt. Soc. Am. **61**, 203 (1971)
2.68 J. Upatnieks, C. D. Leonard: Appl. Opt. **10**, 2365 (1971)
2.69 K. Biedermann, S. Johansson: J. Opt. Soc. Am. **64**, 862 (1974)
2.70 H. Hannes: Optik **26**, 363 (1967)
2.71 H. M. Smith: J. Opt. Soc. Am. **58**, 533 (1968)
2.72 R. L. Lamberts: J. Opt. Soc. Am. **60**, 1389 (1970)
2.73 R. L. Lamberts: Appl. Opt. **9**, 1345 (1970)
2.74 R. L. Lamberts: Appl. Opt. **11**, 33 (1972)
2.75 H. M. Smith: J. Opt. Soc. Am. **59**, 1492 (1969)
2.76 H. T. Buschmann: Phot. Sci. Eng. **16**, 425 (1972)
2.77 R. L. Lamberts, F. C. Eisen, F. G. Kaspar: J. Opt. Soc. Am. **63**, 480A (1973)
2.78 A. A. Friesem, J. L. Walker: Appl. Opt. **8**, 1504 (1969)
2.79 K. S. Pennington, J. S. Harper: Appl. Opt. **9**, 1643 (1970)
2.80 M. Lang, G. Goldmann, P. Graf: Appl. Opt. **10**, 168 (1971)
2.81 M. Lang: Opt. Commun. **3**, 141 (1971)

2.82 B. Hill: Appl. Opt. **11**, 182 (1972)
2.83 G. Goldmann: Optik **34**, 254, 312 (1971)
2.84 A. Bardos: Appl. Opt. **13**, 832 (1974)
2.85 W.-H. Lee, M.O. Greer: J. Opt. Soc. Am. **61**, 402 (1971)
2.86 M. Chang, N. George: Appl. Opt. **9**, 713 (1970);
M. Chang: Thesis. California Institute of Technology 1969
2.87 R. L. van Renesse, N. J. van der Zwaal: Opt. Laser Technol. **3**, 41 (1971)
2.88 R. L. van Renesse, F. A. J. Bouts: Optik **38**, 156 (1973)
2.89 R. L. van Renesse: Appl. Opt. **14**, 1763 (1975)
2.90 J. H. Altmann: Appl. Opt. **5**, 1689 (1966)
2.91 V. Russo, S. Sottini: Appl. Opt. **7**, 202 (1968)
2.92 K. S. Pennington, J. S. Harper, F. P. Laming: Appl. Phys. Lett. **18**, 80 (1971)
2.93 G. Liebmann, F. Schmidt, A. Storch: J. Signalaufzeichnungsmat. **6**, 471 (1975)
2.94 J. N. Latta: Appl. Opt. **7**, 2409 (1968)
2.95 J. Upatnieks, C. Leonard: Appl. Opt. **8**, 85 (1969)
2.96 C. B. Burckhardt, E. T. Doherty: Appl. Opt. **8**, 2479 (1969)
2.97 K. S. Pennington, J. S. Harper: Appl. Opt. **9**, 1643 (1970)
2.98 P. Barlai: Z. Naturforsch. **27a**, 544 (1972)
2.99 D. H. McMahon, W. T. Maloney: Appl. Opt. **9**, 1363 (1970)
2.100 M. Lehmann, J. P. Lauer, J. W. Goodman: Appl. Opt. **9**, 1948 (1970)
2.101 F. P. Laming, S. L. Levine, G. Sincerbox: Appl. Opt. **10**, 1181 (1971)
2.102 S. L. Norman: Appl. Opt. **11**, 1234 (1972)
2.103 T. Inagaki, J. Nakajima, Y. Nishimura: Fujitsu Sci. Techn. J. 135 (1974)
2.104 N. Nishida: Appl. Opt. **13**, 2769 (1974)
2.105 H.-G. Weide, E. Lau, S. Dähne: Z. Wiss. Phot., Photophysik Photochem. **61**, 145 (1967)
2.106 H. Thiry: Appl. Opt. **11**, 1652 (1972)
2.107 A. Graube: Appl. Opt. **13**, 2942 (1974)
2.108 R. L. Lamberts, C. N. Kurtz: Appl. Opt. **10**, 1342 (1971)
2.109 H. T. Buschmann: Optik **34**, 240 (1971)
2.110 P. Hariharan, C. S. Ramanathan: Appl. Opt. **10**, 2197 (1971)
2.111 C. Clausen, H. Dammann: Opt. Commun. **2**, 263 (1970)
2.112 G. Lippmann: Compt. Rend. Séanc. Acad. Sci. Paris **112**, 274 (1891);—**114**, 124 (1892);—**115**, 575 (1892);—**118**, 92 (1894)
2.113 Yu. N. Denisyuk: Opt. Spectrosc. **14**, 279 (1963)
Yu. N. Denisyuk, I. R. Protas: Opt. Spect. **14**, 381 (1963)
2.114 E. N. Leith, A. Kozma, J. Upatnieks, J. Marks, N. Massey: Appl. Opt. **5**, 1303 (1966)
2.115 V. V. Aristov, V. Sh. Shektman: Soviet Phys. Uspekhi **14**, 263 (1971)
2.116 M. R. B. Forshaw: Opt. Laser Technol. **6**, 28 (1974)
2.117 L. H. Lin, K. S. Pennington, G. W. Stroke, A. E. Labeyrie: Bell. Syst. Tech. J. **45**, 659 (1966)
2.118 J. Upatnieks, J. Marks, R. J. Fedorowicz: Appl. Phys. Lett. **8**, 286 (1966)
2.119 L. H. Lin, C. V. LoBianco: Appl. Opt. **6**, 1255 (1967)
2.120 A. A. Friesem, J. L. Walker: Appl. Opt. **9**, 201 (1970)
2.121 H. M. Smith: J. Opt. Soc. Am. **62**, 802 (1972)
2.122 F. G. Kaspar: J. Opt. Soc. Am. **63**, 37 (1973)
2.123 S.-I. Ragnarsson: Opt. Commun. **14**, 39 (1975)
2.124 Y. Belvaux: Nouv. Rev. Optique **6**, 137 (1975)
2.125 D. H. R. Vilkomerson, D. Bostwick: Appl. Opt. **6**, 1270 (1967)
2.126 O. Bryngdahl: Appl. Opt. **11**, 195 (1972)
2.127 N. Douklias, J. Shamir: Appl. Opt. **12**, 364 (1973)
2.128 N. Brousseau, H. H. Arsenault: Appl. Opt. **14**, 1679 (1975)
2.129 K. Biedermann, S.-I. Ragnarsson, P. Komlos: Opt. Commun. **6**, 205 (1972)
2.130 S.-I. Ragnarsson, P. Komlos, K. Biedermann: Preparation of Volume Holograms by Increasing the Thickness of Photographic Silver Halide Emulsions. Technical Report 1973-08-03, Institute of Optical Research, S-100 44 Stockholm, Sweden

2.131 M. R. B. Forshaw: Opt. Commun. **8**, 201 (1973)
2.132 H. Kiemle, W. Kreiner: Phys. Lett. **28A**, 425 (1968)
2.133 G. D. Mintz, D. K. Morland, W. M. Boerner: Appl. Opt. **14**, 564 (1975)
2.134 P. Hariharan: Opt. Commun. **6**, 377 (1972)
2.135 E. G. Ramberg: RCA Rev. **33**, 5 (1972)
2.136 S.-I. Ragnarsson: Appl. Opt. **16**, (in printing)
2.137 K. A. Stetson: private communication

# 3. Dichromated Gelatin

D. Meyerhofer

With 5 Figures

Dichromated gelatin is an important holographic material, because it possesses almost ideal properties for phase holograms. When properly used, it forms a clear film that exhibits very little absorption and optical scattering. Furthermore, a large index change can be produced in the interior of the gelatin, making possible thick phase holograms of close to 100% efficiency. Both transmission and reflection holograms of this kind can be generated.

Dichromated gelatin and other dichromated colloids are among the oldest photographic materials. They record information either as a variation of the index of refraction or as a thickness variation, or as a combination of the two. Thus, these materials may be used to record phase holograms. Conventional processing, in the photographic or printing industry, emphasizes the thickness variation. Similar techniques can be used for surface relief holograms, but there are no special advantages in using dichromated gelatin in this manner.

The significance of dichromated gelatin as a holographic material is in use as thick phase holograms produced by modulation of the refractive index. This use depends on an effect discovered by *Shankoff* [3.1]. He used hardened gelatin layers as the starting material so that there would be no, or only small, variation in thickness. Instead of developing the exposed layer in a single liquid, usually water, the layer is sequentially treated in two different liquids, the first one being of small molecular size and a solvent for gelatin such as water, the second one of larger size such as isopropyl alcohol. The second step causes a very rapid drying of the material of the first liquid. The stresses induced cause the gelatin to crack or form voids in regions which are softer than the average, due to the lack of optical exposure. The cracks and voids produce large differences in phase shifts between exposed and unexposed areas. These can be 100 times larger than phase shifts produced by water wash alone. In this way, holograms can be produced which exhibit high efficiency and also very low noise, a very desirable combination.

The main reasons that dichromated gelatin holograms have not been widely used, despite their high promise, are the difficulty of obtaining reproducible results and problems related to the distortion of the photosensitive layer from exposure to developed image. The problems can be overcome, but considerable care is required.

This chapter discusses the photochemical processes taking place in the dichromated gelatin layer and how they are used for photographic and holographic purposes. The use of this material as a resist will be briefly described.

This is followed by a detailed consideration of the *Shankoff* technique, based on hardened gelatin layers. After the techniques for making and processing thick phase holograms are listed, the holographic parameters are discussed. Finally, there is a consideration of some of the applications that dichromated gelatin holograms have been used for.

## 3.1 Basic Photochemistry

### 3.1.1 Dichromated Colloid Layers

It has been known since the early 1800's that the addition of water-soluble dichromates to certain biological organic colloids causes these materials to become photosensitive [3.2]. The areas exposed to light are hardened and become less soluble than the unexposed areas. This differential solubility can be converted into a variation of the thickness or the density of the material by washing it in water. The resulting phase images can be made visible by dying the layer. Amplitude images have also been obtained through the chemical reaction of dichromate ions with a supporting layer. Many variations of this process were used in early photographic experiments. They are considerably less sensitive than silver halide emulsions, but are attractive because of their very high quality of tone rendition.

The most important way of dying these layers has been by dispersing carbon particles in the colloid. The particles are retained in the hardened part of the layers and are washed away in other areas during development. The process was optimized by transferring the gelatin layer to a second substrate after exposure so development would proceed from the softer less exposed side. This formed the basis for the "carbon tissue" process used extensively in the printing industry for over 100 years.

Many different kinds of colloids have been used to form photosensitive layers: gelatin, albumen, fish glue, polyvinyl alcohol, etc. They have been sensitized with ammonium, sodium, and potassium chromates and dichromates. The most attractive combination of these materials is ammonium dichromate and gelatin. One of the reasons is the high solubility of $(NH_4)_2Cr_2O_7$ in water. This permits a high concentration to be introduced into the colloid without crystallization taking place upon drying of the coating. Regions in which the dichromate has crystallized cause severe optical scattering in a hologram, even though the crystallites have been washed away in the development process.

The gelatin has the practical advantage that it can absorb large quantities of liquids, and that the liquids can penetrate it rapidly, thus facilitating all processing steps. Also, it can be produced in uniform sheets of good optical quality, and its ability to gel makes film formation easy. The gel can be hardened to various degrees by chemical cross linking to change the mechanical properties. Well hardened, it remains insoluble to high temperatures.

### 3.1.2 The Properties of Gelatin

Gelatin is derived from the natural protein collagen which is obtained from animal sources [3.3, 4]. In a lengthy chemical procedure, the collagen molecules are broken into irregularly shaped polypeptide chains of amino acids which constitute the gelatin. The chemical structure may be represented as

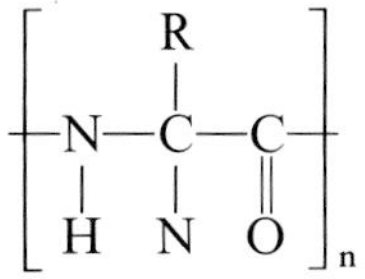

$$\left[ \begin{array}{ccccc} & & \mathrm{R} & & \\ & & | & & \\ -\mathrm{N} & - & \mathrm{C} & - & \mathrm{C}- \\ | & & | & & \| \\ \mathrm{H} & & \mathrm{N} & & \mathrm{O} \end{array} \right]_n$$

where R represents one of many different amino acid residues. The amino acids are ordered in such a way that there are both polar and nonpolar regions in each gelatin chain. Because of the natural origin, the chemical composition and structure of the gelatin varies from one sample to another.

When the gelatin is first formed, it is in aqueous solution and the molecules exist as single chains surrounded by water molecules. Upon standing at temperatures below 30° C, solutions of more than 1% gelatin become rigid and exhibit rubber-like mechanical properties. This gel phase consists of a three-dimensional network of gelatin molecules which are partially aligned and loosely bound together at the nonpolar parts of the chains. The bound sections are considered to be polymeric crystallites and consist of triple helices of the nonpolar parts connected by hydrogen bonds. Gelation occurs more rapidly at higher gelatin concentrations and at lower temperatures. It does not depend much on the pH of the solution. The final degree of binding depends strongly on the origin of the material.

Because the bonding of the gel is relatively weak, it can be reversed by heating. Above a well-defined melting point, the gel turns into a solution. The melting point depends on the microstructure and, therefore, on the previous history of the gel. The melting point is typically 10° higher than the gelation point.

Both in solution and in the gel, the gelatin can take up large quantities of water. At the lower concentrations, the water molecules are hydrogen-bonded to the polar parts of the gelatin molecule. As the humidity increases, more and more water is incorporated until a layer of water is sandwiched into each space between the protein layers. The nonpolar helix structures expand and water can be absorbed between the helices. If the gel is placed in water, it swells up even further until the gelatin dissolves.

At low relative humidity, thin layers of both solutions and gels dry to clear films. The two kinds of films are very different, the former being amorphous, the latter, crystalline. This leads to significant differences in the mechanical properties. The crystalline films are considerably stronger and less brittle.

Gelatin can be hardened and rendered insoluble by a process similar to the tanning of leather. This can be caused by heat or by ultraviolet radiation, in which case it must involve the breaking of some bonds and their rearrangement into a more stable three-dimensional network. More important is the hardening by chemicals. They produce cross linking of the polar parts of different gelatin molecules by strong chemical bonds [3.4]. For example, very small quantities of formaldehyde can render the gelatin insoluble. The photographic process depends on photochemical hardening. The chromic ion $Cr^{3+}$ forms a coordinated complex with the carboxylate group $-COO^-$. This complex constitutes the cross link between the gelatin chains. Chemical hardening becomes more effective and rapid if the gelatin molecules are closer to each other, e.g., in dried layers, or in high-concentration solutions.

As the hardening is increased, the rigidity of the gel increases, the solubility decreases, and the melting point rises. The swelling of the layer and the amount of water it can take up also decrease. For any given application, the amount of hardening must be controlled to produce optimum results.

### 3.1.3 Photosensitivity of the Dichromate Ion

Despite the widespread use of chromates in photochemical reactions, the detailed processes are not well understood and a number of different mechanisms have been proposed [3.2]. However, the general outline is reasonably clear. In sensitizing the gelatin by dissolving $(NH_4)_2Cr_2O_7$, or some other dichromate, in it, the chromium ion is incorporated as $Cr^{6+}$. The ion is deposited in the gelatin from water solution and does not react directly with the gelatin.

The absorption of light by hexavalent chromium ion in the presence of the oxidizable gelatin and some water initiates the photochemical process. This was demonstrated by *O'Brian* [3.5] who showed that the sensitivity of dichromated resins used in photoengraving has the same spectral response as the absorption curve of dissolved ammonium dichromate. He was able to obtain good quantitative agreement between the number of absorbed quanta and the threshold for hardening as function of wavelength.

After the absorption of a light quantum, the Cr ion is reduced to some lower ionization state. A number of processes have been proposed leading to various ionization states between 3+ and 6+. Most likely, the processes are different in different colloids [3.2]. Some of the intermediate reactions require the presence of water and gelatin. In any case, the end product is chromium oxide $(Cr_2O_3)$ with the trivalent Cr ion. It reacts with the polar part of the gelatin and causes the cross linking of the gelatin molecules, as discussed in the previous subsection. Since all the chemical reactions take place in the interior of the gelatin layer, the $Cr^{3+}$ is formed in the vicinity of the polar segments with which it must react. Thus, this hardening process can be more effective than the hardening by chemicals introduced from the outside, which must first penetrate to the locations where the reaction can take place.

Because the photochemical process involves, at most, a few dichromate ions and the adjacent gelatin molecules, it exhibits a very high spatial resolution (spatial periods much smaller than a light wavelength). This resolution may be degraded in subsequent processing steps which provide an amplification of the recorded information.

## 3.2 Dichromated Gelatin as a Photosensitive Medium

### 3.2.1 Dichromated Gelatin as a Photoresist

In photographic and printing applications, dichromated gelatin has generally been used as a negative-working photoresist. This means that the areas exposed to light form a "resist" which prevents the substrate from being attacked by a subsequent processing step. The nonexposed areas either are washed away or transmit the etchant. This procedure can be used to produce holograms in the form of relief images and, while it is not widely used, it is included in this chapter to complete the description of dichromated gelatin.

For use as a resist, the gelatin layer is formed on the substrate and is, at most, lightly hardened prior to use. The exposed regions then become cross linked and relatively insoluble. Development takes place in warm water where the unexposed regions are fully or partially washed away. When the substrate with the gelatin layer of variable thickness is immersed in an etching bath, the transmission rate of the etchant through the gelatin layer varies with thickness and, consequently, the etch rate of the substrate varies in the same fashion. Thus, continuous tone images are produced in the substrate, a feature that is taken advantage of in gravure printing. This is in contrast to other kinds of resists which form a complete barrier to the etchant so that etching takes place only where the resist is removed entirely. This latter procedure is restricted to forming half-tone images.

Relief phase holograms can be formed in unhardened layers of dichromated gelatin [3.6] in the same way as in other resist layers that depend on decreasing the solubility by photopolymerization. The variation in the thickness of the layer after development produces the main contribution to the optical phase shift on readout. There is also a change in refractive index with exposure, but this will make a much smaller contribution to the diffraction. Holograms formed in this way can be duplicated in the usual manner by replicating the thickness profile [3.7].

Unhardened dichromated gelatin layers may be produced by drawing a substrate out of a gelatin solution, or by doctor-blading the solution onto the substrate. The sensitizing material, ammonium dichromate, is dissolved in the solution. To obtain a high sensitivity, the concentration should be as high as possible. It is limited to about 20% by weight of the dried gelatin, if the dichromate is not to crystallize out. The layers are air dried in the dark before exposure.

The properties of thin unhardened gelatin layers were investigated by *Meyerhofer* [3.6]. Sinusoidal gratings were exposed with a He–Cd laser

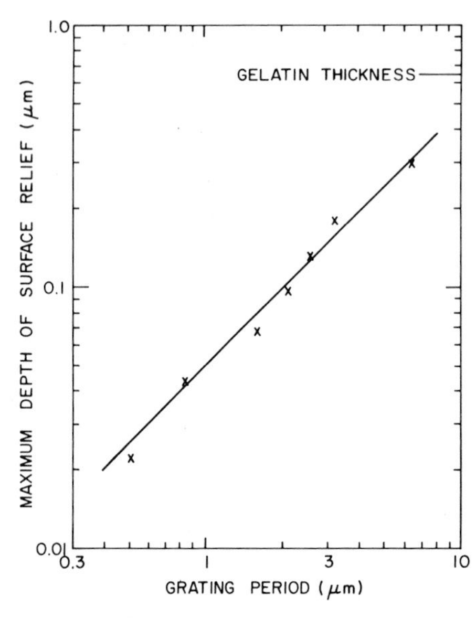


Fig. 3.1. Maximum depth of modulation to which a surface relief grating can be developed in dichromated gelatin [3.6]

(441 nm). The diffraction efficiency was measured with a He–Ne laser before development. The values were very small, but independent of grating frequency, as expected.

When an exposed plate was developed in water to form a relief image, the response was strongly frequency dependent. Figure 3.1 shows the maximum relief between peak and valley of the sinusoidal grating that could be produced at any exposure. In the range shown, this value is proportional to the grating period, which means that the shape of the grating under peak conditions is constant, or that the slope of the relief surfaces is limited to a maximum angle at all periods. For Fig. 3.1, the maximum angle is found to be 9°. These observations are similar to what has been observed in other negative-working resists. They suggest that the softness of the gelatin causes it to flow during the development step when it is strongly swollen. Another explanation is that the chemical cross linking spreads out by migration of the $Cr^{3+}$, e.g., this would be similar to effects observed in tanning of gelatin [3.3]. However, all observations on hardened layers that will be discussed in the following sections lead to the conclusion that the chemical reactions remain localized.

### 3.2.2 Photochemistry of Prehardened Dichromated Gelatin Layers

In thick, or Bragg-type, holograms, the phase changes must take place in the interior of the gelatin layers. In order that the gelatin is not washed away during sensitizing or development, it must be prehardened and rendered insoluble. The hardening can take place in any of the conventional ways (Sect. 3.1.2). The degree of hardening will determine the subsequent characteristics of the layer. If the gelatin is hardened too much, it will no longer deform and become insensitive.

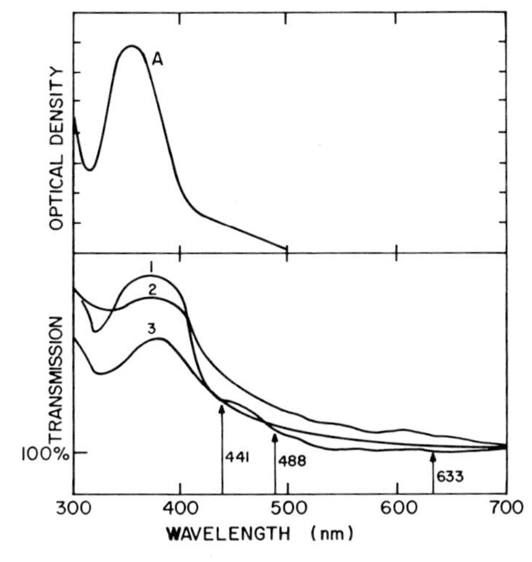


Fig. 3.2. Absorption spectra of dichromated gelatin [3.1] compared to the absorption of ammonium dichromate solution (Curve *A*) [3.5]. Curve 1 is the spectrum before exposure, Curve 2 after exposure, and Curve 3 after development. Note that the two scales are different

The hardened gelatin layer is sensitized by soaking in dichromate solution. For obtaining maximum sensitivity, ammonium dichromate solutions of 5% by weight are typically used. The hardened gelatin in the solution swells up and absorbs a large volume of water, about five or six times its own volume. Next, the layer is dried in air. The water evaporates and the dichromate remains distributed throughout the gelatin. This leads to dichromate concentrations as high as 25 to 30%.

The absorption spectrum of a dichromated gelatin layer, as measured by *Shankoff* [3.1] is shown by Curve *1* in Fig. 3.2. For comparison, Curve A is the absorption of ammonium dichromate in solution [3.5]. As in the case of other colloids [3.5], there is close correspondence between the two curves. This confirms that the absorption of light by the $Cr^{6+}$ ion initiates the hardening process.

During the exposure, some of the Cr is reduced and this changes the absorption spectrum somewhat (Fig. 3.2, Curve *2*). This, in turn, is reflected in a small change in the index-of-refraction, so that a phase hologram is recorded. *Shankoff* [3.1] has measured the index of dichromated gelatin under a number of different conditions. His results are listed in Table 3.1. Here, exposure was

Table 3.1. Refractive Indices at 589 nm under various conditions [3.1]

| Material | Refractive Index |
|---|---|
| *1*) Gelatin | 1.5426 |
| *2*) Gelatin + $(NH_4)_2Cr_2O_7$ | 1.5486 |
| *3*) Gelatin + $(NH_4)_2Cr_2O_7$, exposed | 1.5572 |
| *4*) Gelatin + $(NH_4)_2Cr_2O_7$, exposed, developed in isopropyl alcohol | 1.5515 |
| *5*) Gelatin + $(NH_4)_2Cr_2O_7$, unexposed, developed in isopropyl alcohol | 1.5488 |

uniform and not in the form of holograms. The change of index due to exposure alone is 0.009.

To determine the index change under holographic recording conditions, it is simplest to record a two-dimensional sinusoidal grating with two symmetric laser beams. For linear recording, the variable part of the index is then given as $n=n_0+n_1\cos Kx$. If the grating frequency $K$ is made high enough so that the recording falls in the thick hologram limit, then the first-order diffraction efficiency is given by (1.51)

$$\eta=\sin^2(\pi n_1 d/\lambda\cos\theta_0), \tag{1.51}$$

where $d$ is the thickness of the layer, $\lambda$ the vacuum wavelength, and $\theta_0$ the angle of incidence. Typical grating diffraction efficiencies obtained by long exposures of gelatin layers of the order of 5 µm thick are $10^{-3}-10^{-2}$. From (1.51), a maximum index variation of $2n_1=(2-6)\times10^{-3}$ is derived, which is consistent with the value given above.

The next step in processing an exposed plate is to wash it in water. This removes all the unreacted $Cr^{6+}$ ions, as well as the intermediate reaction products, and makes the layer insensitive to light. After a washed layer is dried, there is usually an increase in hologram efficiency to a few percent. This is probably due to differential retention of water in the gelatin with higher concentrations in the unexposed regions [3.8]. However, the effect is not large enough to make this a practical holographic recording process.

### 3.2.3 The Shankoff Technique of Phase Shift Enhancement

*Shankoff* [3.1] discovered that the optically induced phase shift of an exposed and developed hardened dichromated gelatin plate could be vastly increased by changing the development process. The revised process is described by: "all films are mildly agitated in water maintained at 35 °C for 30 s, dipped in isopropanol for 30 s, and air-dried with a stream of dry air under pressure" [3.1]. The important points are that the gelatin is still saturated with water and fully swollen at the time it is immersed in the isopropyl alcohol, and that the alcohol rapidly replaces the water in the layer by mixing with it. In effect, it dries the layer extremely rapidly because the isopropyl alcohol does not dissolve the gelatin and so does not wet it.

The same effect can be obtained with other solvents which do not dissolve gelatin, but are very miscible with water, such as acetone, ethanol, and butanol. Similarly, other solvents can be used for the wash, as long as they dissolve gelatin, swell it, and wash out the unreacted chemicals. Glacial acedic acid and ethylene glycol are among the solvents used [3.9].

With this technique, *Shankoff* produced 90% efficient sinusoidal holograms in 1.3 µm-thick layers of dichromated gelatin. If (1.51) applies, this means that the peak index variation is $2n_1=0.26$, which is a remarkably large number. He further observed that, if the developed hologram was reimmersed in water, the large index change would disappear and the hologram efficiency returned to a small value. In fact, the entire process is reversible if the wet layer is again dried in isopropyl alcohol. This cycle can be repeated a number of times before the hologram gradually degrades. It indicates that, within limits, the photochemical cross linking that is produced during exposure remains intact during subsequent steps and that it controls whatever process occurs during the drying step.

This is also consistent with the observation that both the low-efficiency and the high-efficiency holograms have very high frequency response. Gratings with periods smaller than 200 nm can be recorded in both the water-washed and the dried state, with efficiencies comparable to those obtained at longer periods.

These observations raise the question of the origin of the large phase shift or change in index. Consider first a uniformly exposed layer of gelatin. Curve *3* of Fig. 3.1 shows that the absorption spectrum of a layer developed by this technique does not differ significantly from the spectrum of the exposed layer before development. Similarly, the index-of-refraction at 589 nm changes by less than 0.006 during development, and the difference between exposed and unexposed layers is about half that large (Table 3.1). Because most of the $Cr^{6+}$ is removed during washing, this change must be due to $Cr^{3+}$. It appears that isopropyl alcohol molecules are bound to reduced chromium sites where the gelatin is cross linked. When the absorption spectra of exposed layers developed with isopropyl alcohol are compared with those of layers developed in water only, the former exhibit an increase in absorption between 340 and 410 [3.10]. This peak must be due to the $Cr^{3+}$-gelatin-isopropyl alcohol complex. However, the index change related to this absorption is much too small to account for the effective index change that is observed.

*Curran* and *Shankoff* [3.8] investigated the development process in great detail in order to try to explain the large refractive index changes. It is quite clear that there must be some sort of voids in the interior of the developed gelatin to account for the observed effects. *Curran* and *Shankoff* suggest that these voids are cracks between the Bragg planes, i.e., the planes of high exposure or great hardness. The dehydration of the gelatin by isopropyl alcohol causes a rapid shrinkage of the gelatin. It sets up large stresses in the material which are relieved by cracking at the weakest points—those that have not been cross linked. The authors have observed such splitting in microscopic observations. By producing simple reflective holograms with Bragg planes in the plane of the gelatin layer, they could observe under certain conditions what appeared to be an onion-peel effect of these 200 nm-thick layers breaking apart. They also made extensive measurements with index-matching fluids which showed that, if the fluid can be introduced into the gelatin, then a fluid of the appropriate index completely wipes out the hologram. They interpret this as meaning that the fluid fills up the cracks. A curious part of this observation is that the fluid can be introduced from the top of the gelatin layer only if the gelatin is first soaked with methanol. If the gelatin is separated from the substrate, the index-matching fluid extinguishes the hologram by itself. This suggests that the cracks do not extend to the top surface.

There are a number of problems with this explanation. As has been mentioned, holograms formed by this technique can be made with very high quality and low noise. Cracks the size of a wavelength or longer would, however, cause a great deal of scattering and noise. Furthermore, high efficiencies are observed as well in diffuse holograms, where there are no clear-cut Bragg planes, but a continuous variation of phase shift. Any cracks in such layers would have to be qualitatively different. Secondly, *Curran* and *Shankoff* have shown by band pass measurements that, for both high and low exposures, it is possible to produce gratings which are equal in depth to the film thickness. This means that, if there are cracks, they would have to be very narrow at low exposure and increase in width as the exposure is increased. But, at low exposure, the stresses must be small and one would not expect a complete crack to form. Finally, the experiments with index-matching fluids led to the conclusion that the front surface of the gelatin is defect-free and impenetrable to the liquid, while their model for gratings with planes perpendicular to the layer suggests that the cracks should start at the surface and penetrate into the layer.

It, therefore, seems likely that the voids produced in development have a somewhat different character. To explain the excellent optical properties, we must assume that the voids are much smaller than an optical wavelength. This would also explain the efficiency of diffuse holograms where phase variations are more localized. It would not contradict the ability of a grating hologram to split into Bragg planes when further stressed. The concentration of voids would be high throughout the plane of low exposure, allowing cracks to propagate. Small voids do not make the experiments with index-matching fluid any easier to explain.

Assuming that the voids are indeed very small, we can calculate the volume that they must occupy. If the index of the gelatin in this region is $n_g$ and it occupies a fractional volume $v$, then the effective index of the layer $n$ is given by [3.11]

$$(n^2-1)/(n^2+2)=v(n_g^2-1)/(n_g^2+2)\,. \qquad (3.1)$$

This formula has been found to apply well for gases and mixtures of liquids and should, therefore, represent this case as well. If $n_g=1.53$ and the peak index variation is 0.26, then $n=1.27$ and (3.1) gives $v=0.55$. When this is averaged over all the gelatin, this means that over 20% of the volume is empty. In thicker gelatin layers where the maximum index variation obtainable is somewhat smaller, this can still amount to voids taking up 10% of the volume. It appears to be related to the observation of *Curran* and *Shankoff* [3.8] that 7 to 8% of the gelatin is removed during development.

When developed holograms are immersed in water, the gelatin swells up and the voids disappear. If the drying takes place in air, the water is removed relatively slowly and the gelatin has a chance to relax the stresses set up by the difference between hardened and unhardened regions. No voids are formed and the hologram remains at low efficiency. Inserting such a layer in isopropyl alcohol has no effect, since the alcohol does not dissolve or swell up the gelatin. Only when the gelatin is saturated with water does the rapid drying produce the high stress condition needed.

It might be expected that a process which depends on such a complicated process as the rapid dehydration and formation of voids would not be very linear. In fact, it will be seen later that the index change is proportional to the exposure at low exposures. The index-of-refraction changes linearly with the volume of voids for small changes, as can be seen from (3.1), and, therefore, the volume of voids must be proportional to exposure for any given development condition.

### 3.2.4 Variation in Emulsion Thickness During Processing

One of the characteristics of gelatin is that its volume can change by a large amount as it absorbs and desorbs water. When a gelatin layer is coated on a glass surface, the lateral dimensions are fixed and the layer thickness is the parameter that changes. If the moisture content is changed too rapidly, the gelatin may not change uniformly and the layer may become distorted. This condition must, of course, be avoided if the optical quality is to be retained. In case of hardened layers, the gelatin molecules are tied together into a continuous three-dimensional network. During swelling and shrinking, this network is not changed and so the basic photochemical information recorded in the gelatin is not destroyed during development in water, and remains in the volume independent of the thickness of the layer. This is the reason why faded holograms can be reconstituted. However, the changes in thickness do affect the properties of holograms recorded in the volume of the gelatin, especially thick three-dimensional holograms [3.12]. It causes planes of constant index change in the emulsion to change their angle of tilt. For a simple holographic grating, therefore, both the angle of the readout beam and the image beam must be changed to obtain maximum diffraction. For a picture hologram, which can be thought of as a superposition of many gratings, it is clear that each grating will have a different angular change resulting in reduced diffraction efficiency and/or distortion of the image beam. A useful holographic material must, therefore, have the thickness of the fully developed emulsion approximately equal to that during the exposure.

This property of holograms can be used to determine the thickness of the gelatin layer at various stages during the processing [3.13]. Simple gratings are recorded in the gelatin such that the Bragg plane forms an acute angle with the plane of the gelatin. Then the angle of maximum diffraction efficiency on readout determines the thickness [3.12]. The results obtained in this way by *Meyerhofer* [3.13] are listed in Table 3.2. The thickness of the emulsion during the first exposure is normalized to 1.0. After the first complete development, the layer thickness has increased by 30 to 50%. This value is larger than the 6% observed by *Lin* [3.14] for reflection holograms, probably due to the different recording geometry. To check the permanence of these changes, we sensitized the

Table 3.2. Relative thickness of the gelatin layer at various stages of the processing of dichromated gelatin [3.13]

| | |
|---|---|
| During exposure | 1.0 |
| After developing | 1.3–1.4 |
| During resensitizing | 3.0 |
| After resensitizing (dry) | 0.98 |
| After renewed developing | 1.3 |
| In the wash | 4.1 |
| Dry | 1.01 |
| Redevelop | 1.3 |
| After heating | 1.0–1.3 |

plate for the second time and found the thickness at the exposure time to be the same as the first time. The same was true after another development. It is known that, if the developed plate is immersed in water and then dried, most of the hologram disappears. At the same time, the thickness returns to the value after sensitizing. This demonstrates that it is the water wash, followed by drying, that determines the original thickness, not the dichromate (this, despite the high concentration of dichromate which can be 30% by weight). The table also shows that the thickness of the gelatin during immersion in water increases by a factor of 3 to 4.

It should be noted that the diffraction efficiencies varied greatly during these measurements, in agreement with previous discussion. It was always possible, however, to measure the angles of maximum diffraction and so determine the layer thickness.

The last entry in Table 3.2 shows a very interesting result. If the developed layer is baked at temperatures of about 150 °C, the thickness gradually reduces to the relative value 1, or even further. Often, this reduction takes place without any change in efficiency. This means that the large variations in refraction index and/or density of the layer must remain, even though the average density is increased by the shrinkage. This implies that the shrinkage is not a disappearance of voids, which would greatly reduce the efficiency, but probably a combination of some reductions in voids and a concentration of the gelatin. All this confirms the considerable stability of the cross-linked gelatin.

Similar, more extensive, measurements were made by *McCauley* et al. [3.15]. They recorded reflection holograms with 514 nm light and measured the wavelength of peak efficiency. The changes in thickness are much smaller than for transmission holograms. They observed a gradual decrease of thickness with drying time at room temperature, as might be expected. More interestingly, there was an approximately linear decrease of emulsion thickness after development with increasing dichromate concentration in the sensitized layer. Typical values range from an increase in thickness after development of 5 to 10% for 1% ammonium dichromate solution to a decrease of the same amount for 10% solution. Again, this shows how dichromated gelatin can be tailored to the desired properties.

## 3.3 Preparation and Processing of Hardened Dichromated Gelatin Plates for Thick Holograms

The application of dichromated gelatin to holograms is most significant for thick three-dimensional holograms because of the ability to obtain close to theoretical values of maximum efficiency with low noise. Consequently, this chapter concentrates on these holograms. In the present section, we shall discuss the sensitization and the development of thick holographic plates.

### 3.3.1 Preparation of Plates

The preparation of hardened dichromated gelatin layers is described in detail by *Brandes* et al. [3.9]. They used both dip-coating and doctor-blading techniques for applying the gelatin to the glass substrate. They found that commonly available gelatin of 125 bloom strength was most suitable for holographic purposes. Among the many variables that characterize gelatins, bloom strength appeared to be the only one that was related to the unique holographic process under discussion.

The recommended dip-coating procedure is listed in Table 3.3. The thickness of the gelatin layer produced in this way depends on the gelatin suspension concentration, the pulling rate, and the size of the plate being pulled. For small plates, layers of 1 to 15 μm thickness can be produced with good uniformity. For larger plates (> 10 cm in dimension), the thickness is limited to about 4 μm. For thicker layers, doctor-blading is a suitable alternative. The recommended procedure is listed in Table 3.4. The thickness again depends on the gelatin suspension concentration and on the height of the doctor-blade. Uniform layers of 6–15 μm can be produced reliably.

The hardening agent used should not change the properties of the gelatin, such as its melting point. Many commercial hardeners behave in this way. *Brandes* et al. found that 0.5% ammonium dichromate with heating is an excellent solution because it does not cause cross linking until the layer is completely dried out and heated.

Hardened films produced in either of these ways are stable and can be stored for many months. These films are sensitized by dipping in 4% aqueous solution of $(NH_4)_2Cr_2O_7$ for 2 min. A small amount of wetting agent improves the uniformity. The films are then dried in the dark and stored at 20 °C or lower for no more than 12 h before exposure.

A different way of preparing dichromated gelatin plates was used by *Lin* [3.14]. He started with Kodak 649F spectroscopic plates. Because of its simplicity, this procedure has been widely used. It also has the advantage of producing uniform and repeatable layers relatively easily. The process for preparing plates that has been used by the author [3.13] is listed in Table 3.5. It is based on the results of numerous trials throughout an extended period of time. This sequence of steps was found to produce optimum results with fair

Table 3.3. Procedure for producing gelatin layers by dip-coating (after *Brandes* et al. [3.9])

1) Prepare 12 to 18% by weight gelatin (J. T. Baker Chemical Company, U.S.P. grade, 125 bloom) suspension in water (initial mixing at 20 °C, final stirring at 70 °C). Filter with heated filter.

2) Thermostat at 40 °C and add $(NH_4)_2Cr_2O_7$ to give a ratio of 0.5% to the weight of gelatin.

3) Insert glass plate and withdraw vertically at rates of 1 to 5 cm/min.

4) Air dry plates in vertical position for 1 h.

5) Harden by baking at 150 °C for 2 h.

Table 3.4. Procedure for producing gelatin layers by doctor-blading (after *Brandes* et al. [3.9])

1) Prepare 4% by weight gelatin suspension as in Table 3.3.

2) Apply at 50 °C.

3) Apply excess quantity of suspension to a leveled glass plate.

4) Wipe doctor-blade across plate with 0.5–1.0 mm spacing from plate.

5) Air dry plates in dust-free atmosphere for 24 h. Plates must be accurately horizontal.

6) Harden by baking at 150 °C for 1 h.

Table 3.5. Preparation of sensitized material from spectroscopic plates [3.13]

1) Preparation of 649F plates.
   1.1) 15 min in Kodak fixer with hardener.
   1.2) 10 min wash in running water (20–25 °C).
   1.3) 10 min in methyl alcohol or until plate is clear (agitate).
   1.4) 10 min in clean methyl alcohol.
   1.5) Dry in vertical position.

2) Preparation of sensitizing solution.
   2.1) 50 g/l of Baker, Merck, or equivalent ammonium dichromate fine crystal (use distilled water).
   2.2) Filter solution before use. There is no limit as to how many plates one can sensitize with each solution, but, after storing one week, a new solution should be used. The use of Kodak FotoFlow 0.2 oz/gal is optional.

3) Sensitizing plates.
   3.1) Soak plates 5 min in dichromate solution in flat pan with emulsion side up.
   3.2) Tilt plate ~10° and allow excess to run off (about 3 min). Remove residue at edge with a paper towel.
   3.3) Store in light tight box with same tilt as in 3.2 until ready to use.
   3.4) Expose between 15 h and 40 h after sensitizing.

reproducibility. There are occasional variations in the results that appear to be caused by variations in the gelatin obtained from the spectroscopic plates. The thickness of the gelatin after processing is about 12 μm.

The first series of procedures in Table 3.5 is designed to produce as clean a gelatin plate as possible. All the previously incorporated chemicals must be washed out. The sensitizing solution is chosen to incorporate the highest concentration of dichromate in the gelatin without causing undesired side effects. During the first stage of the drying process, the solution is allowed to run off the almost horizontal plates. This results in a uniform concentration except for the lowest region of the plate (about 1 cm wide). If the plates are exposed too soon after sensitizing, they tend to "crystallize" during development. By this, we mean that regions of the gelatin appear disturbed as if one of the components in the layer had crystallized. Such regions cause extra scattering and, therefore, noise in the final image. If the sensitized plates are kept too long before exposing, they start to lose sensitivity.

One may ask why 649F plates are used for this procedure. Experimentally, it is found that this particular gelatin formulation does give the best result. Holograms can also be made from Kodak high resolution plates which are about 7 μm thick, but the total phase shift is considerably lower (in both cases, only the first 4–5 μm take part in the information storage if 441 nm light is used [Sect. 3.4.1]). Many of the other commercial emulsions are not suitable for this procedure at all. On the other hand, the 649F emulsion can be used on a film base as well as on glass plates, as long as the film is mounted sufficiently stably during exposure.

It is frequently desirable to produce holograms on other kinds of substrates which can have nonplanar shapes. *McCauley* et al. [3.15] have shown how dichromated gelatin holograms can be produced on Plexiglass substrates by gel casting or by film transfer, in addition to doctor-blading. This allows much more flexibility in hologram generation.

### 3.3.2 Development Procedure

The most critical step in producing high efficiency, low noise holograms is the development step. A number of variations of the procedure have been used successfully. The one favored by the author [3.13] is listed in Table 3.6.

The water wash serves two purposes. First, it softens the gelatin for the following drying step. Secondly, it removes all the unreacted dichromate. This is essential to preventing crystallization and the introduction of scattering centers upon drying. For $10 \times 12$ cm plates, 10 min at room temperature seems to be adequate.

The washed plate is immersed wet into the isopropyl alcohol. There seems to be some advantage to making the first step a mixture of alcohol and water, but we have not been able to document this conclusively. In any case, at least two steps must be used so that the last bath remains as free of water as possible. We

Table 3.6. Development procedure for exposed dichromated gelatin plates [3.13]

1) Wash 10 min in running water at 20 °C.

2) Soak 2 min in mixture 50% isopropyl alcohol and 50% distilled water, agitated.

3) Repeat in 90% isopropyl alcohol—10% water mixture.

4) The hologram is mounted on a stainless-steel plate with gelatin layer away from the plate, and inserted vertically into fresh isopropyl alcohol with agitation for 10 to 20 min.

5) The plate is pulled out of the container at a rate of about 1 cm/min. At the same time, a flow of hot air is directed against the gelatin.

have found that dehydration is complete after two min for $10 \times 12$ cm plates. However, for larger plates ($20 \times 25$ cm), the process takes a much longer time, as it appears that the reaction proceeds from the edges of the plate to the interior in a diffusion-controlled mechanism. This can be observed visually, but we cannot offer an explanation for it.

The final step of removing the alcohol is more critical than had first been suspected. When plates are removed from the alcohol and dried in air, the results are very sensitive to the humidity in the air. Low humidity encourages crystallization, while high humidity reduces the hologram efficiency. The problem was overcome by pulling the plates slowly out of the bath and drying them with a stream of warm air directed at the liquid interface. When the plate is well cleaned, the interface forms a uniform straight line, and the dried plate has a uniform appearance and efficiency. After the drying is complete, the plates are no longer affected by relative humidity up to 90%.

As has been mentioned in Section 3.2, the high efficiency of this holographic process depends on producing voids in the gelatin at the appropriate locations. To fully develop these, it is necessary to dehydrate the washed layer very rapidly. However, if this is done too vigorously, the plates may take on a milky-white appearance which produces heavy light scattering. It has been attributed to a precipitation of gelatin from aqueous solution when water-miscible solvents are added [3.1]. Another explanation [3.16] is that the very rapid dehydration induces such large strains that there is large-scale tearing and cracking rather than just formation of voids.

*Brandes* et al. [3.9] showed that both the light scattering and the sensitivity increase with increasing wash water temperature and decrease when the sensitized plates had been hardened more extensively. Their best compromise calls for plates hardened, as in Table 3.3, which are developed in a water wash at 25 to 40 °C followed by isopropyl alcohol at 70 °C.

*Lin* [3.14] used two additional processing steps to prevent a milky appearance being generated. He treated the exposed plate in 0.5% ammonium dichromate solution for 5 min and then in a reducing agent such as Kodak rapid fixer with hardener. He proposed that the complete reduction of hexavalent chromium to trivalent chromium caused the improved appearance.

A different approach was taken by *Chang* [3.16]. Based on the properties of gelatin (Sect. 3.1.2), he argued that the stronger and less brittle crystalline films produced from gel layers would leave less tendency to whiten, particularly if the crystallites were large enough. He converted 649F gelatin into this state by washing the plates in water which started at 21 °C and then was gradually raised to 32 °C. The plates were then soaked for 15 min at 32°. This eliminates all small crystallites and allows larger crystallites to form on subsequent cooling. Hardening, sensitizing, and developing steps follow, but all take place at room temperature.

It is clear that a number of different tradeoffs among the many variables must be considered by each experimenter in designing his particular process.

### 3.3.3 Stability of Dichromated Gelatin Holograms

Because the holograms are destroyed by washing the plates in water, it would seem likely that they are also sensitive to moisture. This is, however, the case only at very high relative humidities (over 80%). Under ordinary room atmosphere conditions, the holograms are very stable. We have had many holograms lying around unprotected for 5 years without any noticeable change in efficiency. If such holograms are wiped out in water, they can again be redeveloped.

In addition, it is relatively easy to protect dichromated gelatin holograms from humidity by covering them with an appropriate organic adhesive [3.14, 15]. Since the top surface of the developed gelatin is impervious to nondissolving liquids, the hologram will not be affected by the overlayer.

The fact that each exposure produces a permanent cross-linking network means that one can resensitize a previously developed hologram and superimpose a second exposure. Both holograms will then be developed simultaneously in the isopropyl alcohol step [3.14].

## 3.4 Holographic Properties of Hardened Dichromated Gelatin

### 3.4.1 Phase Shift Available

Because the available phase shift $\Delta\phi$ in a phase-recording material may be many times $2\pi$, it allows the storage of a number of high efficiency holograms superimposed. To determine the maximum phase shift, one needs to know how much material is exposed, how many photosensitive centers there are, and how much phase shift each exposed center produces after development. The phase shift is, of course, related to the index variation in a layer of thickness $d$ by

$$\Delta\phi = 2\pi\,\Delta n d/\lambda_0\,. \tag{3.2}$$

Dichromated gelatin plates produced according to the prescriptions of Section 3.3 typically contain 20 to 30% of dichromate by weight, or $1\text{–}2\times10^{21}$ molecules/cm$^2$. Typical absorption coefficient values for unexposed layers are

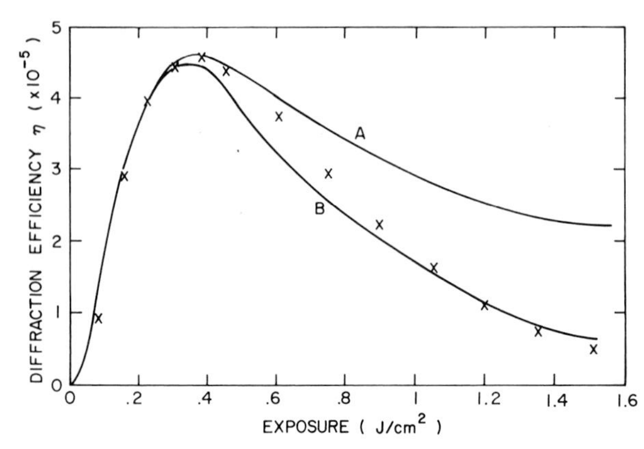


Fig. 3.3. Diffraction efficiency of a dichromated gelatin layer, 0.7 μm thick, during exposure with 441 nm light [3.6]. The points are experimental, the curves are calculated for a sinusoidal spatial variation of exposure. Curve *B* includes a small dc bias due to unequal energy in the two exposing beams

0.2–0.4 $(\mu m)^{-1}$ at 441 nm and 0.04–0.06 $(\mu m)^{-1}$ at 488 nm (Sect. 3.2.2). The calculated absorption cross sections are approximately $2 \times 10^{-18}$ cm$^2$ at 441 nm and $4 \times 10^{-19}$ cm$^2$ at 448 nm. This means that, at 448 nm, the exposure is relatively uniform in the most commonly used plates (up to 15 μm thickness), while at 441 nm, only about the first 4 μm are exposed.

To determine the amount of phase shift produced by the exposure alone, the grating can be probed with nonabsorbed red light, while being exposed with blue light. A typical experimental result is shown in Fig. 3.3 [3.6]. Measured efficiency values are independent of the grating period to the smallest values used (0.5 μm). The data of Fig. 3.3 were taken on a very thin layer of dichromated gelatin (0.7 μm), so that the exposure is uniform in 441 nm light. Then, the expected efficiency can be calculated exactly [3.6] and fit to the data. Two curves are shown, the lower of which represents the experimental situation in which the two interferring beams were not exactly equal. From the fit to the data, the maximum phase shift at complete exposure is determined to be 0.05 and the maximum index-of-refraction change is found to be 0.006, in good agreement with Table 3.1. The change in index per molecule of dichromate per cm$^3$ is therefore $5 \times 10^{-24}$ cm$^3$.

In thicker layers, similar results apply. If the exposing light is heavily absorbed, then this must be taken into account in the calculations. For example, for a plate based on 649F gelatin, a maximum obtainable phase shift of 0.3 is measured in 441 nm light [3.13], while for lightly absorbed light, a phase shift of about 1 is expected. An exposure of about 1 J/cm$^2$ is required to reach saturation.

Of much greater practical interest is the phase shift that can be produced by the water-alcohol development procedure. In Section 3.2.3, it was shown that in a 1.3 μm thick gelatin layer, a peak refractive index variation of 0.26 had been obtained. Thus, this development procedure has caused a 30- to 40-fold increase in phase shift. Similar results were attained by the author in 649F plates. Using 441 nm light (approximately 4 μm effective grating thickness), the efficiency versus exposure relationship was followed through the first peak (90% efficiency), through the first minimum ($<10$%), and to near the second peak at 80% efficiency.

Other experimenters, using 488 nm exposure in thicker layers (649F plates, 12 to 13 μm thick) were able to reach maximum values of index change of 0.04 only at 1 J/cm$^2$ exposure [3.17]. The differences must be due to differences in processing conditions. It does not appear to be due to a partial development of the gelatin layer, such as would be the case for 441 nm exposure. *Curran* and *Shankoff* [3.8] showed, by band pass measurements, that gelatin layers up to 15 μm thick can have high efficiency gratings which penetrate the entire layer. This is confirmed for 649F plates and double grating exposures by *Alferness* and *Case* [3.18].

### 3.4.2 Sensitivity

From what has been said so far, it is clear that there can be no unique measurement of sensitivity. There are many different ways of processing dichromated gelatin films and they lead to different measured sensitivities. As discussed in Section 3.3.2, there is a tradeoff between sensitivity and scattering. However, it is possible to make some general statements based on our previous discussions.

The initial photochemical process depends primarily on the density of dichromate ions available in the emulsion. There are limits, both in the solubility of the ions and in the thickness of the layer that can be used because of the absorption of light. Thus, the basic exposure process is limited to the point where all the photons are absorbed.

The next step in which the $Cr^{3+}$ reacts with the gelatin depends on the nature of the gelatin, but, again, there is not much that can be changed about that. The major opportunity for varying the sensitivity is in the development step (Sect. 3.3.2). The harder the original gelatin layer is, the lower will be the sensitivity. Conversely, the development can be increased by increasing the bath temperatures in the washing and dehydration steps. Thus, in order to have a reliable photographic system, it is necessary to control the mechanical properties of the gelatin, such as was done by *Chang* [3.16], as well as to tightly monitor the processing.

The problem of determining the sensitivity is most apparent in Fig. 3.4 taken from *Curran* and *Shankoff* [3.8]. The exposures were made in 488 nm light with two equal beams symmetrically incident on the plate. The curves are for different thicknesses of gelatin. From Section 3.4.1, we expect an absorption length of 20 μm, so that the total absorbed light increases proportional to the gelatin thickness. Thus, one would also expect a monotonic decrease of exposure needed for a given diffraction efficiency. This is clearly not the case for the thinner layers which exhibit strong oscillatory behavior.

These data are summarized in Table 3.7, where various measured sensitivities are listed for comparison. There are many possible definitions of sensitivity. To be able to make appropriate comparisons, it is defined as the inverse of the exposure required to reach 20% efficiency, for simple symmetric gratings with equal exposing beam powers.

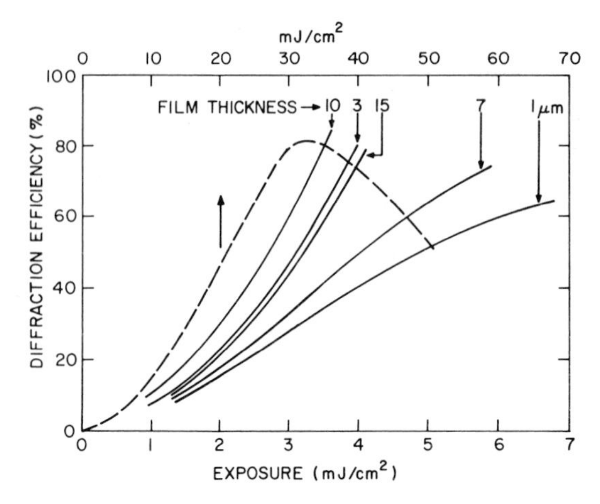


Fig. 3.4. Diffraction efficiency of simple gratings versus exposure for various film thicknesses at 488 nm. The solid curves (grating periods 488 nm) were measured by *Curran* and *Shankoff* [3.8] on films of various thicknesses, while the dashed curve (grating period 282 nm) from *Lin* [3.14] is for a sensitized 649F plate (upper exposure scale)

Figure 3.4 also shows the efficiency-exposure characteristic measured by *Lin* [3.14] on holographic films derived from 649F plates. Exposures were at 448 nm as for the other curves. The grating spacing is somewhat different, but, as shown in Table 3.7, this difference is not significant. The sensitivity is substantially lower (note the different exposure scales) for the films derived from the 649F plates than that for the films formed directly from gelatin. This may be due to the initial hardness difference as well as due to the added processing by *Lin* to reduce the noise. Experimental results similar to *Lin*'s have been reported by *Fillmore* and *Tynan* [3.17] with even lower sensitivities (Table 3.7). The author has made sensitivity measurements at 441 nm where all light is absorbed in a sensitized 649F plate, and, as expected, somewhat higher sensitivities are obtained (typically 160 $cm^2/J$).

Table 3.7. Sensitivity of dichromated gelatin plates (reciprocal of exposure needed to obtain 20% diffraction efficiency)

| Experimentor and conditions | Gelatin thickness [μm] | Grating period [μm] | Sensitivity [$cm^2/J$] |
|---|---|---|---|
| *Curran* and *Shankoff* [3.8] | 1 | 0.5 | 420 |
| 448 nm | 3 | 0.5 | 560 |
| | 7 | 0.5 | 430 |
| | 10 | 0.5 | 650 |
| | 15 | 0.5 | 520 |
| *Lin* [3.14] | 12 | 1.4 | 150 |
| 448 nm, 649F plates | 12 | 0.28 | 80 |
| | 12 | 0.16 | 62 |
| *Fillmore* and *Tynan* [3.17] 448 nm, 649F plates | 14 | 0.4–10 | 40 |
| *Meyerhofer* [3.13] 441 nm, 649F plates | 13 | ~0.5 | 200 |

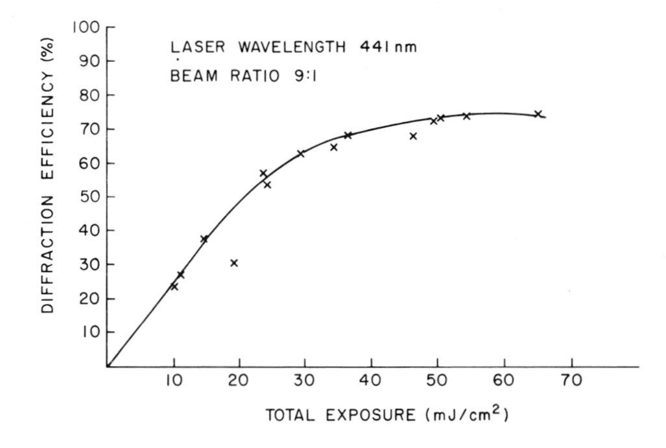


Fig. 3.5. Diffraction efficiency of a diffuse hologram [3.13]. The objects consisted of a square random diffuser in a black background

It is worth noting that the curves of Fig. 3.4, which are all taken with two equal beams, follow (1.51) reasonably well. By plotting the square root of the efficiency, a substantial linear recording range is demonstrated [3.14].

These measurements were also extended to diffuse image holograms. Figure 3.5 shows the amount of light diffracted into the real image of a diffuse hologram recorded in dichromate 649F plates with a reference to object beam ratio of 9 to 1 and with 441 nm laser light [3.13]. The sensitivity, as defined above, is 125 $cm^2/J$, but if account is taken of the beam ratio, that value increases to about 200 $cm^2/J$ for the comparison in Table 3.7. The curve does not appear to follow (1.51).

### 3.4.3 Resolution

The high resolution of the holographic recording process in hardened dichromated gelatin has been described in earlier sections. The data in Table 3.7 further demonstrate this fact. In the case of *Lin*'s results, there is a drop-off of sensitivity from a grating period of 1.4 μm to one of 0.16 μm (in a reflection grating). However, the drop is small compared to any effect caused by a real cut-off frequency, and it may be due to experimental differences rather than due to a fundamental limit. In fact, the results of *Fillmore* and *Tynan* show no change of sensitivity over the wide range of grating periods from 0.4 μm to 10 μm. This is strong evidence that the voids in the gelatin that produce the large phase shift must be very small and continuously distributed compared to dimensions of optical wavelength.

### 3.4.4 Image Quality and Noise

The almost ideal character of dichromated gelatin as a holographic recording medium is especially well demonstrated in fabricating holographic optical elements. As striking examples are the diffraction gratings produced by *Shankoff* and *Curran* [3.19] which exhibit properties close to those expected theoretically.

These gratings were produced in 3 μm thick gelatin layers with 3 to 8 $mJ/cm^2$ exposure of 488 nm light. Exposure was by two systematically arranged beams, so that the grating planes are perpendicular to the layer. However, it is just as easy to "blaze" these gratings for other angles by changing the angles of the recording beams. The responses of two different gratings were shown, with periods of 0.5 and 0.9 μm. The peak diffraction efficiency was over 90% and the spectral response was as expected for gratings of this thickness. The resolving power of the gratings was 85 to 90% of the theoretical value and limited by their optical system. The background scatter was less than $10^{-4}$ of the signal and there were no detectable ghosts.

The properties of diffuse holograms have been studied by *Meyerhofer*. A typical exposure curve was shown in Fig. 3.5. For noise measurements, the image geometry of *Upatnieks* and *Leonard* [3.20] was used so that the data could be compared with theory. The object consisted of a 32 mm square diffuser with a $3 \times 5$ mm opaque rectangle in the center. The power density was measured in the reconstructed real image; the value at the center of the opaque spot representing the noise, the value just outside the opaque spot representing the signal plus noise. Reasonable agreement was obtained with the predicted signal-to-noise ratios. For example, for an object to image beam ratio of 1, and an exposure of 8 $mJ/cm^2$, an efficiency of 50% and a ratio $(s/n)_c = 1.7$ was measured. Since the values are at least as good as the calculated ones, one can conclude that the image noise is dominating over excess noise in the hologram.

The image quality was determined by recording holograms of diffusely illuminated resolution charts. Care was taken to shrink the emulsion after development by heating it at 150 °C. With equal image and object beam on recording, and with a $f/6$ optical aperture on readout, a resolution of 40 cyc./mm was obtained. Of course, much higher resolution is possible recording non-diffused objects.

### 3.4.5 Storage Capacity

The storage capacity of a phase recording medium, such as dichromated gelatin, is directly related to the maximum possible phase shift that can be induced. Since a very high efficiency hologram can be formed in a 1.3 μm thick layer [3.1], three separate such holograms should be feasible in 4 μm, ten in 13 μm.

These considerations were confirmed by *Meyerhofer* [3.13] by recording multiple holographic elements in 649F emulsion with 441 nm light. The reference and object beams were changed for each exposure so the holograms could be read out individually by turning to the Bragg frequency. More than 10 holograms with over 20% efficiency each could be superimposed. The efficiency was the same for all holograms, and the cross-talk was negligible. The plates had been shrunk to eliminate aberrations as was confirmed by enlarging the focal points of the lenses.

In a related experiment, a multiple beam splitter was produced by superimposing a number of exposures in which the reference beam was kept constant and the object beam changed for each exposure. Then, on readout with the reference beam, all image beams are produced simultaneously. In this case, equal beams were used during exposure. The exposure was similar to that required for an efficient single hologram. Up to 30 exposures could be made before the material saturated, and all could be read out with an efficiency between 10 and 20%.

## 3.5 Applications

### 3.5.1 Holographic Optical Elements

One of the objects of developing dichromated gelatin as a holographic recording material was to make use of its high efficiencies for holographic optical elements. Efficiency is particularly important in this use, because one usually wants to work with the lowest possible exposures and because getting rid of the undiffracted beam without increasing the background illumination is generally a problem.

The simplest holographic optical element is the grating, the analog of a prism. The fact that some of the highest quality gratings have been recorded in dichromated gelatin [3.19] has already been discussed in Section 3.4.4. The exposing condition and the gelatin thickness can be chosen to control the grating parameters. This technique has been extended to beam splitters and beam combiners by *Case* [3.21]. He obtained experimental results which closely approached the theoretical optimum values.

A somewhat more complicated element is the holographic lens, which combines a change in wavefront curvature with a deflection. It can be formed by a plane and spherical beam or by two spherical beams. When it is used in reconstruction at or near the recording directions, it is capable of forming high-resolution images, but only if the emulsion thickness does not vary from exposure to readout. By using the final baking step for dichromated gelatin, lenses were formed with over 90% efficiency and high quality imaging at the same time [3.13].

This technique was applied to a special beam splitter-lens combination required for a holographic computer memory by *Stewart* et al. [3.22]. The single holographic element they used is much simpler than the array of beam splitter lenses and masks that would be required by a conventional optical system.

If the hologram thickness is not the same as during recording, there will be aberrations in the holographic optical elements. *McCauley* et al. [3.15] have measured these effects in dichromated gelatin holograms on plain and spherical substrates and indicated how the aberrations can be compensated for.

Dichromated gelatin has particular advantages for recording multiple elements because of its large phase change. Multiple lenses have been of interest

[3.23], but they have been difficult to produce because of the complicated object beam required. If each lens is recorded separately, that problem can be alleviated considerably. Up to 20 lenses have been recorded in this way in dichromated gelatin [3.13]. Since larger numbers of lenses are usually required, thicker layers would be necessary and less heavily absorbed light (longer wavelengths) would have to be used. This is no serious handicap, since the sensitivity does not change much, as long as most of the incident light is absorbed in the layer.

### 3.5.2 Duplicate Holograms in Dichromated Gelatin

Dichromated gelatin holograms combine the advantage of high efficiency and low noise with the disadvantage of the low sensitivity. This is particularly important for diffuse objects, where most of the incident light goes into illumination of the object. Various schemes have been investigated for overcoming this problem [3.13]. First, diffuse object holograms were formed in bleached 649F plates, which have very much higher sensitivity than dichromated gelatin. The diffraction efficiency of these holograms was deliberately kept at 10% or lower. The holograms were then contact-printed onto dichromated gelatin. The bleached hologram was located in exactly the same position as during recording and was illuminated with the reference beam only. Because of the low diffraction efficiency of the original holograms, the undiffracted "new" reference beam for the dichromated gelatin exposure was adequately uniform. The diffracted image beam became the "new" object beam. The procedure produced excellent copies with the high efficiencies characteristic of dichromated gelatin.

The low sensitivity of dichromated gelatin is much less of a problem for copies than in making originals; first because all the light is used to expose the dichromated gelatin and second because much longer exposures can be tolerated in a contact printing system without encountering any coherence problems due to vibration of thermal variations. The procedure is quite similar to the hybrid technique of *Pennington* et al. [3.24] who essentially combine the silver and dichromate solutions in the same emulsion.

## References


3.1 T.A.Shankoff: Appl. Opt. **7**, 2101 (1968)
3.2 J.Kosar: *Light-Sensitive Systems* (John Wiley and Sons, New York 1965) Chap. 2
3.3 C.E.K.Mees, T.H.James: *The Theory of the Photographic Process*, 3rd ed. (Macmillan, New York 1966) Chap. 3
3.4 J.E.Jolley: Phot. Sci. Eng. **14**, 169 (1970)
3.5 B.O'Brian: J. Opt. Soc. Am. **42**, 101 (1952)
3.6 D.Meyerhofer: Appl. Opt. **10**, 416 (1971)
3.7 R.A.Bartolini, N.Feldstein, R.J.Ryan: J. Electrochem. Soc. **120**, 1408 (1973)
3.8 R.K.Curran, T.A.Shankoff: Appl. Opt. **9**, 1651 (1970)
3.9 R.G.Brandes, E.E.Francois, T.A.Shankoff: Appl. Opt. **8**, 2346 (1969)
3.10 D.L.Matthies: Private communication

3.11 M.Born, E.Wolf: *Principles of Optics*, 2nd ed. (Macmillan, New York 1964) pp. 87–90
3.12 D.H.R.Vilkomerson, D.Bostwick: Appl. Opt. **6**, 1270 (1967)
3.13 D.Meyerhofer: RCA Rev. **33**, 110 (1972)
3.14 H.Lin: Appl. Opt. **8**, 963 (1969)
3.15 D.G.McCauley, C.E.Simpson, W.J.Murbach: Appl. Opt. **12**, 232 (1973)
3.16 M.Chang: Appl. Opt. **10**, 2550 (1971)
3.17 G.L.Fillmore, R.F.Tynan: J. Opt. Soc. Am. **61**, 199 (1971)
3.18 R.Alferness, S.K.Case: J. Opt. Soc. Am. **65**, 730 (1975)
3.19 T.A.Shankoff, R.K.Curran: Appl. Phys. Lett. **13**, 239 (1968)
3.20 J.Upatnieks, C.Leonard: J. Opt. Soc. Am. **60**, 297 (1970)
3.21 S.K.Case: J. Opt. Soc. Am. **65**, 724 (1975)
3.22 W.C.Stewart, R.S.Mezrich, L.S.Cosentino, E.M.Nagle, F.S.Wendt, R.D.Lohman: RCA Review **34**, 3 (1973);
W.C.Stewart, L.S.Cosentino: Appl. Opt. **9**, 2271 (1970)
3.23 S.Lu: Proc. IEEE **56**, 116 (1968)
3.24 K.S.Pennington, J.S.Harper, F.P.Laming: Appl. Phys. Lett. **18**, 80 (1971)

# 4. Ferroelectric Crystals

D. L. Staebler

With 13 Figures

Hologram storage in ferroelectric (FE) media began in 1968 with the work of *Chen* et al. on undoped $LiNbO_3$ [4.1]. They found that an "optical damage" effect, although undesirable for nonlinear optical applications [4.2], could be used to record thick phase holograms. Their model for the effect, which in its basic parts has been substantiated by others, is that the incident light allows charge to migrate within the illuminated area. This sets up a space charge field that modifies the refractive index through the electro-optic effect. Since then, subsequent work on $LiNbO_3$ and other FE materials has led to a number of advances in this field. These include the improvement of the storage and erasure sensitivity by orders of magnitude, development of techniques for hologram fixing, the storage of many thick holograms at different angles within the same volume of a crystal, and electronic means for storage and readout enhancement. As a result, FE materials (also called electro-optic, pyroelectric, or photorefractive in the literature) have now become major contenders for a wide variety of optical storage and display applications.

The advantages of FE materials are excellent resolution, readout efficiency, reversibility, storage capacity, and sensitivity. In addition, they are highly flexible, being useful in both read/write and read-only systems. The read/write operation is particularly simple because the as-recorded holograms can be immediately read out without processing, and then erased with the same wavelength used for storage. The prime limitation of these materials at present is that most of the holographic storage properties are strongly interrelated; in enhancing one property (e.g., sensitivity) by proper choice of material or material treatment one finds a tradeoff in others (e.g., storage capacity). Future research, both in materials and in methods of use, is needed before FE crystals can fulfill their potential.

This chapter describes the operation and applications of FE materials. The major ones considered here are $LiNbO_3$, SBN, $BaTiO_3$, and PLZT ceramics. It begins with a discussion of the storage, fixing, and erasure mechanism in the various materials. Then follow sections on the methods of preparations, and methods of use. Finally, the characteristics of these materials directly related to their use for hologram storage are considered—sensitivity, dynamic range, resolution, storage time, and noise.

## 4.1 Theory of Mechanisms

Holograms in FE materials consist of bulk space charge patterns. They are formed when an interference pattern of light generates a pattern of electronic charge carriers that are free to move. Migration of the carriers toward areas of low optical illumination, and subsequent capture in localized sites called traps, produces patterns of net space charge. This sets up a corresponding electric field pattern. Since FE crystals have strong electro-optic properties, the result is a pattern of variations in the refractive index, a phase hologram.

The as-recorded holograms are immediately visible (i.e., there is no latency period), and are stable (or at least metastable), because the charges are bound to the localized traps. Nevertheless, they can be erased by exposure to a *uniform* beam of light of a wavelength that can release the trapped charge. In some cases, e.g., during readout with such light, this is an undesirable feature. Here, one can fix the charge patterns by converting them to patterns of ions which are not sensitive to light. This is done via thermally activated ionic conductivity, as in $LiNbO_3$, or through electrically initiated ferroelectric reversal, as in SBN and $BaTiO_3$.

As an introduction to the theory of storage, erasure, and fixing phenomena, it is instructive to consider the special case of Fe-doped $LiNbO_3$. This material is one of the best FE storage media known to date, and its mechanisms have been well established. Where appropriate, mechanisms in other materials will also be described.

### 4.1.1 Storage

The essential features of hologram storage in Fe doped material are shown in Fig. 4.1. The crystal contains a single species of photosensitive electron traps (introduced by Fe doping) and storage comes about from the spatial redistribution of electrons in these traps. The occupied traps are $Fe^{2+}$ ions and have an absorption band in the visible region that corresponds to the excitation of the trapped electron into the conduction band,

$$Fe^{2+} \underset{h\nu}{\rightarrow} Fe^{3+} + e^- .$$

The $Fe^{3+}$ ions are the empty traps, and have no absorption. To allow for net redistribution of the electrons, a certain percentage of the traps must be empty.

Initially, the trapped electrons are distributed uniformly throughout the crystal, as shown in Fig. 4.1a. Then the crystal is exposed to an interference pattern. Figure 4.1a shows a sinusoidal pattern formed by two plane beams. This light, upon absorption by the trapped electrons, produces a corresponding pattern of free electrons. The electrons can then move toward regions of low light intensity, usually in successive steps of excitation, migration, and retrapping. The charge migration can come about through 1) diffusion due to the periodic gradient in free electron density [4.3]; 2) drift due to the presence of an applied

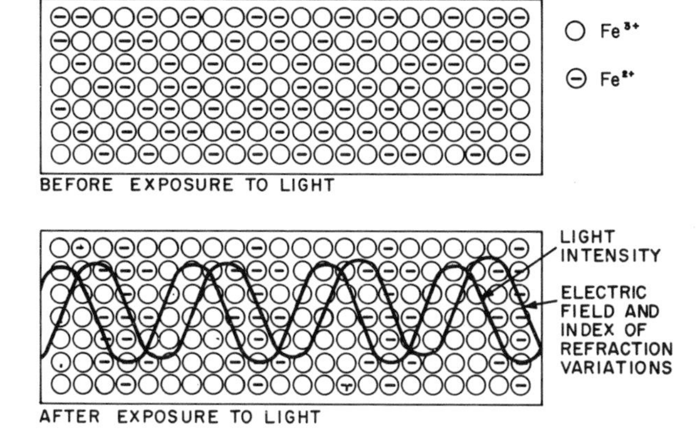


Fig. 4.1. Hologram storage process in an FE material. The light redistributes electrons (– signs) trapped in localized defects (circles). The resulting field pattern produces an index pattern via the linear electro-optic behavior of the crystal

electric field [4.1, 3]; or 3) a bulk photovoltaic effect [4.4] caused by the asymmetry of the lattice of all FE materials. As is discussed later in this section, the latter effect can be represented, at least mathematically, as drift in an internally generated electric field. By either process, the electron transport produces space charge, positive where the electrons came from, negative where they accumulate. Figure 4.1b shows an extreme case; nearly all trapped electrons have been displaced. The result is an electric field pattern that produces the desired variations in refractive index. Calculations [4.5] show that one can store a hologram with 100% diffraction efficiency in a 1-cm-thick crystal (assuming ~1 μm grating spacing) with a trapped electron density of only $5\times10^{14}$ cm$^3$.

Theoretical approaches to the storage process have used, out of necessity, a number of assumptions. This is because a complete closed form solution is impossible for the general situation, even for the exposure to simple sinusoidal light patterns. In the following, we give the general equations that must be used in any theory, and then consider some simple cases. The question of whether diffusion, drift, or the photovoltaic effect dominates the storage process will also be considered.

The first consideration in any theory is the free electron density $n$. It is given by the rate equation

$$\partial n/\partial t = G - n/\tau + \nabla \cdot \boldsymbol{J}/e, \tag{4.1}$$

where $G$ is the optical generation rate of the free carriers, $\tau$ is their mean free lifetime before they are captured by an empty trap, $J$ is the net current due to the migrating electrons, and $e$ is the charge of an electron. $G$ represents the information to be stored. Spatial variations of $G$ produce spatial variations in $J$, which in turn builds up a net charge density pattern $\varrho$ from

$$\partial \varrho/\partial t = \nabla \cdot \boldsymbol{J}\,. \tag{4.2}$$

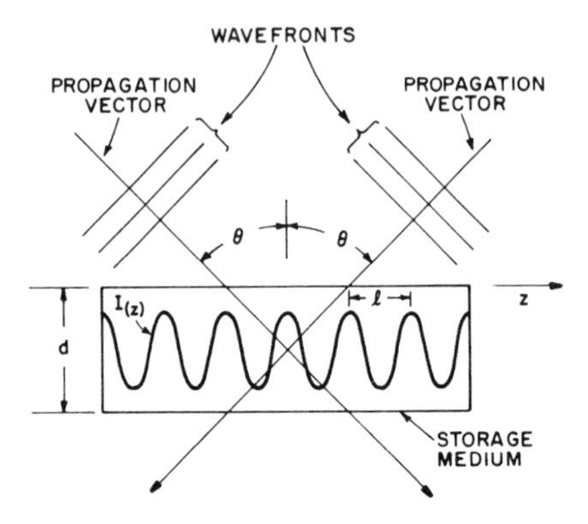


Fig. 4.2. The sinusoidal light pattern produced by two plane collimated beams in a thick FE crystal. $l=\lambda\,(\text{wavelength})/2\sin\theta$

The resulting electric field pattern $E_1$ is found from

$$\nabla\cdot\boldsymbol{E}_1=-\varrho/\varepsilon\,, \tag{4.3}$$

where $\varepsilon$ is the dielectric constant.

*Amodei* [4.3, 6] was the first to develop a quantitative theory for the spatial and time development of $E_1$. *Alphonse* et al. [4.7] have now expanded on his theory, and their approach is given below. Two assumptions are made here, both of which pertain to most practical situations. The first is that the density of occupied and empty traps is large enough so that they remain approximately constant throughout the development of the hologram. This removes the complications of spatial variations in absorption constant and free carrier lifetime. The second is that the mean distance than an electron travels before being retrapped is much shorter than the periodicity of the grating. This is equivalent to saying that $\tau$ of (4.1) is so short that the electron density can be calculated by ignoring the last term, yielding

$$n=G\tau \tag{4.4}$$

for times larger than $\tau$ ($>10^{-9}$ s for most crystals) [4.7]. $G$ is proportional to the incident light intensity which is (see Fig. 4.2):

$$I=I_0(1+m\cos Kz)\,, \tag{4.5}$$

where $m$ is the modulation ratio and depends on the relative intensity of the two beams, $K$ is $2\pi/l$, and $l$ is the grating spacing (Fig. 4.2). Thus the free carrier density becomes

$$n=n_0(1+m\cos Kz)\,. \tag{4.6}$$

From this the current density $\boldsymbol{J}$ is found using

$$\boldsymbol{J}=e\mu n\boldsymbol{E}+eD\nabla n\,, \tag{4.7}$$

where $\mu$ and $D$ are the mobility and diffusion constant of the free carriers, and $\boldsymbol{E}$ is the total electric field. $E$ is now separated into $E_1$, the unknown hologram field, and $E_a$, the known or applied field, both along the axis of the hologram (the $z$ axis in Fig. 4.2):

$$\boldsymbol{E}=(E_a+E_1)\boldsymbol{z}\,. \tag{4.8}$$

From (4.2), (4.3), and (4.6)–(4.8) one finds then for $E_1$

$$E_1(z,t)=[m/f(z)]E_m\{1-\exp[f(z)t/T_0]\}\cos(Kz-\phi)\,, \tag{4.9}$$

where $f(z)=1+m\cos Kz$, $T_0$ is the dielectric relaxation time given by

$$T_0=\varepsilon/e\mu n_0\,. \tag{4.10}$$

$E$ is the saturation field (for $t\to\infty$) given by

$$E_m=(E_D^2+E_a^2)^{1/2}\,, \tag{4.11}$$

and $E_D$ is an effective field for diffusion given by

$$E_D=K(kT/e) \tag{4.12}$$

($k$ is Boltzmann's constant and $T$ is the temperature). $\phi$ is a spatial phase factor, given by $\tan^{-1}(E_D/E_a)$.

The salient results from these equations are the following:

1) for short storage times, $t\ll T_0$, the field is sinusoidal and builds up linearly with time, i.e.,

$$E_1(z,t\ll T_0)=mE_m(t/T_0)\cos(Kz-\phi)\,, \tag{4.13}$$

2) the storage rate is proportional to the light intensity and to the value of $E_m$. The first can be seen from

$$T_0=C/I_0\,, \tag{4.14}$$

where $C$ is a material constant that conveniently lumps all of the parameters in (4.4) and (4.10). The second shows that for storage dominated by drift, the rate is simply proportional to the applied field. For storage by diffusion, $E_D$ is the important parameter, and this has been seen to vary with the recording angle [4.7] as expected from (4.12),

3) the position of the grating depends on the storage mechanism. If drift dominates, $\phi=0$, i.e., the grating is simply proportional to the light pattern. If diffusion dominantes, $\phi=90°$, i.e., the grating is shifted with respect to the light pattern. This effect has been verified by coupled wave experiments [4.8],

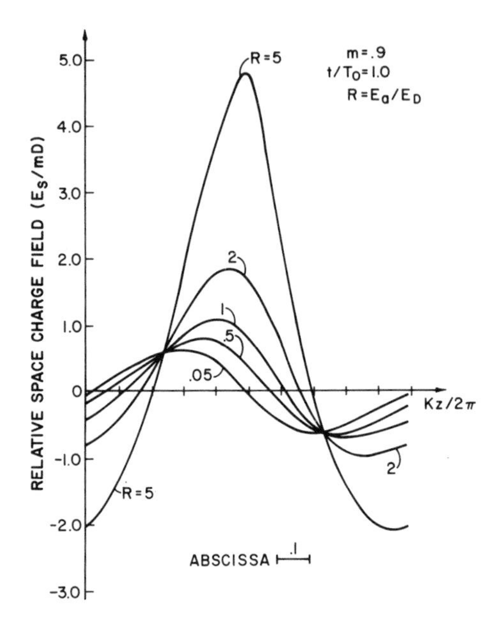


Fig. 4.3. Spatial distribution of stored space charge field near saturation. The magnitude increases with increasing values of applied field (from [4.7])

4) when the storage time approaches the dielectric relaxation time $T_0$, the field of the hologram retards the storage process, leading to a saturation field

$$E_1(z, \tau \to \infty) = [mE_m/f(z)] \cos(Kz - \phi). \tag{4.15}$$

Because $f(z) = 1 + m \cos Kz$, $E_1$ is highly nonsinusoidal when $m \leqq 1$. This is expected to introduce higher orders of diffration [4.9].

Many of the above features can be seen in Fig. 4.3. In particular, $E_1$ increases and shifts to the right, for a given exposure, as drift begins to dominante. The nonsinusoidal character of the field pattern is also quite apparent.

The question of charge transport mechanism will now be considered. Drift will dominate the storage process only when $E_a \gg E_D$. For a typical case of $\lambda \cong 500$ nm, $\theta \cong 15°$, and $T = 300$ K, $E_D \cong 10^3$ V/cm. Thus quite high fields must be applied. A drift-like storage process can also occur without applied fields because of the presence of a bulk photovoltaic effect [4.4]. The model for this effect is based on the work of *Clark* et al. [4.10] who suggested that photoionization of a $Fe^{2+}$ ion occurs by intervalence transfer of the electron to a neighboring Nb ion from where it is transferred into the conduction band. Since the crystal lattice is asymmetric, this transfer has a preferential direction. Thus, each photoactivated electron has superimposed on its normal motion an extra shift ($\Delta$). For $LiNbO_3$, it occurs in the $+C$ direction. The net current due to this effect is

$$J = eG\Delta, \tag{4.16}$$

where, as before, $G$ is the generation rate of free electrons. Comparing this to (4.4) and (4.7) we find that the effect can be represented as drift in an equivalent photovoltaic field [4.11] given by

$$E_a = \Delta/\mu\tau\,, \tag{4.17}$$

where $\Delta$ is the electron shift and $\mu$ and $\tau$ are the mobility and lifetime of the free electrons. Of all these, only the last can be adjusted in a given material. It varies inversely with the concentration of empty traps [4.12]. In $LiNbO_3$ crystals with a high concentration of $Fe^{3+}$ ions (e.g., $\sim 10^{19}\,cm^3$) one expects [4.12] a $\tau$ of $\sim 10^{-12}$ s, which gives, using $\mu \cong 1\,cm^2/V\text{-}s$ [4.12] and $\Delta \cong 0.1$ nm [4.4], an equivalent photovoltaic field of $\sim 10^4$ V/cm. In fact, equivalent fields of nearly $10^5$ V/cm have been measured in highly doped crystals [4.13, 14]. Thus, the photovoltaic effect will predominate. For crystals with a low concentration of $Fe^{3+}$ ions (e.g., $\sim 10^{17}\,cm^3$), the effective field will be only $\sim 10^2$ V/cm. In this case, diffusion effects can be easily observed [4.8, 12] because $E_D > E_a$.

A theory that considers arbitrary electron mean free paths was developed by *Young* et al. [4.15]. They showed that for a free electron generation rate

$$G = G_0(1 + m\cos Kz) \tag{4.18}$$

the initial buildup of space charge is given by

$$E_1(z,t) = teG_0 mL\varepsilon^{-1}(1 + K^2L^2)^{-1}(KL\sin Kz + \cos Kz) \tag{4.19}$$

for drift and

$$E_1(z,t) = teG_0 mL'\varepsilon^{-1}(1 + K^2L'^2)^{-1}(KL'\sin Kz) \tag{4.20}$$

for diffusion.

In both equations $L$ is the mean free path of the electrons, and is given by

$$L = \mu E_a \tau \tag{4.21}$$

for drift and

$$L' = (D\tau)^{1/2} \tag{4.22}$$

for diffusion.

For mean free paths much shorter than the periodicity (i.e., $LK \ll 1$), one obtains the equations predicted from the Amodei-Alphonse theories [see (4.13)]. The new contribution is for the case $LK \gg 1$. Here both (4.19) and (4.20) become

$$E_1 = teG_0 m(\varepsilon K)^{-1}\sin Kz\,. \tag{4.23}$$

This equation gives the maximum charge-up rate possible in FE crystals and will be used later to calculate the ideal sensitivity. Note that it does not pertain to the case where the photovoltaic effect is the dominant storage mechanism. For this effect, $L=\Delta$, which is only $\sim 0.1$ nm.

### 4.1.2 Erasure

A hologram is erased by exposing the crystal to a uniform light beam. The simplest way to do this is with one of the beams used for storage, as is done during readout. The light uniformly excites all electrons out of the traps. They then can redistribute evenly throughout the volume to eventually remove the field pattern. This erases the hologram by bringing the sample back to the original condition—a completely uniform distribution of trapped electrons.

The theory of erasure is straightforward. Following the approach of *Amodei* [4.6], one assumes that the free electron density produced by the erasing exposure is completely uniform, i.e., $n=n_0$. Thus, from (4.7) we see that

$$\boldsymbol{J}=e\mu n_0(E_\mathrm{a}+E_1)\boldsymbol{z}\,. \tag{4.24}$$

From (4.2), this becomes, since $E_\mathrm{a}$ is uniform,

$$\partial\varrho/\partial t=e\mu n_0\nabla\cdot\boldsymbol{E}_1 \tag{4.25}$$

and using (4.3) one finds

$$\partial(\nabla\cdot\boldsymbol{E}_1)/\partial t=-(e\mu n_0/\varepsilon)\nabla\cdot\boldsymbol{E}_1\,. \tag{4.26}$$

The solution is

$$E_1=E_1(t=0)\exp(-t/T_0)\,. \tag{4.27}$$

The field pattern decays exponentially with a time constant given by the dielectric relaxation time [(4.10)].

The important points to remember here are:

1) the time required to decay to a certain fraction of the original field strength is independent of the original value, e.g., decay to 10% or the original takes $\sim 2T_0$ s.

2) The decay time is unaffected by $E_\mathrm{a}$, even if it is due to the photovoltaic effect [(4.17)]. $E_\mathrm{a}$ creates a uniform current density which cannot contribute to the erasure process.

Although drift does not affect erasure, it does affect storage, and this can lead to a large asymmetry in the relative times for storage and erasure. Consider the time to store a hologram with a space charge field magnitude of $E_\mathrm{sc}$. From (4.13) this is, for $E_\mathrm{sc}\ll E_\mathrm{m}$,

$$t_\mathrm{s}=mT_0(E_\mathrm{sc}/E_\mathrm{m})\,. \tag{4.28}$$

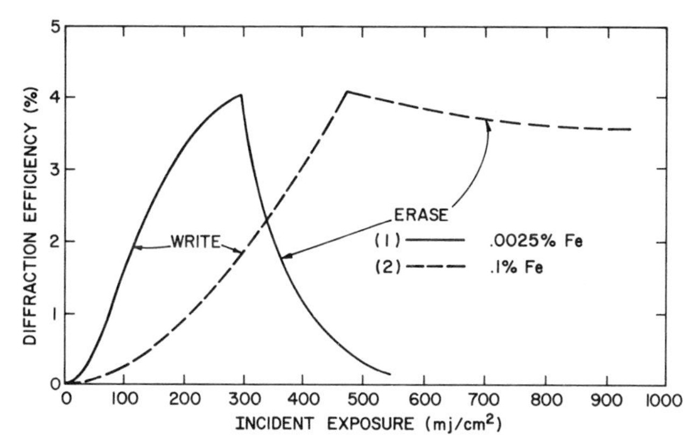


Fig. 4.4. Write/erase behavior for two doping levels of Fe in $LiNbO_3$. The crystals were prepared so as to have similar absorption strengths at 488 nm (see Sect. 4.2.1). The difference is due to the concentration of $Fe^{3+}$ ions [4.12, 31]. Crystal thickness for both is ~2 mm

Comparing this to the erase time $T_e = 2T_0$ for the same incident optical power, we find

$$t_s/t_e = (m/2)(E_{sc}/E_m). \tag{4.29}$$

Thus, if $E_m$ is much larger than $E_{sc}$, the erase time will be much longer than the storage time. Figure 4.4 shows this effect in a Crystal 2) that has a high $E_m$ due to the photovoltaic effect; erasure requires at least a factor of 100 more time than does storage. The other Crystal 1) has a lower $E_m$, and thus its write/erase behavior is more symmetrical. It also has a higher sensitivity for erasure. The differences are due to the free carrier lifetime; it is longer for the second crystal than the first [4.12]. The increase in lifetime is accomplished by material preparation techniques (Sect. 4.3.1) that decrease the concentration of empty traps.

### 4.1.3 Fixing

Although crystals can be prepared in which erasure takes much longer than storage, it is clear that continued readout of a hologram will eventually erase it. It is possible to use a different wavelength for readout and writing, but for thick holograms this will introduce distortions in the diffracted beam unless multi-photon absorption techniques are employed (see Sect. 4.3.4). Thus a method for fixing (making the holograms insensitive to the recording wavelength) is highly desirable, particularly for archival and read only applications.

Fixed holograms in FE crystals are optically stable ionic charge patterns. They are formed by the electric field pattern of the original hologram. In materials such as $LiNbO_3$ the fixing process uses thermally activated ionic conductivity [4.16, 17]. In others, such as $BaTiO_3$ and SBN, it involves field-assisted reversals of ferroelectric domains [4.18, 19]. It can be done after the

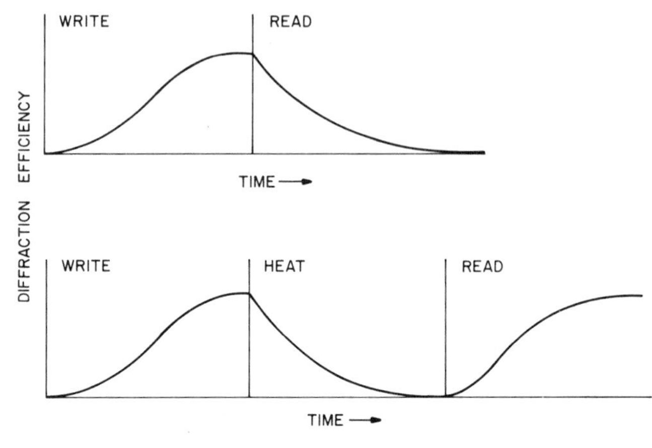


Fig. 4.5. Diffraction efficiency during readout for a normal hologram (top) and a hologram fixed by heating (bottom)

hologram is recorded, or during the recording process itself. $LiNbO_3$ will be discussed first.

Hologram fixing is accomplished in $LiNbO_3$ by heating the crystal during or after storage. At temperatures above 100° C, $LiNbO_3$ has an ionic conductivity that quickly neutralizes the space charge patterns of the hologram. Since the electrons are so deeply trapped that they remain stable during this process, the result is an ionic charge pattern that perfectly mirrors the pattern of the recorded hologram. Upon cooling to room temperature, the ionic pattern is frozen in. The crystal is then exposed to uniform light with a laser or an incoherent light source. The light tends to redistribute the electrons so as to reveal a certain fraction of the electric field pattern due to the ions. The hologram can then be read out without erasure.

Figure 4.5 schematically shows what happens to the diffraction efficiency during the above steps. The optical erasure of a normal hologram is shown for comparison. The fixing process also leads to decay of the diffraction efficiency. It neutralizes the space charge field, and thus removes the index modulation, but it does not erase the ionic pattern. That erasure has not occurred is shown by the reappearance of the hologram upon redistribution of the electrons.

The theory for the ionic relaxation process closely follows that given in Sect. 4.1.2 for optical erasure. The field pattern decays exponentially with a time constant given by a dielectric relaxation time $T_i$. The difference is that ionic conductivity, instead of photoconductivity, is used to calculate it, i.e.,

$$T_i = \varepsilon/\sigma_i , \tag{4.30}$$

where $\varepsilon$ is the dielectric constant and $\sigma_i$ is the ionic conductivity. The variation of $T_i$ with temperature for $LiNbO_3$ is shown in Fig. 4.6. At $\sim 160°$ C, $\sigma \cong 10^{-13} (\sim \text{cm})^{-1}$, giving $T_i \cong 30$ s. If the density of ionic defects is high enough ($> 10^{18}\,\text{cm}^{-3}$), this relaxation process will completely neutralize the field pattern [4.17]. That is, the ionic charge pattern will be exactly equal (and opposite) to the charge pattern of trapped electrons. The ionic defects involved in this process have not been positively identified.

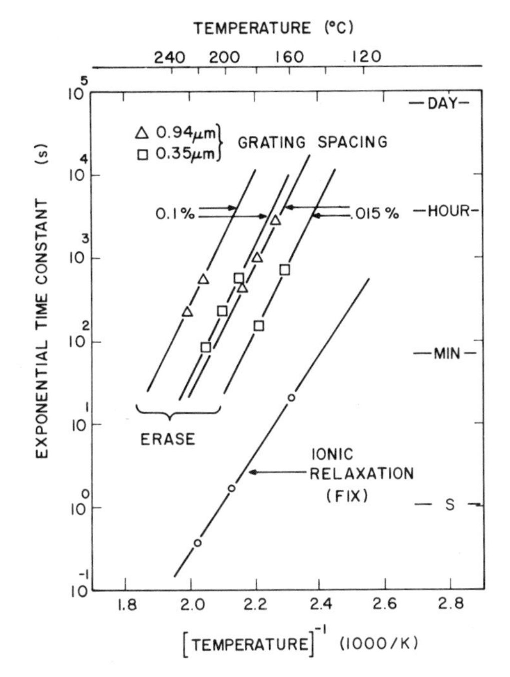


Fig. 4.6. The time constants for thermal fixing (lower line) and thermal erasure (upper lines) of holograms in Fe-doped $LiNbO_3$. Erasure time depends on the grating spacing and doping level, the fixing time does not [4.21]

An important consideration here is the amount of recovery of the diffraction efficiency during the last step shown in Fig. 4.5—exposure to light. In principle, one can obtain a final efficiency equal to that of the recorded hologram. In practice, however, this is not true; the photoconductivity induced by the light partially neutralizes the ionic charge pattern. The fraction of the original field that is recovered depends on the concentration of trapped electrons in the crystal ($N_0$). The specific effect of $N_0$ depends on which mechanism dominates electron transport—diffusion or drift.

Diffusion is expected in crystals with a low concentration of empty traps (see Sect. 4.1.1). The theory for this case was given by *Staebler* and *Amodei* [4.17]. They assumed an ionic charge pattern given by

$$\varrho_i = \varrho_0 \cos Kz \tag{4.31}$$

and considered the equilibrium situation of $J=0$ produced by a sufficiently long exposure to uniform light. Assuming a net density of trapped electrons $N_0$, and an equilibrium modulation on this density of $N_1 \cos Kz$, one has a free carrier density of

$$n = \alpha(N_0 + N_1 \cos Kz)\,, \tag{4.32}$$

where $\alpha$ is a factor that depends on the light intensity, absorption cross section, and free carrier lifetime. From (4.7), this gives for $N_1 \ll N_0$

$$e\mu N_0 E_f = eDN_1 K \sin Kz\,, \tag{4.33}$$

where $E_f$ is the final or equilibrium field and is the difference between the fields due to the ionic and electronic charges, i.e.,

$$E_f = (\varrho_0 - eN_1/\varepsilon K) \sin Kz\,. \tag{4.34}$$

From (4.33) and (4.34), using $D = \mu(kT/e)$, one finds

$$E_f/E_0 = (1 + e^2 N_0/kTK^2\varepsilon)^{-1}\,, \tag{4.35}$$

where $E_0$ is the maximum possible value of $E_f$ (for $N_1 = 0$) and is equivalent to the field of the original hologram. At room temperature and grating spacings of $\sim 1\ \mu m$, this becomes

$$E_f/E_0 = (1 + N_0/1.6 \times 10^{15}\ \mathrm{cm}^{-3})^{-1}\,. \tag{4.36}$$

Thus high recovery is expected only when the density of trapped electrons is less than $\sim 10^{15}\ \mathrm{cm}^{-3}$. Such behavior has been observed in undoped crystals [4.17]. Typically, however, crystals with a useful absorption coefficient contain $\sim 10^{17}\ \mathrm{cm}^{-3}$ trapped electrons. In this case, one must rely on drift for redistributing the electrons. This can be done by applying a field, or it can occur via the equivalent field of the bulk photovoltaic effect as suggested by *Von der Linde* and *Glass* [4.20].

The theory for drift-assisted recovery of the ionic field is done as above with the exception that the equilibrium condition is $\nabla \cdot \boldsymbol{J} = 0$. Thus one finds from (4.7), ignoring diffusion,

$$(\boldsymbol{E}_a + \boldsymbol{E}_f) \cdot \nabla n = -n \nabla \cdot \boldsymbol{E}_f\,, \tag{4.37}$$

where $E_a$ is the drift field and $n$ is the free carrier density. In this case the trapped electron density can be shifted by the field so that one must use

$$n = \alpha[N_0 + N_1 \cos(Kz + \phi)] \tag{4.38}$$

instead of (4.32). Proceeding as before, assuming $N_1 \ll N_0$ and $E_f \ll E_a$, we find

$$E_f/E_0 = [X^2 + 1]^{-1/2}\,, \tag{4.39}$$

where

$$X = eN_0/KE_a\varepsilon\,. \tag{4.40}$$

Thus, if $E_a \geqq eN_0/K\varepsilon$, a significant fraction of the hologram is recovered. Consider the case for $N_0 \cong 10^{17}\ \mathrm{cm}^{-3}$ and grating spacings $(2\pi/K)$ of $1\ \mu m$. A value of $\sim 10^5$ V/cm or larger is needed for $E_a$ and it can be provided entirely by the photovoltaic effect if the crystals contain a large concentration of empty traps (see Sect. 4.1.1). This explains the good fixing results found in heavily doped crystals [4.21].

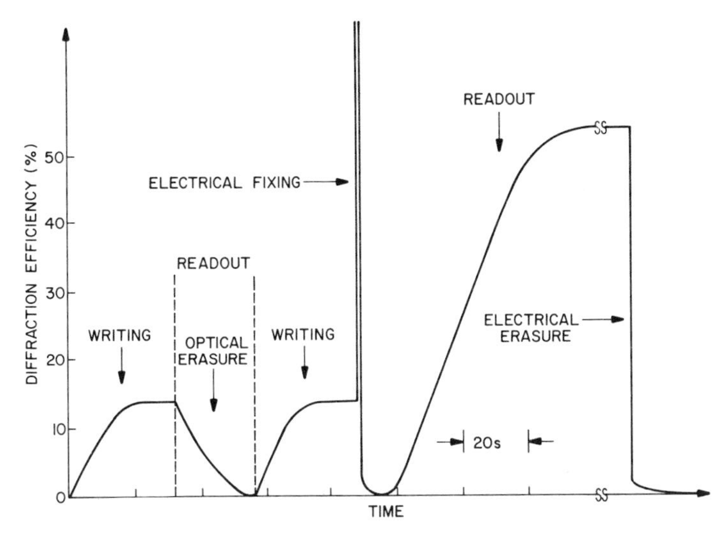


Fig. 4.7. Diffraction efficiency during readout in SBN of a normal hologram and a hologram fixed by an electric field (adapted from [4.19])

The fixed holograms, although quite stable at room temperature, can be erased. This is done by heating them to the fixing temperature and exposing them to uniform light. The same effect is obtained by heating the crystal to an even higher temperature, one that thermally activates the trapped electrons (see Fig. 4.6). In either case, the electrons redistribute while the ions follow them in an effort to maintain charge neutrality. Finally, both patterns, ionic and electronic, are erased [4.21].

FE materials with a low coercive field ($E_c$) can be used for an electrical fixing technique [4.18, 19]. This has been done in $BaTiO_3$ ($E_c = 1.1$ kV/cm) and SBN ($E_c = 970$ V/cm). A coercive field is the critical value of applied electric field required to reverse the internal polarization of an FE crystal. Application of such a field after the recording of a hologram produces a spatial pattern of domain reversal. This is an ionic pattern that can be read out without erasure.

A typical result of this process is shown in Fig. 4.7 for SNB [4.19]. The crystal is first poled to remove all domains. This is done with a field of $\sim 2$ kV/cm, much larger than $E_c$. Then the hologram is recorded as described for $LiNbO_3$. As in that case, readout of the hologram erases it. After poling the crystal again, another hologram is recorded. This hologram is then fixed by applying for 0.5 s a 1.25 kV/cm field antiparallel to the original poling field. Polarization reversals occur only in regions where the net field (applied field plus the space charge field of the hologram) is greater than $E_c$. The result is a domain pattern that replicates the recorded pattern of trapped charges. The domain pattern tends to neutralize the trapped charge pattern; thus the efficiency becomes quite low. During exposure to light, however, the trapped charges redistribute, and the hologram reappears. The high efficiency is attributed to overcancellation of the electronic space charge fields by local polarization switching. Erasure is done with another poling field. It wipes out all domain reversals associated with the fixed hologram.

A thermal fixing technique has also been devloped for SBN [4.22]. The holograms are recorded in crystals heated above the Curie temperature ($T_c \cong 50\,°C$). This is the critical temperature at which the material switches from a ferroelectric to a paraelectric state. The sample is then cooled and as it passes through $T_c$ the field patterns of the hologram induce corresponding polarization domains. The result, at room temperature, is a thermally and optically stable hologram.

Fixing can also be done by recording holograms in SBN with high light intensities [4.23]. This may be due to a heating effect, or to spontaneous reversals of polarization.

### 4.1.4 Electrical Enhancement

There are two enhancement effects that can be produced in FE materials with applied fields. The first is an enhancement of the recording process. This is due to a speed-up of the charge transport, and is described in Section 4.1.1. The second is an enhancement of the diffraction efficiency of a given hologram. It is due to an enhancement of the electro-optic coefficients and is described below. It occurs only in materials such as PLZT [4.19] and SBN [4.23]—those whose Curie point is close to room temperature.

To understand this effect we begin with the equation for the diffraction efficiency for a thick phase hologram [4.24]

$$\eta = \sin^2(\pi \Delta n d/\lambda \cos\theta)\,, \tag{4.41}$$

where $d$ is the thickness, $\lambda$ is the optical wavelength, $\theta$ is the recording angle shown in Fig. 4.2, and $\Delta n$ is the fundamental component of the modulation on the refractive index, i.e., $n = n_0 + \Delta n \cos Kz$. For the simple case of a sinusoidal electric field pattern, $E_1 = \Delta E \cos Kz$ [see e.g., (4.13)], and a linear electro-optic coefficient, we find

$$\Delta n = B \Delta E\,, \tag{4.42}$$

where $B$ depends on the electro-optic tensor and the orientation of the crystal [see (4.47) for example].

For a material whose Curie point is below room temperature, such as PLZT ceramics, the electro-optic properties are quadratic, i.e.,

$$n = n_0 + (A/2)(E_a + \Delta E \cos Kz)^2 \tag{4.43}$$

thus a first order diffraction does not occur unless a net electric field $E_a$ is applied. The reason is seen in Fig. 4.8. The field provides a bias that gives a first-order component

$$\Delta n = A E_a \Delta E\,. \tag{4.44}$$

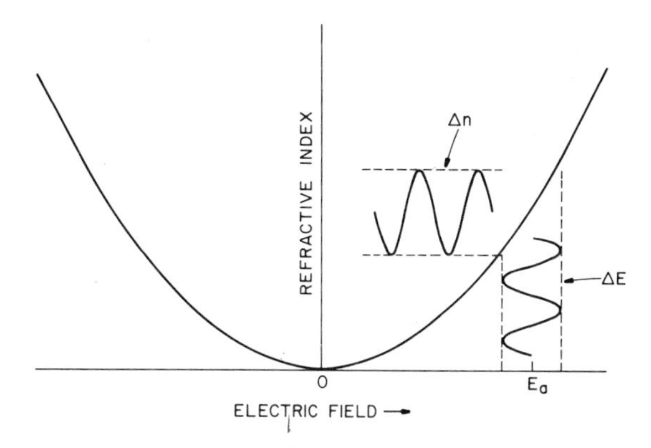


Fig. 4.8. Use of an applied field to give a first-order diffraction grating in a quadratic material

Thus the value of $\Delta n$, and thus the efficiency, can be altered with an applied field without affecting the space charge pattern. When storage occurs during the application of an electric field, both enhancements occur—charge transport and refractive index. In this case, one must consider the effect of the uniform space charge field that builds up across the illuminated region during storage [4.19].

Similar effects have been observed in SBN [4.23]. This material has, in addition to a strong linear coefficient, a second-order coefficient. Following the same argument as above, one finds a net linear coefficient of $B + AE_a$. At an applied field of $-B/A$, the net coefficient becomes zero, and the hologram is not read out. This is the origin of the latent-to-active operation discussed in Section 4.3.5.

## 4.2 Materials and Preparation

### 4.2.1 Lithium Niobate ($LiNbO_3$)

Single crystals of $LiNbO_3$ are grown using the Czochralski method [4.25]. Starting materials of lithium carbonate and niobium pentoxide, usually in the congruent composition [4.26], are melted in a crucible. Dopants are added to the melt as the appropriate transition metal oxide [4.27]. Growth is begun on a seed crystal lowered into the top of the melt. The seed is then rotated and, at the same time, slowly raised up from the melt to increase the size of the growing crystal. Crystals up to 6 cm diameter can be prepared. Poling of the crystals during growth, using a ~2 mA current from melt to seed, decreases the domain structure [4.28]. In furnaces with a low vertical temperature gradient, however, the crystals remain above their Curie temperature for some time after growth is terminated. These crystals must be poled later with a low current while the crystal is maintained near 1200 °C [4.28]. After growth, a crystal is annealed at 1150 °C for 4 or 5 h, then slowly cooled to room temperature. They are then cut and polished into the desired geometry (see Sect. 4.3). Both doped and undoped crystals grown as described above have excellent optical quality and are suitable for high resolution hologram storage.

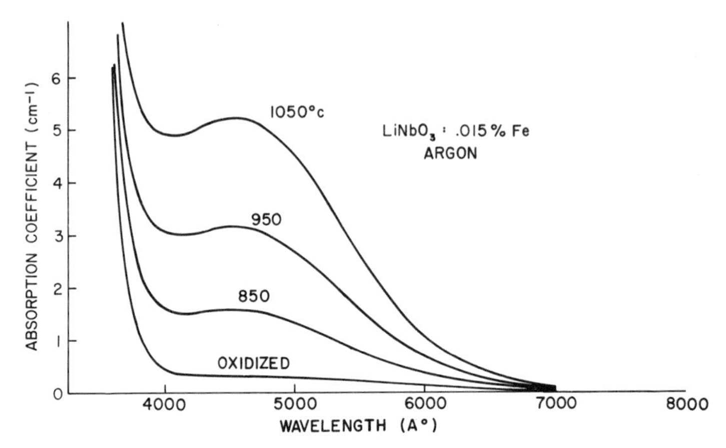


Fig. 4.9. Absorption spectra of $Fe^{2+}$ in $LiNbO_3$ for various reduction treatments. The net transmission is given by $\exp(\alpha d)$ where $\alpha$ is the absorption coefficient and $d$ is the thickness

The dopant most commonly used is Fe [4.27, 29]. It is a common impurity in undoped crystals [4.29], and introduces the traps required for hologram storage. It enters the lattice as either $Fe^{2+}$ or $Fe^{3+}$ ions [4.30]. In the hologram storage process, the $Fe^{2+}$ ions are occupied traps, and the $Fe^{3+}$ ions are empty traps. The concentrations of these two species determine the holographic behavior of the crystals and thus must be controlled. The means for control is 1) the amount of dopant added to the melt; and 2) oxidation-reduction heat treatments done after growth. The doping determines the total concentration of Fe ions, while the heat treatments determine the fractions that are in the $Fe^{2+}$ or $Fe^{3+}$ valence state. Heating in an argon gas atmosphere tends to reduce $Fe^{3+}$ ions into the $Fe^{2+}$ ions (reduction); heating in oxygen (oxidation) reverses the process [4.27, 29, 31]. The treatments also introduce or remove charge compensators to keep the crystal neutral, but there is no direct evidence that they are directly involved in the storage process.

One of the first considerations in the thermal treatment of an Fe-doped $LiNbO_3$ crystal is the absorption at the wavelength of interest. This absorption is due to the $Fe^{2+}$ ions (the occupied traps) and it determines (in part) the sensitivity of the crystal to incident light. Of course, if too much incident light is absorbed, very little is transmitted through the crystal for readout. A reasonable compromise is a net absorption (determined by the density of $Fe^{2+}$ ions and the crystal thickness) of $\sim 67\%$ or less; two-thirds is absorbed for storage and one-third is transmitted for readout [4.20]. Figure 4.9 shows the increase in the $Fe^{2+}$ absorption spectrum with higher degrees of reduction. For a thickness of 2 mm, a total absorption of 67% at 488 nm requires a $Fe^{2+}$ concentration of $\sim 10^{17}$ $cm^{-3}$ [4.31].

The second consideration is the concentration of empty traps ($Fe^{3+}$ ions) and this depends on the desired application. In applications requiring an erasure sensitivity much lower than for storage, a high equivalent photovoltaic field is

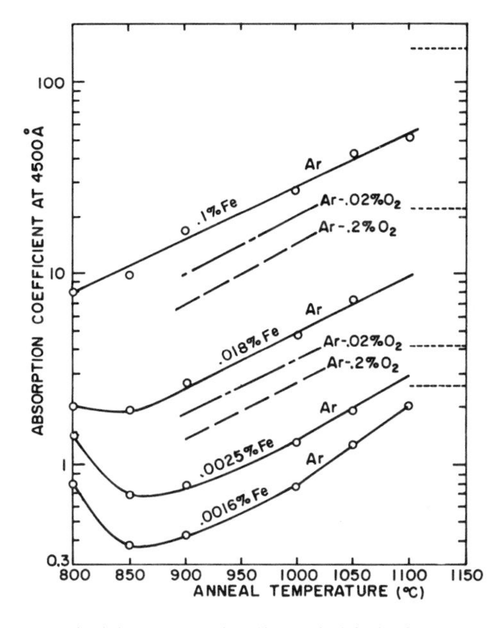


Fig. 4.10. Absorption coefficient of Fe-doped $LiNbO_3$ after heat treatments in pure argon, and in argon-oxygen mixtures. The dotted line to the right is the coefficient expected for full reduction (all ions reduced to the $Fe^{2+}$ state) for each of the doping levels [4.31]

needed (see Sect. 4.1.1), and this is found in crystals with an $Fe^{3+}$ concentration of $10^{18}$ $cm^{-3}$ or larger. Here one uses a doping level of 0.01 mol-% of more of Fe, and a reduction treatment that converts only a small fraction of the ions into the divalent state. In applications requiring a symmetrical write/erase behavior, such as the read/write application discussed in Section 4.3.1, the concentration of $Fe^{3+}$ ions must be low ($\sim 10^{17}$ $cm^{-3}$). In this case, crystals with a low doping level ($\sim$0.001 mol-% of Fe) are used, and a very large fraction of the ions is reduced into the divalent state. Figure 4.4 shows the resulting write/erase behavior for both cases. The absorption coefficient, and fractional amount of reduction, can be determined from Fig. 4.10 for crystals of different doping levels and various heat treatments. Very large degrees of reduction (greater than 90%) can be achieved in these crystals with another type of treatment. The crystals are packed in $Li_2CO_3$ powder, and then annealed in oxygen at $\sim$550 °C [4.31].

Other dopants have also been used in $LiNbO_3$. These are Cu [4.27], Mn [4.27], Rh [4.32], and U [4.33]. These show no significant improvement over Fe doping, with the exception of an improved thermal stability for U-doped crystals [4.33]. The excitation spectra of the dopants vary [4.34, 35] and this may allow some flexibility in the choice of storage wavelength.

### 4.2.2 Strontium Barium Niobate (SBN)

Holograms were first recorded in SBN by *Thaxter* [4.36]. This material has a coercive field and Curie temperature much lower than for $LiNbO_3$. Thus it has some interesting effects, e.g., electrical fixing and erasure, not exhibited by $LiNbO_3$. The electron traps, however, are not well understood.

Single crystals are grown with the Czochralski technique [4.37]. The starting materials are strontium and barium carbonates, and niobium pentoxide. Crystals can be grown over the composition range $Sr_{0.75}Ba_{0.25}Nb_2O_6$ to $Sr_{0.25}Ba_{0.75}Nb_2O_6$. The former, SBN (75/25), is best for hologram storage. Single crystals are grown from oriented seeds by pulling from the melt. Crystals of good optical quality have been grown as large as 1 cm in diameter and 7 cm in length. A modified Czochralski technique produces strain-free crystals with improved homogeneity [4.38].

The crystals are dark amber as grown and bleach to a pale yellow or amber after annealing in oxygen at 1400 °C. The nature of the color centers is not known, nor is there a published study of doped crystals.

### 4.2.3 Barium Titanate ($BaTiO_3$)

Storage in $BaTiO_3$ was first seen in $BaTiO_3$ by *Townsend* and *Lamacchia* [4.39]. This material, like SBN, has a low coercive field and Curie temperature. Crystals are flux grown by the Remeika technique [4.40]. The melt is composed of ~30% barium titanate covered with ~70% potassium fluoride. The crystals are grown on the bottom of the crucible, and normally form in thin plates (butterfly twins) that are convenient for use. Crystals have been grown with Fe impurities [4.41], and these are better for storage than undoped crystals [4.42].

### 4.2.4 Barium Sodium Niobate

Holograms were recorded in Fe, Mo-doped crystals of barium sodium niobate by *Amodei* et al. [4.43]. The crystals were grown from the melt by slowly cooling the crucible to room temperature. They were of poor optical quality. As yet, no reports of storage in high quality material have been published.

### 4.2.5 Lithium Tantalate ($LiTaO_3$)

Single crystals of $LiTaO_3$ are grown as in the case of $LiNbO_3$, by the Czochralski method [4.25]. Both holograms [4.44] and optical damage patterns [4.45, 46] have been recorded in the material. Various transition elements have been used as dopants [4.46]. As yet, however, doped $LiTaO_3$ has not been well characterized for hologram storage.

### 4.2.6 Potassium Tantalate Niobate (KTN)

The first studies of KTN involved optical damage effects [4.47, 48]. Recent experiments on optical damage via two-photon absorption show the promise of very high holographic sensitivities [4.49]. The crystals used in these studies were flux grown. The composition determined the Curie temperature. For the work

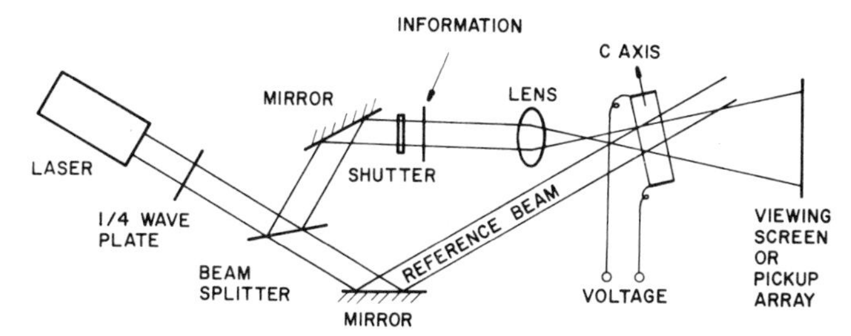


Fig. 4.11. Hologram storage setup for FE crystals. The 1/4 wave plate is used to rotate the polarization vector of the laser beam. The applied voltage, if necessary, enhances the storage process or the readout efficiency. For some crystals it is not required

described in [4.49] the composition was $KTa_{0.65}Nb_{0.35}O_3$ which gave a ferroelectric Curie temperature of ~10 °C. Thus the electro-optic behavior of this material is quadratic (Kerr effect) at room temperature. KTN is difficult to grow in large single crystals with good optical quality.

### 4.2.7 PLZT Ceramics

Hologram storage was first demonstrated in a PLZT ceramic by *Micheron* et al. [4.50]. This material is fabricated by classical ceramic techniques [4.51], i.e., the powder of the desired composition is cold pressed, then sintered. Large diameters are possible since the sample is not a single crystal.

The composition used for hologram storage is PLZT 9 (9% substituted), given by $Pb_{0.91}La_{0.09}(Zr_{0.65}Ti_{0.35})_{0.9775}O_3$. Its electro-optic effect at room temperature is quadratic. The photosensitivity is relatively low [4.52] and is only slightly increased by transition metal substitutions [4.53].

## 4.3 Methods of Use

The basic setup for hologram storage in FE materials is shown in Fig. 4.11. The *c*-axis (the optic axis) of the crystal is parallel to the polarization plane of the beams, and both are normal to the plane that bisects the angle between the beams. This maximizes the observed refractive index modulations, and thus maximizes the net sensitivity. Electric fields, where desired, are applied along this direction. Information to be recorded can be an image, such as a line drawing or printed matter, or it can be a page composer if the system is to interface with a digital system. Since this is a bulk storage mechanism, one can store many holograms, each one at a different angle, within the same volume. If this 3D approach is desired, there must be means for rotating the medium or the light beams. More complete information concerning the specific details of holographic storage systems can be found elsewhere [4.54]. The choice of laser depends on the power and wavelength requirements of the crystal. Most of the work reported on FE storage materials has been done with 488 nm light from an argon ion laser.

### 4.3.1 Read/Write

The read/write application of FE materials is the simplest. A page of digital data is recorded with the object and reference beams directed onto a small portion of the crystal. The reference beam, directed onto the same area, is used for both readout and erasure. The operations are all on-line, i.e., done during real time operation of the system. A readout beam that simultaneously erases the information, as is the case here, is desirable. It shortens the total cycle time by doing two operations at once. If the information is to be retained it can be immediately recycled as done with magnetic core memories, provided that the memory cycle is fast enough.

For a fast cycle time, one requires a material with fast erasure properties such as those found in heavily reduced, lightly doped crystals of Fe-doped $LiNbO_3$ (see Fig. 4.4). *Alphonse* and *Phillips* used this material in a simulated memory having $10^3$ bits per page and 27 dB signal-to-noise ratio [4.55]. They demonstrated the feasibility of a 60 ms cycle time, an improvement of two orders of magnitude over previous experimental prototypes using other optical storage media, but not yet fast enough for a practical system.

### 4.3.2 Read/Write with Nondestructive Readout (NDRO)

In this format, the information can be read out a number of times without erasure. In order to be useful for on-line operation, the total cycle time (the sum of the storage, readout, and erasure times) must be as short as that for the read/write system described above.

An experimental read/write NDRO memory using 3D storage was described by *D'Auria* et al. [4.56]. They took advantage of the thickness of the storage media to record the pages in blocks of superimposed holograms. Access is achieved by rotation of the reference beam. This can, in principle, increase the storage capacity by two orders of magnitude. They used a material (0.5 mol-% Fe-doped $LiNbO_3$) which is not easily erased with light (see Fig. 4.4). This was necessary to prevent the erasure of previously recorded pages during the storage of a new one within the same block. Thus, some readout can be achieved without immediately losing the information. They erased the information by locally heating the blocks to 200 °C with an array of transparent electrodes. The cycle time was several tens of seconds. A method for erasure as fast as storage was recently described [4.57]. It consists of coherently recording complementary refractive index changes to cancel the recorded changes of a single page. The process is selective; only one page out of a block is erased. Registration and alignment problems may limit the practical application of this method.

In principle, SBN can also be used for NDRO operation. The holograms are fixed with an applied field, and then erased with another (see Fig. 4.7). $LiNbO_3$ can also be used in this way, fixing and erasure done thermally as discussed in Section 4.1.3. The major obstacle for these cases is the cycle time; large amounts of light are generally required to recover a fixed hologram before it can be read out.

### 4.3.3 Read-Only

In a read-only system, the information is contained in prerecorded static or archival data files. Therefore, the recording and processing times are not critical. High storage capacity and stability are the most desirable features. In particular, the holograms must not decay during continuous readout or during shelf storage.

A volume read-only storage procedure was demonstrated by *Staebler* et al. [4.21] in Fe-doped $LiNbO_3$. Five hundred holograms were sequentially recorded at different angles in a crystal heated to $\sim 160\,^{\circ}C$, a temperature at which fixing occurs simultaneous with storage. Then the holograms were read out after cooling the crystal to room temperature and exposing it to uniform light (see Sect. 4.1.3). All had an efficiency of more than 2.5%, showing that optical erasure during storage is not a problem in heavily doped crystals. The stability of the holograms is excellent. Holograms stored nearly four years ago and periodically exposed to light have shown no degradation in either image quality or diffraction efficiency.

All holograms can be erased by heating the crystal to $\sim 200\,^{\circ}C$ (see Fig. 4.6). The same crystal can then be reused; no fatigue has ever been observed. From Fig. 4.6 the maximum shelf life of the recorded holograms is $10^5$ years at room temperature.

### 4.3.4 Multiphoton Absorption

A unique means for nondestructive readout has been demonstrated by *Von der Linde* et al. [4.58]. Laser pulses of peak intensity sufficient to create two-photon absorption processes are used to write the hologram. Readout is done with less intense light so that only single photon steps can occur. If the material is insensitive to the single photon wavelength, then the readout is completely nondestructive. Since the interference patterns that create the hologram correspond to the single photon wavelength, no distortion occurs. This technique replaces the fixing techniques possible in FE materials, while maintaining the possibility of fast optical erasure. Erasure is done with two-photon absorption of a spatially uniform light beam. If the cycle times are short enough, it can be used on-line in an NDRO read/write system. If the thermal stability of the recorded hologram is long, it can be used as a read-only memory. Here the sample would have to be protected from ambient light of wavelengths that can erase the holograms.

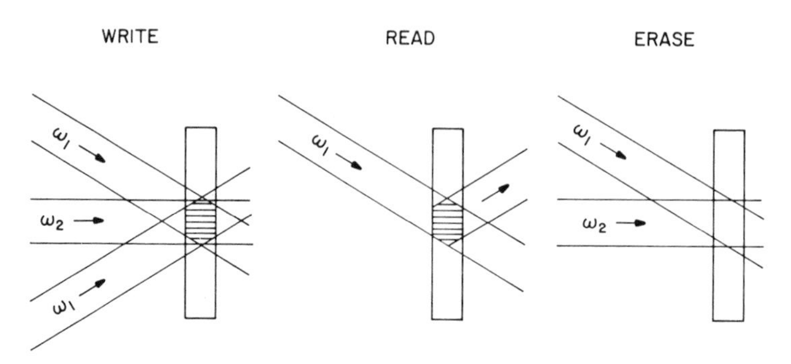


Fig. 4.12. Two-photon holographic steps. Storage and erasure occur via two-photon absorption at an optical frequency of $\omega_1 + \omega_2$. The $\omega_1$ readout beam, by itself, is not absorbed by the crystal. Thus, readout is nondestructive (adapted from [4.20])

This process can also be done with an auxiliary beam, shown in Fig. 4.12. The crystal is sensitive to $\omega_1 + \omega_2$, but insensitive to $\omega_1$. A hologram was recorded in this fashion in undoped $LiNbO_3$ using 1.06 μm for the writing beams, and 0.53 μm for the auxiliary beam. Readout with the full intensity 1.06 μm beam did not erase the hologram [4.58].

This approach has important possibilities, but needs more research. There are, as yet, a number of problems in its practical implementation. The first is that the high peak intensity required for multiphoton processes demands the use of ultrashort pulses. This increases the complexity of the laser and optical setup. In addition, an auxiliary beam may require the use of two lasers. Another problem arises from the requirement of a linear exposure characteristic for practical holographic recording; two-photon absorption via a virtual transition is nonlinear. This can be avoided by using a real two-step operation where the auxiliary beam (for example) excites a localized electron into a long-lived excited state where it remains until excited into the conduction band by the photons of the storage beams. Such a process can also reduce the peak power required for storage. Recent experiments [4.59] on Cr doped crystals of $LiNbO_3$ and $LiTaO_3$ have demonstrated the advantages of this two-step process. Experiments on the relative times for storage and erasure are still needed if these processes are to be evaluated for both read/write and read-only applications.

### 4.3.5 Layered Memory

Another approach to memory storage in FE materials was described by *Thaxter* and *Kestigian* [4.23]. A number of storage planes, each composed of a sheet of SBN, are selectively accessed with applied voltage. The selection is done during storage via the electrical enhancement of the storage process, and during readout via the latent-to-active mode discussed in Section 4.1.4. An experimental model using two lasers was constructed.

### 4.3.6 Holographic Page Synthesis

*Carlsen* [4.60] demonstrated a holographic digital data storage technique in Fe-doped $LiNbO_3$. The main feature of this technique is that the data are stored sequentially, one bit at a time, thus eliminating the need for a page composer. Pages are synthesized by choosing a given angular address for the reference beam, and then sequentially storing the desired information. Readout is parallel, with page address done by varying the angle of the reference beam. The use of a bulk medium allows a high page capacity and reduces intermodulation between the stored elements.

## 4.4 Holographic Properties

### 4.4.1 Sensitivity

The diffraction efficiency produced by a given exposure is a measure of the sensitivity. For FE materials this depends on both the charge transport and electro-optic properties of the particular crystal being used. A theoretical approach to the expected sensitivity is given below. Measured values follow. It will be seen that FE crystals can have extremely high sensitivities, even approaching those of photographic emulsions.

The derivation for the sensitivity follows from the theory given in Section 4.1. The crystal's absorption is taken into account here, using the optimum value [4.20] of two-thirds. Only the early stage of storage, where it is linear with time, is considered. The result holds true only for low diffraction efficiency ($\eta < 40\%$), and is independent of crystal thickness. The sensitivity is defined as

$$S = \eta^{1/2}/\mathscr{E}\,, \tag{4.45}$$

where $\eta$ is the diffraction efficiency (diffracted intensity/incident reatout intensity) and $\mathscr{E}$ is $It$, the product of the irradiance of the incident storage light ($I$) and the storage time ($t$).

First, the expression of *Young* et al. [(4.19)] is used to find the magnitude of the space charge field ($E_1$) for a given free electron generation rate ($G$) and storage time $t$. This is then related to $\mathscr{E}$ from

$$G = 2If/3h\nu d\,, \tag{4.46}$$

where $h\nu$ is the energy of the absorbed photons, $f$ is the probability that an absorbed photon produces a free carrier, and $d$ is the crystal thickness (an arbitrary parameter that cancels out later).

Next, the refractive index change ($\Delta n$) is found from the electro-optic coefficients of the crystal. This is highly dependent on the orientation of the crystal and is maximum for the orientation shown in Fig. 4.11. For this case one has [4.60]:

$$\Delta n = (n_e^3 r_{33}/2)\Delta E\,, \tag{4.47}$$

where $n_e$ is the extraordinary index of refraction and $r_{33}$ is the relevant component of the electro-optic tensor. The diffraction efficiency is now determined from (4.41) giving for the sensitivity

$$S = [m/\cos\theta][2/3\sqrt{3}][fL/(1+K^2L^2)^{1/2}][n_e^3 r_{33}/2\varepsilon][\pi e/hc]\,. \tag{4.48}$$

The bracketed terms, and their maximum values, are discussed below.

The first term depends on the relative intensity and angle ($\theta$) of the beams (see Fig. 4.2). $m=1$ for beams of equal intensity.

The second term reflects the effect of the crystal absorption (2/3) during storage and transmission (1/3) during readout. It is already maximized.

The third term is the effective distance that an absorbed photon displaces a trapped charge. It pertains here to storage via a drift field, but the final conclusions hold equally well for diffusion. When $L$ (the drift length of the electrons) is small, the effective distance can be enhanced by increasing the drift field. The maximum enhancement occurs when $L$ becomes larger than the grating spacing ($l=2\pi/K$) of the hologram. Then the bracketed quantity in (4.48) becomes simply $fl/2\pi$. The largest possible value of $f$ is unity. $l$ is limited by practical considerations. A typical value is $\sim 1\,\mu$m.

The fourth term in brackets reflects the electro-optic and dielectric properties of the particular FE material being considered. It, however, varies little from material to material [4.20]. This is because both $r_{33}$ and $\varepsilon$ represent the ease with which the materials polarization is affected by an electric field. The last bracketed term is composed of the electronic charge ($e$), Planck's constant ($h$), and the velocity of light ($c$).

Equation (4.48) is now used to calculate the theoretical maximum sensitivity. One assumes that $KL \gg 1$, and that $f$, $m$, and $\cos\theta$ are all unity. ($\cos\theta$ is not far from unity in a typical situation, e.g., $\cos\theta = 0.96$ for a 1 μm grating produced by interference of 488 nm beams.) This gives

$$S_{\max} = (l/3\sqrt{3})(n_e^3 r_{33}/2\varepsilon)(e/hc)\,. \tag{4.49}$$

The maximum sensitivity depends on the grating spacing. Consider the case of a 1 μm hologram in $LiNbO_3$ ($n_e = 2.277$, $r_{33} = 3.2 \times 10^{-11}$ m/V, and $\varepsilon_{33} = 2.83 \times 10^{-10}$ F/m). From (4.49) we find that a diffraction efficiency of 1% would require only $10^{-4}$ J/cm$^2$ of incident light. Similar results are expected for any linear electro-optic material. Quadratic materials, which require an applied

Table 4.1

| Material | Storage sensitivity | Erase sensitivity | Index change | Comment |
|---|---|---|---|---|
| | $\eta^{1/2}/\mathscr{E}$ [cm²/J] | $1/\mathscr{E}$ [cm²/J] | | |
| KTN | $2.2\times10^3$ | $10^4$ | $5\times10^{-5}$ | Storage via two-photon absorption with 6.25 kV/cm applied field [4.20, 49] |
| SNB | $3.3\times10^1$ | 5 | $10^{-5}$ | Storage and readout with 3 kV/cm applied field [4.23] |
| $LiNbO_3$:Fe [treated] | 2.5 | $8.3\times10^1$ | $10^{-6}$ | In low doped and heavily reduced crystals [4.12, 31] |
| $LiNbO_3$:Fe | $5\times10^{-1}$ | $1.6\times10^{-1}$ to $4\times10^{-2}$ | $10^{-3}$ | In highly doped and lightly reduced crystals [4.12, 27] |
| $LiNbO_3$:Fe | $5\times10^{-2}$ | 0 | $4\times10^{-4}$ | For fixed holograms in heated crystals [4.21] |
| PLZT | $5\times10^{-3}$ | $4\times10^{-3}$ | $10^{-3}$ | Storage and readout with 10 kV/cm applied field [4.50, 52] |
| $BaTiO_3$:Fe | $3\times10^{-3}$ | Electrical | $3\times10^{-4}$ | [4.42] |

electric field for readout (Sect. 4.1.4), can have a lower electro-optic term in (4.49), and thus would require (in principle) more energy for the same efficiency [Ref. [4.20], Eq. (4.29)].

The erasure sensitivity can be calculated from Section 4.1.3. It does not depend on the strength of the initial hologram, nor is it affected by drift fields. The primary factor is the amount of photoconductivity produced by the erasing light. From (4.4), (4.10), (4.27), and (4.46) one finds that the incident energy needed to erase a crystal (with an absorption of 2/3 and thickness $d$) is given by

$$\mathscr{E} = 3\varepsilon h\nu d/e\mu\tau\,, \tag{4.50}$$

where $\mu\tau$ is the mobility-lifetime product of the free carriers. Clearly, large lifetimes are needed for sensitive erasure, and this has been accomplished in $LiNbO_3$ by decreasing the concentration of empty traps (see Sect. 4.1.2). It is instructive to consider the hypothetical case where $\mu\tau$ has been increased to the point where the maximum storage sensitivity occurs via diffusion, i.e., when the diffusion length [(4.22)] is a factor of $2\pi$ smaller than grating spacing. This gives, for a 1 μm spacing, a $\mu\tau$ of $\sim10^{-8}$ cm²/V, well within the realm of possibility for these materials [4.12, 15]. Now, for a 2 mm thick crystal, and a wavelength of 488 nm, we have

$$\mathscr{E}\,(\text{erase}) = 4\times10^{-4}\ \text{J/cm}^2\,.$$

Erasure can be nearly as sensitive as storage.

Table 4.1 shows the sensitivity for storage and erasure for a number of materials. The largest sensitivities are found in KTN. This material, in fact, exceeds the best storage sensitivity calculated above; perhaps it was measured with a grating spacing much larger than the 1 μm value used for the calculation [see (4.49)]. Both KTN and SBN show promising results but their practical application is severely limited by the problems associated with the application of large electric fields, and the lack of techniques for tailoring the materials storage and erasure behavior. KTN also requires two-photon absorption as discussed in Section 4.3.4, and is difficult to prepare in large crystals of high quality.

$LiNbO_3$, at present, is the most practical material. It can be obtained in large single crystals of excellent optical quality, and material treatments controlling the crystal absorption and relative sensitivities for storage and erasure are well established (see Sect. 4.2). Nevertheless, its sensitivities are still much less than the theoretical maximum. This is due to a combination of a short electron migration length ($L$), and a small quantum efficiency ($f$) for the photogeneration of free carriers. The latter is suggested by the difference between the sensitivity spectrum and absorption spectra of $LiNbO_3$:Fe [4.61]. The migration lengths, and thus the storage sensitivity, can be increased with applied [4.5] or optically generated [4.62] fields, particularly in the treated crystals [4.12], but this requires a complex contact arrangement if large areas are to be used. Clearly more research is needed. A possible approach, suitable for any of these materials, is to use a set of traps with an extremely small capture cross section. This, in principle, could increase the migration length without sacrificing any of the other storage properties dependent on a large trap density. Research on shallow traps [4.63] should also prove fruitful. A tradeoff with dark storage time should be kept in mind here; increasing the migration length can also increase the dark conductivity of the crystal (see Sect. 4.4.4).

### 4.4.2 Dynamic Range

The dynamic range of a phase storage medium is its maximum possible photo-induced change of the refractive index. It determines the largest diffraction efficiency that can be recorded in a crystal of a given thickness, and the number of different holograms that can be recorded in a given volume. Measured values are shown in Table 4.1.

There are two theoretical limits to the dynamic range in FE crystals. The first is the density of empty or occupied traps, whichever is lower. This determines the largest space charge density that can be built up during exposure to light. From (4.3), and assuming sinusoidal gratings, one can see that the charge density required for a given index is inversely proportional to the grating spacing. For a 1 $\mu$m grating spacing, an index change of $2 \times 10^{-5}$ is produced with a trap density of only $10^{15}$ $cm^{-3}$. Impurity-doped crystals do not normally have a problem in this respect because the trap concentrations are in excess of $10^{18}$ $cm^{-3}$. The only case where the traps can limit the diffraction efficiency in a doped crystal is for

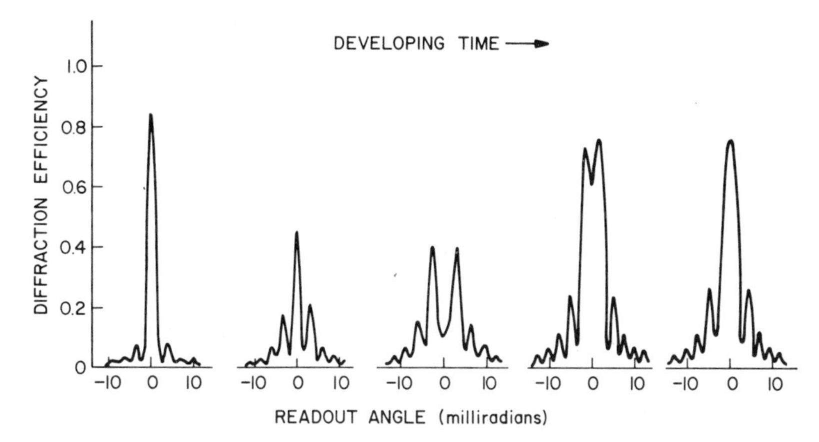


Fig. 4.13. An example of the large dynamic range of a fixed hologram in Fe-doped $LiNbO_3$. Various stages of the developing process (exposure to light) are shown. The final diffraction grating (on the right) has the angular response consistent, for the 2 mm thickness used here, with a refractive index modulation of $4\times10^{-4}$ [4.21]

those prepared so as to have a very low density of empty traps, e.g., the treated crystals of $LiNbO_3$:Fe shown in Table 4.1. This situation enhances the erasure sensitivity but at the expense of a limited storage range [4.12]. Other FE materials (e.g., KTN and SBN) rely on naturally present impurities or defects for electron trapping. Nevertheless, the trap concentration is apparently sufficient for experimentation on these crystals. The large electro-optic coefficients of these materials have little influence here; they are compensated by an equally large dielectric constant [see discussion of (4.48)].

The second limitation on the dynamic range is the equilibrium space charge field. It saturates when the field of the hologram has become comparable to the effective field for storage [see (4.9)]. If diffusion is the dominant process, the equilibrium field is given simply by the effective field for diffusion [(4.12)]. For a grating spacing of $\sim 1\,\mu m$ and a temperature of 300 K, this is $\sim 1.6$ kV/cm. The maximum index change ($\Delta n$) is then determined by the electro-optic coefficient of the material being used. For $LiNbO_3$ $\Delta n$ is $3\times10^{-5}$. It is larger in materials with a larger electro-optic coefficient, e.g., $\sim 10^{-3}$ for SBN. Since this is more than the observed value (Table 4.1), this material is probably trap limited.

The largest possible index charge occurs when storage occurs via a uniform electric field. In $LiNbO_3$ an equivalent field can be generated internally via the bulk photovoltaic effect. The largest field is $\sim 10^5$ kV/cm giving $\sim 2\times10^{-3}$ as the maximum index change, consistent with the measured value. Holograms fixed in $LiNbO_3$ also have a large dynamic range, as shown by Fig. 4.13.

A large dynamic range allows the use of thin crystals (e.g., $\sim 10\,\mu m$) for hologram storage, but it is not necessary for thicker crystals unless one is interested in multiple storage. Fe-doped $LiNbO_3$ is of particular interest here because it has, as normally prepared, the low erasure sensitivity needed to record many holograms, one after the other, at slightly different angles in the same crystal. Its dynamic range would then allow the storage by this procedure of at least $10^3$ holograms in a 1-cm-thick crystal, a capacity well within the limits set by the angular resolution of holograms of this thickness [4.24].

### 4.4.3 Resolution

Since, for the most part, these materials are single crystals, the spatial resolution is in principle limited only by the distance between traps. Consider a trap density of $10^{15}\,\mathrm{cm}^{-3}$, the minimum value required for an index change of $2\times10^{-5}$ at a $\sim 1\,\mu\mathrm{m}$ grating spacing. The distance between traps is, on the average, only 100 nm, an order of magnitude smaller than the grating spacing. Thus, the hologram can clearly be recorded at this spatial frequency, but it may have some scattering noise due to statistical fluctuations of the trapped electrons (see Sect. 4.3.5). The noise would be even more severe at smaller grating spacings. To compound the problem, a smaller grating spacing would lead to a lower refractive index change, assuming a trap limited situation. For these reasons, the use of impurity doping is quite important. Concentrations in excess of $10^{18}\mathrm{cm}^{-3}$ can be easily achieved, giving a distance between traps of only 10 nm. This should be sufficient for any holographic application. PLZT ceramics, which are of particular interest for hologram storage because they can be prepared in quite large dimensions, contain pores as small as $0.5\,\mu\mathrm{m}$ [4.51]. The resolution limits due to these defects have not been studied.

### 4.4.4 Storage Time

The storage time of the recorded holograms is determined by the dark conductivity of the material [see e.g., (4.27) for the erasure due to photoconductivity]. Most FE materials have low conductivity at room temperatures ($\sim 10^{-18}(\Omega\mathrm{cm})^{-1}$). Reported storage times range from 10 h for KTN [4.49], weeks for SBN [4.23] and PLZT [4.52], a few months for $LiNbO_3$:Fe [4.27], and 5 months for $LiNbO_3$:U [4.33]. The conductivity mechanisms are not well understood. There is recent evidence that Si impurities may be involved in $LiNbO_3$ [4.65].

It should be noted, however, that the decay of a hologram's diffraction efficiency does not necessarily mean that the information is lost. The dark conductivity could be primarily ionic, as in the case for $LiNbO_3$. The decay then corresponds in fact to the first step of a fixing process—the formation of an ionic pattern that neutralizes the electronic pattern of the stored hologram. Actual loss of the information occurs only upon thermal activation and subsequent redistribution of the trapped electrons. This process can be quite slow. Storage times for fixed holograms of up to $10^5$ years are predicted for $LiNbO_3$ [4.21]. Holograms in heavily reduced crystals of Fe-doped $LiNbO_3$ decay within weeks, rather than months [4.55]. This may be due to electronic conductivity rather than ionic, caused by the increased free carrier migration lengths [4.63] in such crystals.

### 4.4.5 Noise and Distortion

Due to the recent development of FE storage materials, the problems of noise and distortions are the least studied. Nevertheless, a number of potential problems have been pointed out, and they are discussed here.

One problem is due to beam coupling. Since FE materials have no latency, i.e., the refractive index appears immediately, certain beam coupling effects that can usually be ignored during recording in other media become quite significant. In principle, a phase hologram can act as a periodic array that couples the two interferring beams that record the holograms. This rearranges their amplitude and phase as they travel through the crystal. The result is to limit or distort the hologram that is ultimately recorded. The analysis of this process has been carried out for the storage of sinusoidal gratings [4.8, 63, 67]. A similar effect occurs during optical readout [4.8, 12, 67], even of fixed holograms [4.17]. Although it has not been studied with holograms that contain information, one suspects that it will limit the useful diffraction efficiency. Holograms with low diffraction efficiency produce little coupling.

Another problem associated with the lack of latency is the buildup of optical scattering [4.27, 68]. This occurs when any photosensitive FE crystal is exposed to a coherent beam of light. A defect that scatters light produces a spherical wave that then interferes with the original beam. The resulting interference patterns are recorded as index changes that, in turn, lead to more scattering. The result is a rapid buildup of complicated scattering patterns that produce noise in the readout image. This process occurs during storage, and thus limits the total exposure (and thus the diffraction efficiency) that can be used. It also occurs during readout, and can be a particularly severe problem for holograms fixed in highly sensitive materials.

There are means for avoiding the scattering effects. One is to use a latent storage process. Such a process has already been observed in heated crystals of Fe-doped $LiNbO_3$. At a sufficiently high temperature (e.g., $\sim 160^\circ$ C), fixing occurs simultaneously with storage. This limits the largest index change that can be built up during storage and thus should prevent the buildup of unwanted scattering [4.21]. The fixed holograms are then "developed" by cooling the crystal to room temperature and exposing it to light. The latent-to-active mode in SBN (see Sects. 4.1.4 and 4.3.5) could also be used to prevent scattering. Another method useful only for readout is to render the material insensitive to light without disturbing the recorded hologram. This could be done by removing an applied field required for storage, or by applying a field antiparallel to an internally generated field [4.69]. Another approach is to decrease the crystal absorption after a hologram has been recorded. This has, in fact, been done in photochromic $LiNbO_3$:Fe, Mn [4.70].

A low dynamic range can also prevent the buildup of scattering. This, of course, would be at the sacrifice of a low diffraction efficiency unless thick crystals are used. An important example of the benefit of a low dynamic range is the treated crystals of Fe-doped $LiNbO_3$ shown in Table 4.1. These have the symmetrical storage and erase properties required for read/write operations, and show no buildup of optical scattering during storage or readout. In fact, the Weiner noise at 1000 lines/mm is only $5 \times 10^{-8}$, slightly better than for dichromated gelatin [4.55].

Optical scattering, if produced, does not prevent use of the crystal. It can be erased. One method is to simply shift the coherent beam (or beams) to a new angle. This erases previously stored scattering, but begins to record more at the new angle. Exposure to incoherent light is better. It erases all scattering. This has been used to restore the readout quality of fixed holograms in Fe-doped $LiNbO_3$ [4.21]. Thus, optical induced scattering can be minimized in this archival storage technique—with the latency during storage in heated crystals and with coherent light erasure during readout.

Another factor that can complicate storage and erasure is reflection from the front and back surfaces of the crystal [4.71]. This can be particularly troublesome when the beams heat the crystal. If necessary, it can be prevented with antireflection coatings, an approach that would improve the storage sensitivity and diffraction efficiency by letting more light pass through the crystal.

The ultimate noise sources in FE crystals are crystal defects and the discreteness of the storage process itself. The former is controllable to a certain extent by crystal preparation and polishing treatments. The latter is fundamental; storage is due to localized electrons trapped at discrete defects [4.72]. *Burke* et al. recently investigated noise sources intrinsic to volume phase holograms in iron-doped $LiNbO_3$ [4.73]. These were granularity in the electronic pattern, intermodulation distortion, and crosstalk between holograms stored at different angles. They showed that the only important factor was crosstalk, and that it depended principally on the angle between holograms. A s/n ratio for readout of $\sim$40 dB, a level compatible with high image quality, can be achieved at angular spacings that allow the storage of more than 500 holograms in a $\sim$0.5-cm-thick crystal. The s/n ratio was not studied as a function of information content of the individual holograms, however, and this may be important in the practical application of volume storage [4.72, 74].

## References

4.1 F.S.Chen, J.T.LaMacchia, D.B.Frazer: Appl. Phys. Lett. **13**, 223 (1968)
4.2 A.Ashkin, D.D.Boyd, J.M.Dziedzic, R.G.Smith, A.A.Ballman, H.J.Levinstein, K.Nassau: Appl. Phys. Lett. **9**, 72 (1966)
4.3 J.J.Amodei: Appl. Phys. Lett. **13**, 22 (1971)
4.4 A.M.Glass, D.von der Linde, T.J.Negran: Appl. Phys. Lett. **25**, 233 (1974)

4.5 J.J.Amodei, D.L.Staebler: RCA Rev. **33**, 71 (1972)
4.6 J.J.Amodei: RCA Rev. **32**, 185 (1971)
4.7 G.A.Alphonse, R.C.Alig, D.L.Staebler, W.Phillips: RCA Rev. **36**, 213 (1975)
4.8 D.L.Staebler, J.J.Amodei: J. Appl. Phys. **43**, 1042 (1972).
4.9 S.F.Lu, T.K.Gaylord: J. Opt. Soc. Am. **65**, 59 (1975)
4.10 M.G.Clark, F.J.DiSalvo, A.M.Glass, G.E.Peterson: J. Chem. Phys. **59**, 6209 (1973)
4.11 S.F.Lu, T.K.Gaylord: J. Appl. Phys. **46**, 5208 (1975)
4.12 D.L.Staebler, W.Phillips: Appl. Opt. **13**, 789 (1974)
4.13 Y.Ohmori, Y.Yasojima, Y.Inuishi: Japan. J. Appl. Phys. **14**, 1291 (1975)
4.14 A.M.Glass, D.von der Linde, D.H.Auston, T.J.Negran: J. Electron. Mater. **4**, 915 (1975)
4.15 L.Young, W.K.Y.Wong, M.S.W.Theriault, W.D.Cornish: Appl. Phys. Lett. **24**, 264 (1974)
4.16 J.J.Amodei, D.L.Staebler: Appl. Phys. Lett. **18**, 540 (1971)
4.17 D.L.Staebler, J.J.Amodei: Ferroelectrics **3**, 107 (1972)
4.18 F.Micheron, G.Bismuth: Appl. Phys. Lett. **20**, 79 (1972); — **23**, 71 (1971)
4.19 F.Micheron, C.Mayeux, J.C.Trotier: Appl. Opt. **13**, 784 (1974)
4.20 D.von der Linde, A.M.Glass: Appl. Phys. **8**, 85 (1971)
4.21 D.L.Staebler, W.Burke, W.Phillips, J.J.Amodei: Appl. Phys. Lett. **26**, 182 (1975)
4.22 F.Micheron, J.C.Trotier: Ferroelectrics **8**, 441 (1974)
4.23 J.B.Thaxter, M.Kestigian: Appl. Opt. **13**, 913 (1974)
4.24 H.Kogelnik: Bell System. Tech. J. **48**, 2909 (1969)
4.25 A.A.Ballman: J. Amer. Ceram. Soc. **48**, 112 (1965)
4.26 R.L.Byer, J.F.Young, R.S.Fergelson: J. Appl. Phys. **41**, 2320 (1970)
4.27 W.Phillips, J.J.Amodei, D.L.Staebler: RCA Rev. **33**, 94 (1972)
4.28 K.Nassau, H.J.Levinstein, G.M.Loiacono: J. Phys. Chem. Solids **27**, 989 (1966)
4.29 G.E.Peterson, A.M.Glass, T.J.Negran: Appl. Phys. Lett. **19**, 130 (1971)
4.30 W.Keune, S.K.Date, I.Deźsi, U.Genser: J. Appl. Phys. **46**, 3914 (1975)
4.31 W.Phillips, D.L.Staebler: J. Electron. Mater. **3**, 601 (1974)
4.32 A.Ishida, O.Mikami, S.Miyazawa, M.Sumi: Appl. Phys. Lett. **21**, 192 (1972)
4.33 E.Okamoto, H.Ikeo, K.Muto: Appl. Opt. **14**, 2453 (1975)
4.34 B.Dischler, J.R.Herrington, A.Räuber, H.Kunz: Solid State Commun. **14**, 1233 (1974)
4.35 H.Kunz: Ferroelectrics **8**, 437 (1974)
4.36 J.B.Thaxter: Appl. Phys. Lett. **15**, 210 (1969)
4.37 A.A.Ballman, H.Brown: J. Crystal Growth **1**, 311 (1967)
4.38 J.C.Brice, O.F.Hill, P.A.C.Whiffin, J.A.Wilkinson: J. Crystal Growth **10**, 133 (1971)
4.39 P.L.Townsend, J.T.LaMacchia: J. Appl. Phys. **41**, 5188 (1970)
4.40 J.P.Remeika: J. Amer. Chem. Soc. **76**, 940 (1954)
4.41 H.Arend, P.Coufova: Czech. J. Phys. B **13**, 55 (1963)
4.42 F.Micheron, G.Bismuth: Appl. Phys. Lett. **20**, 15 (1972)
4.43 J.J.Amodei, D.L.Staebler, A.W.Stephens: Appl. Phys. Lett. **18**, 507 (1971)
4.44 J.M.Spinhirne, T.L.Estle: Appl. Phys. Lett. **25**, 38 (1974)
4.45 F.S.Chen: J. Appl. Phys. **40**, 3389 (1969)
4.46 Hedeki Tsuya: J. Appl. Phys. **46**, 4323 (1975)
4.47 F.S.Chen: J. Appl. Phys. **38**, 3418 (1967)
4.48 S.R.King, T.S.Hartwick, A.B.Chase: Appl. Phys. Lett. **21**, 32 (1972)
4.49 D.von der Linde, A.M.Glass, K.F.Rodgers: Appl. Phys. Lett. **26**, 22 (1975)
4.50 F.Micheron, A.Hermosin, G.Bismuth, J.Nicolas: C. R. Acad. S. Paris **274**, 361 (1972)
4.51 G.H.Haertling, C.E.Land: J. Amer. Ceram. Soc. **54**, 1 (1971)
4.52 F.Micheron, C.Mayeux, A.Hermosin, J.Nicolas: J. Amer. Ceram. Soc. **57**, 306 (1974)
4.53 J.M.Rouchon, M.Vergnolle, F.Micheron: IEEE Symposium on the Application of Ferroelectrics, Albuquerque, New Mexico, June 1975
4.54 D.Chen, J.D.Zook: Proc. IEEE **63**, 1207 (1975)
4.55 G.Alphonse, W.Phillips: Final Report for NASA Contract No. NAS8-26808, May 1975 RCA Rev. **37**, 184 (1976)
4.56 L.D'Auria, J.Huignard, C.Slezak, E.Spitz: Appl. Opt. **13**, 808 (1974)

4.57 J.P.Huignard, J.P.Herriau, F.Micheron: Appl. Phys. Lett. **26**, 256 (1975)
4.58 D.von der Linde, A.M.Glass, K.F.Rodgers: Appl. Phys. Lett. **25**, 155 (1974)
4.59 D.von der Linde, A.M.Glass, K.F.Rodgers: J. Appl. Phys. **47**, 217 (1976)
4.60 W.John Carlsen: Appl. Opt. **13**, 896 (1974)
4.61 H.Kurz, E.Krätzig: Appl. Phys. Lett. **26**, 635 (1975)
4.62 J.P.Huignard, F.Micheron: Opt. Commun. **16**, 80 (1976)
4.63 R.Parsons, W.D.Cornish, L.Young: Appl. Phys. Lett. **27**, 654 (1975)
4.64 R.J.Collier, C.B.Burkhardt, L.H.Lin: *Optical Holography* (Academic Press, New York 1971)
4.65 B.F.Williams, W.J.Burke, D.L.Staebler: Appl. Phys. Lett. **28**, 224 (1976)
4.66 D.W.Vahey: J. Appl. Phys. **46**, 3510 (1975)
4.67 R.Magnusson, T.K.Gaylord: J. Appl. Phys. **47**, 190 (1976)
4.68 R.Magnusson, T.K.Gaylord: Appl. Opt. **13**, 1545 (1974)
4.69 Y.Ohmori, Y.Yasojima, Y.Inuishi: Japan. J. Appl. Phys. **14**, 1291 (1975)
4.70 D.L.Staebler, W.Phillips: Appl. Phys. Lett. **24**, 263 (1974)
4.71 W.D.Cornish, L.Young: J. Appl. Phys. **46**, 1252 (1975)
4.72 E.G.Ramberg: R C A Rev. **33**, 5 (1972)
4.73 W.Burke, P.Sheng, H.A.Weakliem: Final Report prepared for Office of Naval Research under Contract No. N00014-75-C-0590, March 1976,
W.Burke, P.Sheng: J. Appl. Phys. **48**, 681 (1977)
4.74 H.Nomura, T.Okoshi: Appl. Opt. **15**, 550 (1976)

# 5. Inorganic Photochromic Materials

R. C. Duncan, Jr. and D. L. Staebler

With 20 Figures

Photochromic materials are materials which undergo reversible changes in color when exposed to radiation of an appropriate wavelength. It is the reversibility of the color change which distinguishes photochromic substances from other photosensitive materials which either photocolor or photobleach only. Photochromic color changes may: a) bleach spontaneously at room temperature when the exciting or "switching" radiation is removed, b) be bleached or "erased" thermally at elevated temperatures, and/or c) be photoerased by radiation of a different wavelength.

Photochromism has been observed and studied in a variety of materials, organic [5.1–3] as well as inorganic [5.3–5]. Because the introduction of the terms photochromic and photochromism is fairly recent [5.6], much of the early literature of historic interest predates their general usage. Reversible radiation-induced coloration of naturally occurring mineral crystals was reported well over a century ago [5.7], and radiation-induced coloration in glass was reported before the turn of the century [5.8]. The early literature on these and related phenomena has been reviewed in detail in a book by *Przibram* [5.9]. Organic compounds exhibiting similar behavior have also been known for at least 100 years [5.10]. The extensive literature on such organic materials, generally quite distinct from that dealing with inorganic materials, has been summarized by *Brown* and *Shaw* [5.1].

Because they require no post-exposure processing and can be erased and reused, photochromic materials appear to be particularly attractive for information storage applications in which all-optical recording, readout and erasure are desired. In particular, they have distinct advantages as hologram storage media [5.11]. They are essentially grain free, resolutions of greater than 3000 lines having been demonstrated. Since no processing is necessary, the holograms stored can be read out during or immediately after recording. This simplification is particularly important in holographic interferometry [5.12]. Finally, the storage process occurs throughout the bulk of the material. Thus, one can store a multiplicity of thick holograms (1 mm to 1 cm) in the same volume by taking advantage of their angular selectivity. It was, in fact, this last capability that spurred the initial interest in photochromic media for hologram applications [5.13–16].

In spite of these advantages, however, these materials have not become widely popular in holography. First, their sensitivity for storage is relatively low, and there appears to be little hope for improvement. Second, the diffraction

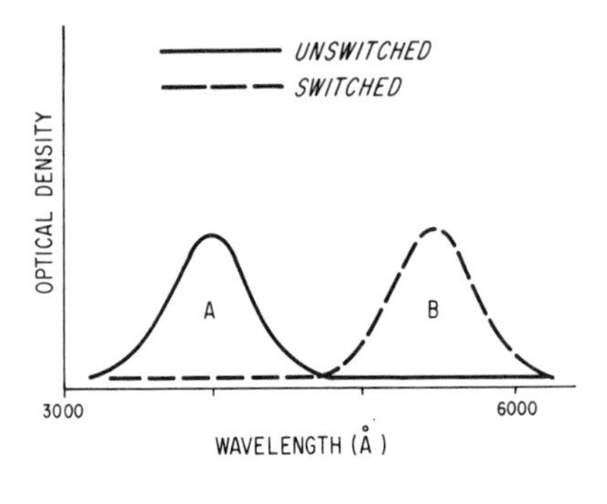


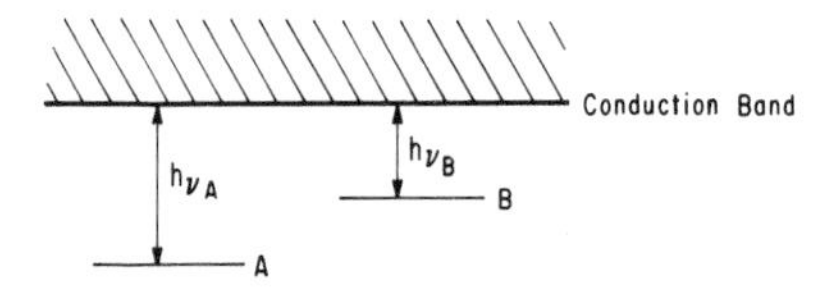


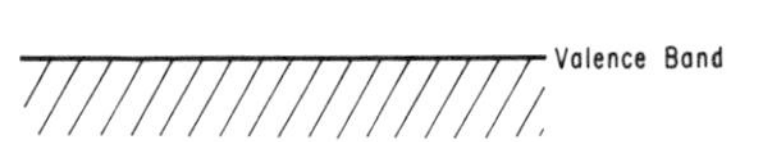


Fig. 5.1. Schematic illustration of the basic model for photochromic behavior in inorganic crystalline materials

efficiency is also low, particularly when many holograms are to be stored in the same volume. Third, readout usually degrades the stored information. This is a severe problem if the same wavelength is used for both recording and readout, a primary requirement for volume holography. An important exception occurs in alkali halides where readout is nondestructive, or at least partially nondestructive, if carried out at temperatures lower than that used for initial storage. Fourth, in-line operation through a complete cycle (record-read-erase) requires at least two wavelengths, one for recording (and perhaps reading), another for erasure and preparation for the next cycle. Fifth, the storage time is fairly short (from hours to weeks) unless the material is refrigerated.

In this chapter, the properties and behavior of state-of-the-art crystalline inorganic photochromic materials are reviewed with the emphasis on quantitative sensitometric characteristics. This review is followed by a section on the applicability of these materials for hologram storage, with emphasis on sensitivity, diffraction efficiency and capacity. A final section deals with photodichroic storage materials. These are similar to photochromics, both in mechanism and application, but are superior in some important respects.

## 5.1 Basic Model

In crystalline inorganic materials, photochromic behavior is usually associated with photoreversible charge transfer between two species of electron traps. This process is illustrated schematically in Fig. 5.1. The two species of traps present, A and B, give rise to characteristic absorption bands, A and B respectively, corresponding to the photo excitation of the trapped electrons into the conduction band. In photochromic materials of practical interest, the trap energies $h\nu_A$ and $h\nu_B$ and their difference $h\nu_A - h\nu_B$ are all large compared to the thermal energy available to the trapped electrons at room temperature. At thermal equilibrium, the trapped electrons preferentially occupy the lower energy A-traps. If the number of such traps is about equal to or greater than the number of electrons available, then the absorption spectrum of this material in its thermally stable or "unswitched" state exhibits band-A only.

When the material is exposed to light in band-A, electrons are excited into the conduction band from which they can be captured by B-traps. This results in a modified absorption spectrum including some band-B and a weakened band-A. If this photo-induced transfer of electrons from A-traps to B-traps is continued to completion, then the new "switched" state absorption spectrum contains only band-B.

When exposure to the band-A "switching" light is ended, the material gradually returns to its original thermally stable or unswitched state spontaneously as the electrons in B-traps are thermally excited into the conduction band and are recaptured by A-traps. This "erase" or bleaching process can be accelerated by elevating the temperature of the photochromic material and/or by exposure to light in band-B.

## 5.2 Photochromic $CaF_2$, $SrTiO_3$, and $CaTiO_3$ [1]

### 5.2.1 Optimum Dopants and Concentrations

A number of recent papers have discussed the photochromic behavior of $CaF_2$ doped with certain rare earths ($CaF_2$ : RE) [5.17–24] and of SrTiO (and $CaTiO_3$) doped with transition metals ($SrTiO_3$ : TM and $CaTiO_3$ : TM) [5.23–28]. At the present time, these are the inorganic photochromic materials which, in the form of single crystal wafers or slabs, offer the greatest promise for applications requiring all-optical recording, readout and erasure.

Of the photochromic $CaF_2$ : RE systems, $CaF_2$ :La, Na exhibits the strongest photochromic effects [5.3, 19]. $CaF_2$ :Ce, Na has also been extensively studied. The Na provides charge compensation and inhibits the growth of a frequently observed nonphotochromic coloration; greater photochromic coloration and correspondingly greater photochromic switching result [5.19]. The La, Ce, and Na concentrations indicated in Table 5.1 lie in the ranges found to give the maximum photochromic absorption changes [5.19]. Appropriately doped $CaF_2$ crystals of excellent optical quality can be grown using a gradient freeze technique [5.19]. The photochromic color centers are generated in thin optically polished wafers by subsequent additive coloration [5.19].

Of the photochromic $SrTiO_3$ :TM systems, $SrTiO_3$ :Ni, Mo is one of the most extensively studied and exhibits among the strongest photochromic switching

[1] Portions of Section 5.2 are taken from [5.24].

Table 5.1. Materials

| Host | Dopants and nominal concentrations | |
|---|---|---|
| $CaF_2$ | $LaF_2$ | 0.05 mol-% |
| | NaF | 0.1 mol-% |
| $CaF_2$ | $CaF_3$ | 0.05 mol-% |
| | NaF | 0.1 mol-% |
| $SrTiO_3$ | NiO | 0.35 mol-% |
| | $MoO_3$ | 0.34 mol-% |
| | $Al_2O_3$ | 0.18 mol-% |
| $CaTiO_3$ | $NiMoO_4$ | 0.19 mol-% |

and the longest switched-state thermal lifetimes [5.5, 23, 28]. With Mo omitted, the $SrTiO_3$:Ni system shows both weaker switching and a shorter lifetime. The Al impurity indicated in Table 5.1 does not affect the photochromic behavior but has sometimes been included to provide additional charge compensation. $CaTiO_3$:TM materials typically exhibit somewhat greater photochromic color changes and somewhat longer switched-state thermal lifetimes than do the corresponding $SrTiO_3$:TM materials [5.28]. The Ni and Mo concentrations indicated in Table 5.1 lie in the ranges found to produce maximum photochromic effects [5.23]. Single crystals of $SrTiO_3$:Ni, Mo, Al and $CaTiO_3$:Ni, Mo, grown by flame fusion, must usually be subjected to mild reduction treatments [5.26, 28] to reduce to a minimum their unswitched-state visible absorption. The optical quality of the $CaTiO_3$:Ni, Mo, Al wafers is generally good, although some signs of strain are frequently evident in crossed-polarizer examination. The optical quality of the $CaTiO_3$:Ni, Mo, Al wafers is significantly poorer, with extensive lamellar twinning evident in spite of extraordinary crystal growth and annealing procedures [5.29].

### 5.2.2 Photochromic Absorption Spectra

Photochromic absorption spectra of wafer samples of each material listed in Table 5.1 are shown in Figs. 5.2–5. The transmission optical density ($D$), i.e., the logarithm of the ratio of the incident ($I_0$) and transmitted ($I$) light intensities, is plotted as a function of the wavelength of light. The various features of these spectra have been discussed in detail elsewhere [5.5, 28]. Here we will only briefly describe the photochromic behaviour indicated and mention some practical considerations related to the attainment of optimum performance.

The spectra of the thermally stable (unswitched) states are indicated by the solid curves (Curves 1 in Figs. 5.2 and 5.3). For materials, particularly $CaF_2$ samples, that have been previously exposed to ultraviolet switching radiation, complete return to this thermally stable state may require some thermal erasure. Material heated to about 150° C for about two minutes and cooled to room temperature in the dark is usually satisfactorily erased. In this state, these materials are all very nearly transparent at visible wavelengths between 500 and 600 nm.

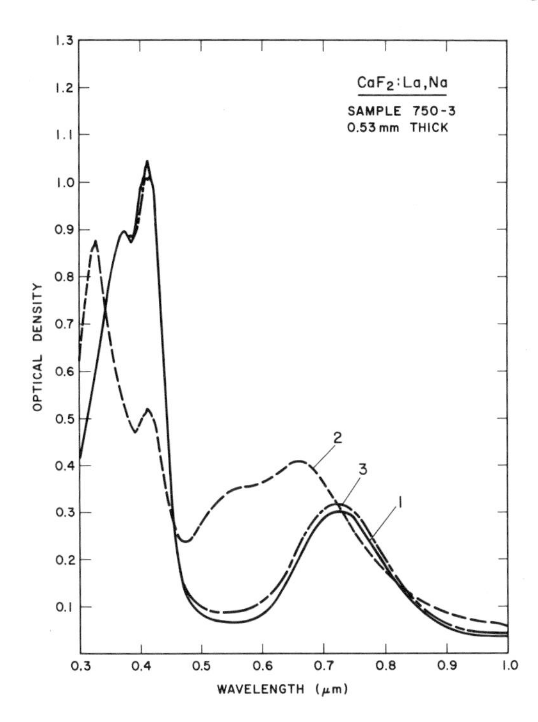


Fig. 5.2. Photochromic absorption spectra of $CaF_2$:La, Na in the unswitched (Curve 1), switched (Curve 2), and optically erased (Curve 3) states

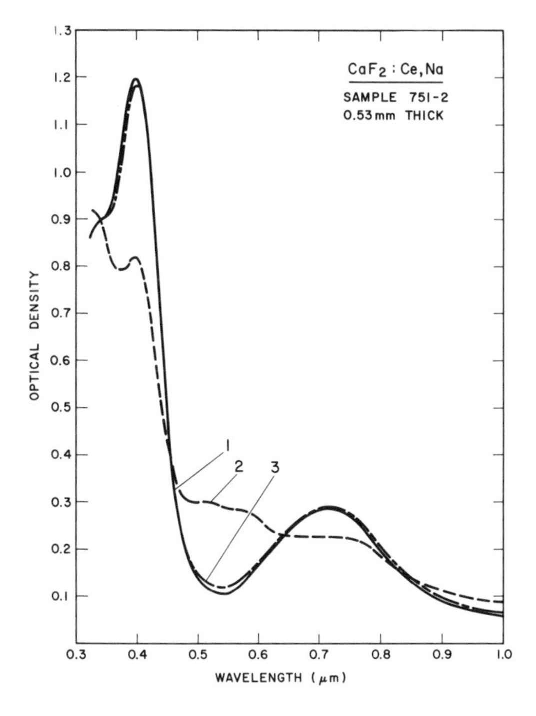


Fig. 5.3. Photochromic absorption spectra of $CaF_2$:Ce, Na in the unswitched (Curve 1), switched (Curve 2), and optically erased (Curve 3) states

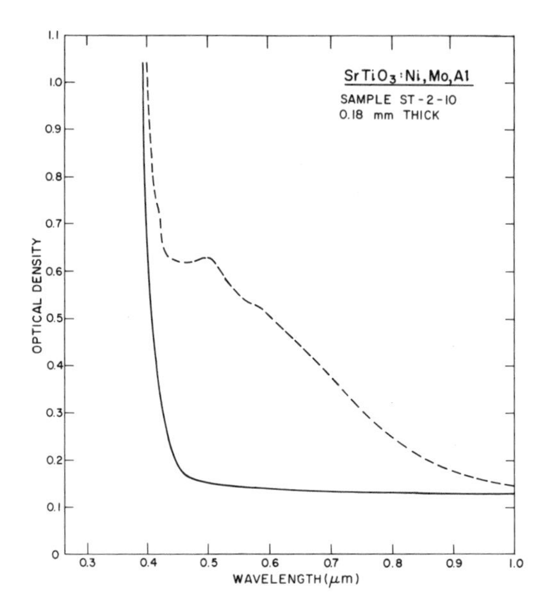


Fig. 5.4. Photochromic absorption spectra of $SrTiO_3$ : Ni, Mo, Al in the unswitched (solid curve) and switched (broken curve) states

After irradiation by the appropriate strong violet or ultraviolet switching radiation, the respective wafers exhibit the switched-state spectra shown by the broken curves (Curves 2 in Figs. 5.2 and 5.3). A two-minute exposure to suitably filtered and focused radiation from a 500 W Hg arc is sufficient to saturate the photochromic changes in absorption.

For the titanate materials, effective switching is produced by radiation in and beyond (shorter wavelengths) the sharp absorption edge at about 400 nm. For maximum photochromic switching, filters should limit the switching radiation to the range 330 nm $\lesssim \lambda \lesssim$ 390 nm.

In the case of the $CaF_2$ materials, the switching radiation is absorbed primarily by the strong unswitched-state absorption band near 400 nm. Absorption by the accompanying band near 700 nm is ineffective in switching. For optimum switching, radiation should be limited to the range 380 nm $\lesssim \lambda \lesssim$ 460 nm. Radiation at wavelengths shorter than about 380 nm can produce up to 50% greater photochromic absorption changes, but this additional switching is not optically reversible, and thermal erasure is required.

Finally, optical erasure is carried out by exposing the wafers to light lying within the visible switched-state absorption bands. The resulting absorption spectra of the $CaF_2$ materials are shown by long-short dashed curves (Curves 3 in Figs. 5.2 and 5.3). The residual switched-state absorption indicated by these curves typically amounts to about 10% of the initial switched state absorption and is *not* cumulative ($<1\%$) with successive photochromic cycling. It can be erased thermally. For the titanates, the spectra produced by optical erasure are indistinguishable from those of the initial thermally stable states.

Because of this erasing effect of visible light, it is clear that switching radiation at wavelengths longer than about 460 nm for the $CaF_2$ materials or

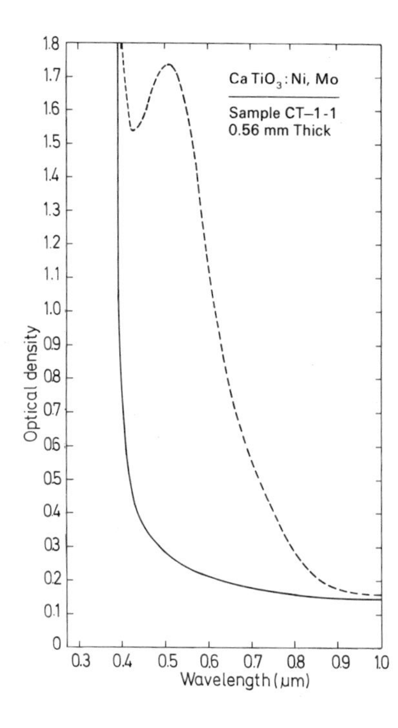


Fig. 5.5. Photochromic absorption spectra of $CaTiO_3$ : Ni, Mo in the unswitched (solid curve) and switched (broken curve) states

430 nm for the titanates is not only ineffective but actually becomes counter-productive as switching proceeds, limiting both the rate of switching and the ultimate saturated level of photochromic absorption change [5.27].

### 5.2.3 Optical Recording Modes

The optical reversibility of these materials clearly makes them subject to two modes of optical recording. In the "write mode", appropriate violet or ultraviolet radiation (as described above) is used to "switch" material initially in its thermally stable or unswitched state. In the "erase mode", appropriate visible "erase" light is used to record on material previously darkened by uniform exposure to switching radiation. In either case, readout of the information thus recorded is obtained by monitoring the absorption in the region of the visible "switched-state" bands shown in Figs. 5.2–5. With $CaF_2$ wafers, readout (of the opposite sign) is also possible at wavelengths lying in the strong photochromic absorption bands near 400 nm.

The photochromic sensitivity of these materials is from two to ten times greater for write-mode recording than for erase-mode recording [5.5, 23]. On the other hand, the only available laser sources with sufficient power output to accomplish the fast addressable or holographic recording required for many applications are those operating at visible (erase-mode) wavelengths.

It appears attractive, therefore, to use the 514.5 nm argon laser line for both erase-mode optical recording and optical readout. From Figs. 5.2–5, it is clear

Table 5.2. Optical readout at 514.5 nm

| Wafer thickness | Maximum absorption changes | | |
|---|---|---|---|
| $d$ [mm] | $\Delta D$ | $R$ | $\overline{\Delta\alpha}$ [cm$^{-1}$] |
| $CaF_2$:La, Na | | | |
| 0.81 | 0.33 | 2.1 | 9.4 |
| 0.53 | 0.23 | 1.7 | 10 |
| 0.29 | 0.12 | 1.3 | 9.5 |
| 0.13 | 0.06 | 1.1 | 11 |
| $CaF_2$:Ce, Na | | | |
| 0.86 | 0.27 | 1.9 | 7.2 |
| 0.53 | 0.18 | 1.5 | 7.8 |
| 0.32 | 0.10 | 1.3 | 7.2 |
| $SrTiO_3$:Ni, Mo, Al | | | |
| 1.02 | 0.86 | 7.2 | 19 |
| 0.53 | 0.73 | 5.4 | 32 |
| 0.28 | 0.60 | 4.0 | 49 |
| 0.18 | 0.45 | 2.8 | 58 |
| $CaTiO_3$:Ni, Mo | | | |
| 0.81 | 2.20 | 158 | 63 |
| 0.56 | 1.44 | 28 | 59 |
| 0.30 | 0.80 | 6.3 | 59 |
| 0.18 | 0.55 | 3.5 | 70 |

that 514.5 nm corresponds very closely to the peaks of the visible switched-state absorption bands of all of the materials shown here. The poorest match occurs for $CaF_2$:La, Na. Even here, however, the photochromic absorption change at 514.5 nm is still about 80% of that at 575 nm where the maximum change does occur. This fortunate correspondence makes possible near maximum utilization (absorption) of the optical recording radiation from a high intensity, coherent, and addressable source, and, in addition, the option of readout with coherent light at a near optimum readout wavelength.

### 5.2.4 Optical Readout at 514.5 nm

The ranges of optical density change available for 514.5 nm optical readout are illustrated in Table 5.2. Saturated photochromic optical density changes at 514.5 nm, $\Delta D$, for wafer samples of various thickness, $d$, are given in the first two columns. The corresponding maximum obtainable transmission contrast ratios, $R = 10^{\Delta D}$, and average bulk absorption coefficient changes, $\overline{\Delta\alpha} = 2.3\, D/d$(cm), are also indicated.

#### Thickness Effects

The $CaF_2$ materials exhibit saturation optical density changes which are nearly directly proportional to wafer thickness, or alternatively, maximum changes in average bulk absorption coefficient that are independent of wafer thickness.

This is an indication that, at saturation, the photochromic coloration is essentially uniformly distributed throughout the bulk of these materials.

For the $SrTiO_3$:Ni, Mo, Al, on the other hand, while wafer thickness varies by a factor of six, the maximum $\Delta D$ varies by less than a factor of two, and the maximum $\overline{\Delta\alpha}$ varies by a factor of almost three. These wafers absorb strongly at wavelengths shorter than 400 nm (see Fig. 5.4), and photochromic coloration is confined to the relatively limited penetration depth of the switching radiation in this wavelength range. The effective penetration depth is about 0.1 to 0.2 mm.

In spite of its sharp UV absorption edge (Fig. 5.5), the $CaTiO_3$:Ni, Mo appears from the data shown in Table 5.2 to color more uniformly throughout its thickness than does the $SrTiO_3$:Ni, Mo, Al. This results from the fact that the $CaTiO_3$ absorption edge occurs about 20 to 30 nm further toward the UV than that of $SrTiO_3$ [5.28] so that some of the switching radiation used ($330 \lesssim \lambda \lesssim 390$ nm) can penetrate more deeply into the bulk material. In addition, the longer (by a factor of more than 10) switched-state lifetime of the $CaTiO_3$:Ni, Mo permits this penetrating radiation, even if fairly weak, to induce relatively larger saturated photochromic switching.

**Contrast Ratio**

The thickest $CaF_2$ wafers exhibit maximum transmission contrast ratios, $R$, at 514.5 nm of only about 2:1. Because of the depth of photochromic coloration in these materials, larger values of $R$ can be obtained with thicker wafers. For non volume-holographic applications, however, increased wafer thickness means reduced image resolution and reduced storage capacity. In cases where coherent readout light is not necessary, a better readout wavelength for these materials is near 400 nm (see Figs. 5.2 and 5.3). Here the maximum values of $\Delta D$ are somewhat more than twice those at 514.5 nm, and the maximum values of $R$ are approximately the squares of those at 514.5 nm.

Markedly greater transmission contrast ratios are attainable with the titanate wafers, as shown in Table 5.2. Because of the penetration limitations discussed above, $SrTiO_3$:Ni, Mo, Al wafers thicker than about 1 mm provide little increase in $R$. The values of $R$ available with $CaTiO_3$:Ni, Mo wafers even under 0.5 mm in thickness are probably adequate for most applications. Unfortunately, the usefulness of $CaTiO_3$ is limited by its relatively poor optical quality.

### 5.2.5 Erase-Mode Sensitivity at 514.5 nm

Families of erase-mode sensitivity curves for wafer samples of each of the four materials of Table 5.1 are shown in Figs. 5.6–9. The photochromic change in optical density at 514.5 nm is shown as a function of time of exposure to 514.5 nm argon laser erasing radiation of various intensities. Optical density change ($\Delta D$) plotted here is given by $\Delta D = \log(T_u/T_s)$, where $T_u$ and $T_s$ represent the transmission of the wafer in its unswitched state and its (partially or completely)

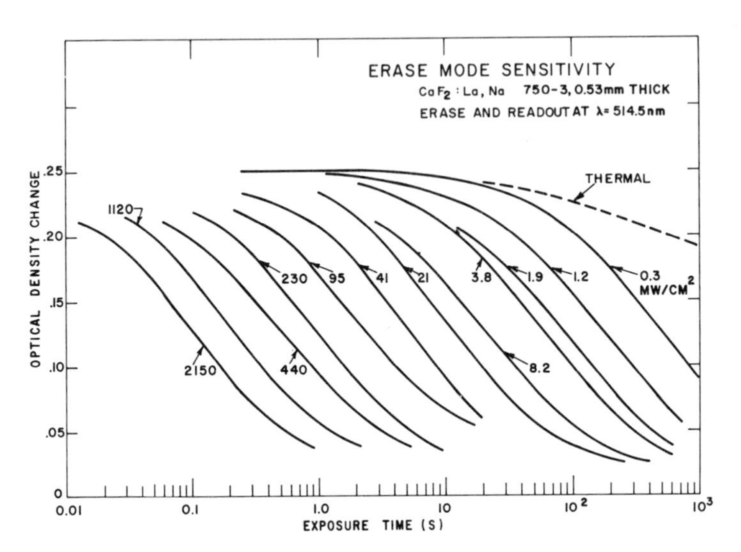


Fig. 5.6. Erasure of a $CaF_2$ : La, Na wafer as a function of erase exposure time for various erase beam intensities

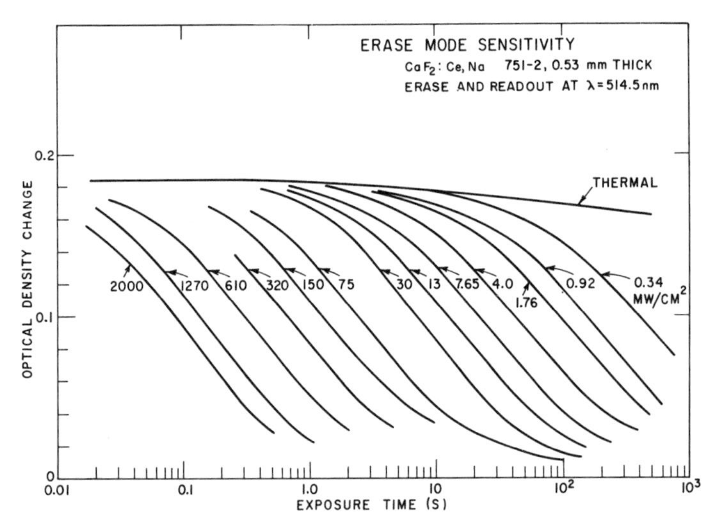


Fig. 5.7. Erasure of a $CaF_2$ : Ce, Na wafer as a function of erase exposure time for various erase beam intensities

switched state, respectively. The uppermost curve in each family, that showing the slowest change in optical density, corresponds to room temperature thermal erasure [5.23]. The $CaF_2$ materials exhibit the longest thermal switched-state lifetimes, the optical density dropping by about 25% in 1000 s for $CaF_2$ :La, Na (Fig. 5.6) and about 10% to 15% in 1000 s for $CaF_2$:Ce, Na (Fig. 5.7). The $SrTiO_3$ :Ni, Mo, Al (Fig. 5.8), on the other hand, exhibits the shortest switched-

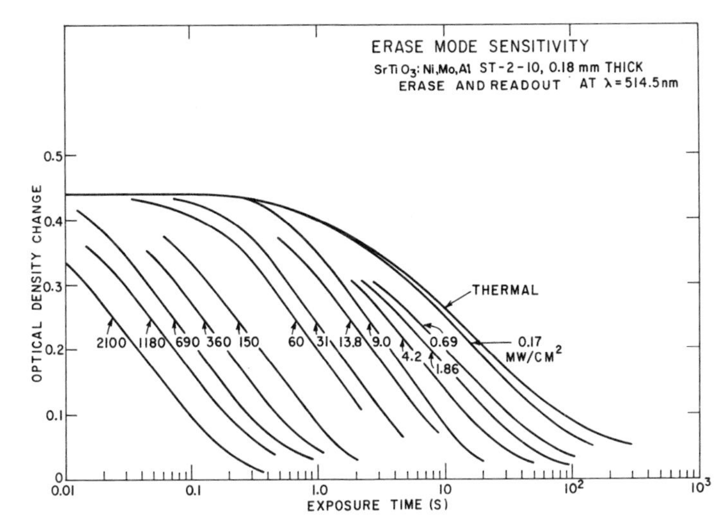


Fig. 5.8. Erasure of a $SrTiO_3$ : Ni, Mo, Al wafer as a function of erase exposure time for various erase beam intensities

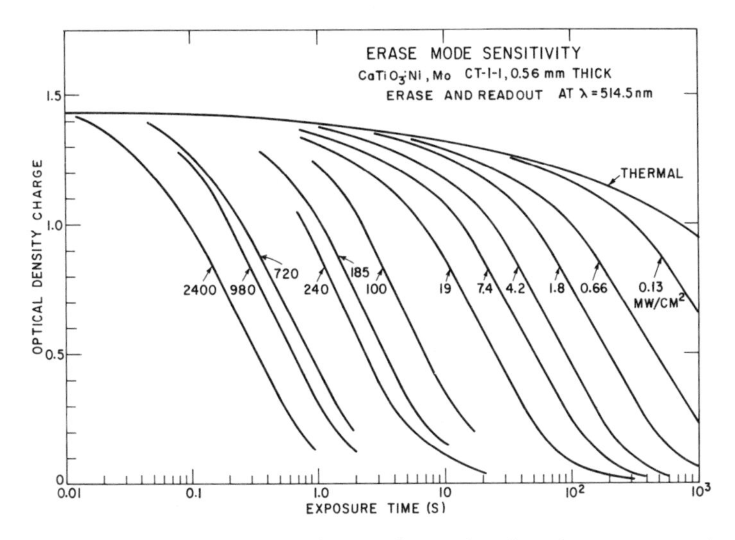


Fig. 5.9. Erasure of a $CaTiO_3$ : Ni, Mo wafer as a function of erase exposure time for various erase beam intensities

state lifetime, the optical density falling by a factor of two in about 25 s. The corresponding decay times for $CaTiO_3$:Ni, Mo (Fig. 5.9) are of the order of 20 min, almost two orders of magnitude longer.

As erase beam intensity is increased, the switched state optical density falls away from the thermal decay curve at progressively shorter erase exposure times.

Table 5.3. Erase-mode recording at 514.5 nm

| Wafer thickness $d$ [mm] | $\Delta D_{max}$ at 514.5 nm | $\gamma$ | | $E_{1/2}$ at 1 W/cm$^2$ [mJ/cm$^2$] |
|---|---|---|---|---|
| | | $\bar{\gamma}$ | $\bar{\gamma}/D_{max}$ | |
| $CaF_2$:La, Na | | | | |
| 0.81 | 0.36 | 0.18 | $0.50 \pm 0.04$ | 290 |
| 0.53 | 0.25 | 0.12 | $0.48 \pm 0.02$ | 230 |
| 0.29 | 0.13 | 0.058 | $0.45 \pm 0.03$ | 270 |
| 0.13 | 0.076 | 0.034 | $0.45 \pm 0.03$ | 190 |
| $CaF_2$:Ce, Na | | | | |
| 0.86 | 0.27 | 0.14 | $0.52 \pm 0.03$ | 270 |
| 0.53 | 0.18 | 0.095 | $0.53 \pm 0.04$ | 220 |
| 0.32 | 0.10 | 0.057 | $0.57 \pm 0.06$ | 220 |
| $SrTiO_3$:Ni, Mo, Al | | | | |
| 1.02 | 0.90 | 0.57 | $0.63 \pm 0.05$ | 120 |
| 0.53 | 0.79 | 0.46 | $0.58 \pm 0.04$ | 110 |
| 0.28 | 0.48 | 0.31 | $0.65 \pm 0.02$ | 77 |
| 0.18 | 0.44 | 0.26 | $0.59 \pm 0.03$ | 65 |
| $CaTiO_3$:Ni, Mo | | | | |
| 0.81 | 2.06 | 1.34 | $0.65 \pm 0.06$ | 530 |
| 0.56 | 1.43 | 0.97 | $0.68 \pm 0.08$ | 420 |
| 0.31 | 0.96 | 0.65 | $0.68 \pm 0.04$ | 270 |
| 0.18 | 0.52 | 0.33 | $0.63 \pm 0.02$ | 200 |

At the higher erase beam intensities, above about 1 mW/cm$^2$, the influence of thermal erasure is small and the erase curves for each material approach a nearly fixed shape with a central region of constant slope or gamma. The existence of this linear central region is important for many potential applications, and its slope, $\gamma = \Delta D / \Delta(\log \text{exposure})$, is an important sensitometric quantity.

Erase-mode recording characteristics at 514.5 nm of a number of photochromic wafer samples are summarized in Table 5.3. These characteristics include $\Delta D_{max}$, the saturated photochromic optical density change at 514.5 nm, and $\bar{\gamma}$, the average $\gamma$ observed for erase intensities greater than 1 mW/cm$^2$. For each material, the values of $\bar{\gamma}/\Delta D_{max}$, listed in the fourth column of Table 5.3, appear to be a characteristic of a given photochromic material. Even the variations in $\bar{\gamma}/\Delta D_{max}$ among the four different materials are not large. In practical terms, $\gamma$ can be adjusted somewhat, as might be desired for a particular application, by varying wafer thickness, and thus $\Delta D_{max}$. Similar adjustments in $\Delta D_{max}$ and $\gamma$ can be achieved by varying the readout wavelength.

Dependence of $\gamma$ on $\Delta D_{max}$ is not unexpected. At smaller values of $\Delta D_{max}$, a smaller fraction of the incident radiation is absorbed and the erasure would be expected to proceed more slowly. An equation for the transmission of a photobleachable material as a function of bleaching exposure has been developed by *Kessler* [5.30] (and, independently, by *Staebler*). The model on which the equation is based is not applicable in complete detail to the situation

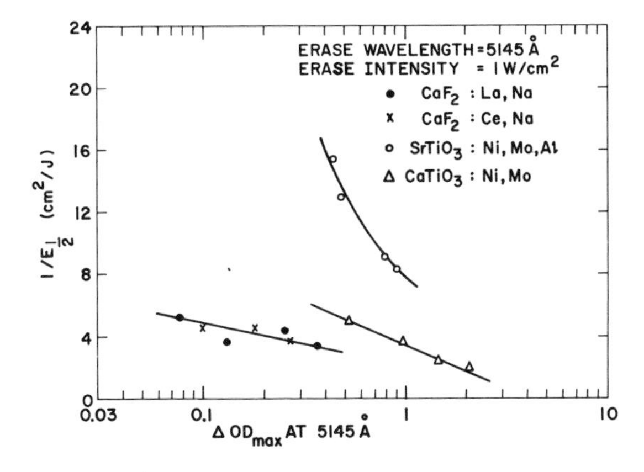


Fig. 5.10. Reciprocal of erase exposure $E_{1/2}$ (to $1/2\Delta D_{max}$) as a function of $\Delta D_{max}$ for several wafer samples of each of four materials

under consideration here. The equation can be used, however, to show a dependence of $\gamma$ on $\Delta D_{max}$ which, in a limited regime, might be approximated by $\gamma/\Delta D_{max}$ = constant. The fact that this constant varies little among the four materials examined is indicative of the fundamental similarities in the photochromic processes.

The right-hand column of Table 5.3 shows the erase exposures $E_{1/2}$ required, at an erase beam intensity of 1 W/cm$^2$, to reduce the readout optical densities to half their initial (saturated) values. Clearly the reciprocal of $E_{1/2}$ is a measure of erase-mode sensitivity. (It should be noted that these are "incident" exposures, not "absorbed" exposures.) For each material, these values of $E_{1/2}$ exhibit a dependence on $\Delta D_{max}$, although not a consistent one. For $SrTiO_3$ : Ni, Mo, Al the two are very nearly proportional, while for the $CaF_2$ materials, the dependence is much weaker. Qualitatively, this dependence is again consistent with the arguments given above for the variation of $\gamma$ with $\Delta D_{max}$ and with the equation derived by *Kessler* [5.30].

For each material, Fig. 5.10 shows $1/E_{1/2}$ plotted as a function of $\Delta D_{max}$ (see Table 5.3). For a particular application requiring a specific contrast ratio of $\Delta D_{max}$, this graph gives an indication of the relative sensitivities of the four materials. Thus, wafers of $CaF_2$ : La, Na and $CaF_2$ : Ce, Na of comparable $\Delta D_{max}$ exhibit comparable sensitivities. For $\Delta D_{max} = 0.4$, the $SrTiO_3$ : Ni, Mo, Al is about twice as sensitive as the $CaTiO_3$ : Ni, Mo and five times as sensitive as the $CaF_2$ materials.

### 5.2.6 Sensitometric Characteristics and Time-Intensity Reciprocity

To the extent to which time-intensity reciprocity holds for the erase-mode operation of these photochromic materials, the entire family of sensitivity curves (Figs. 5.6–9) for each material should coalesce into a single sensitometric characteristic when replotted as a function of erase exposure energy density (i.e., the product of erase beam intensity and exposure time) rather than of exposure time alone.

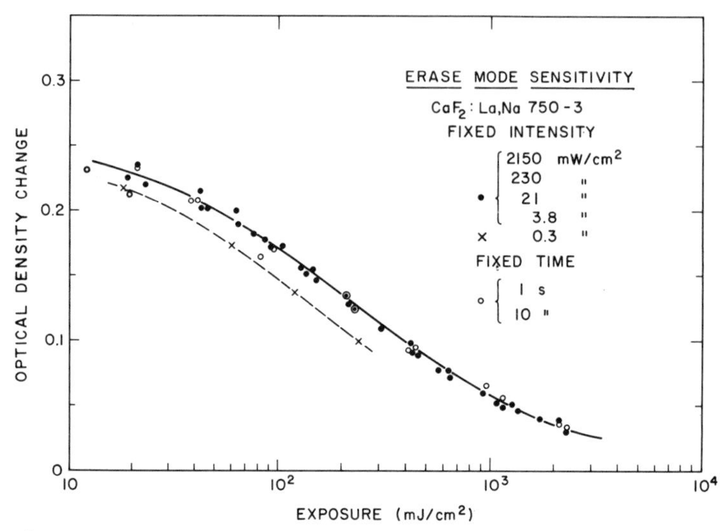


Fig. 5.11. Erase-mode sensitometric characteristic of a $CaF_2$ : La, Na wafer, showing extensive validity of time-intensity reciprocity

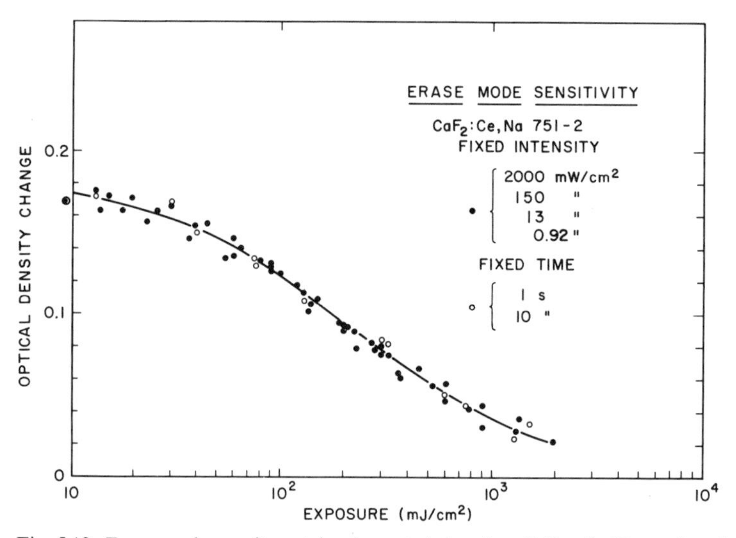


Fig. 5.12. Erase-mode sensitometric characteristic of a $CaF_2$ : Ce, Na wafer, showing extensive validity of time-intensity reciprocity

## $CaF_2$ : La, Na and $CaF_2$ : Ce, Na

Such sensitometric characteristics for photochromic $CaF_2$ are shown in Figs. 5.11 and 5.12. The solid points indicate the erasures induced in two $CaF_2$ wafers by exposure for various times to erase beams of selected *fixed intensities*. The abscissae are the corresponding erase exposure energy densities. These points

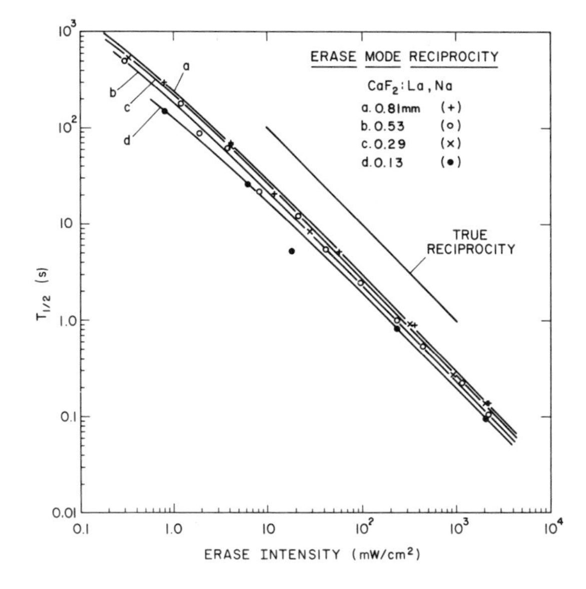


Fig. 5.13. Erase time $T_{1/2}$ (to $1/2\,\Delta D_{max}$) versus erase beam intensity for four wafers of $CaF_2$ : La, Na

consitute a straightforward replotting of selected sensitivity curves from Figs. 5.6 and 5.7, respectively, with beam intensities spanning approximately three decades. In each case, the several separate curves are indeed well represented by single sensitometric characteristics. With few exceptions, the plotted points along the central linear regions of these characteristics deviate from the curves drawn by less than $\pm 10\%$ in exposure or $\pm 5\%$ in optical density. Thus, in this intensity range, time-intensity reciprocity does hold, and these single characteristic curves provide condensed displays of the sensitivity data contained in the families of curves in Figs. 5.6 and 5.7.

The open-circle points in Figs. 5.11 and 5.12 indicate the erasures induced in selected *fixed times* by erase beams of various intensities. These points, also directly transposed from Figs. 5.6 and 5.7, sample ten or more different intensities in each case. Their close agreement with the previously determined sensitometric characteristics is a further demonstration of the validity of time-intensity reciprocity for these wafers in this intensity range.

Because of the effects of room temperature thermal erasure, time-intensity reciprocity cannot be expected to hold for arbitrarily low-erase beam intensities. The primary effect is that, as thermal processes begin to contribute, a given integrated erase energy density appears to induce greater erasure. The $x$-points in Fig. 5.11 were determined in the same manner as were the solid points, but for an erase intensity smaller by another factor of ten, 0.3 mW/cm$^2$. It is clear from Fig. 5.6 that the time required for appreciable erasure at this beam intensity is also that required for thermal erasure to become significant. The accompanying reciprocity failure is evident in the $x$-point curve in Fig. 5.11.

The "reciprocity characteristics" of Figs. 5.13 and 5.14 show time-of-erasure as a function of erase beam intensity for each of the $CaF_2$ sample wafers listed in

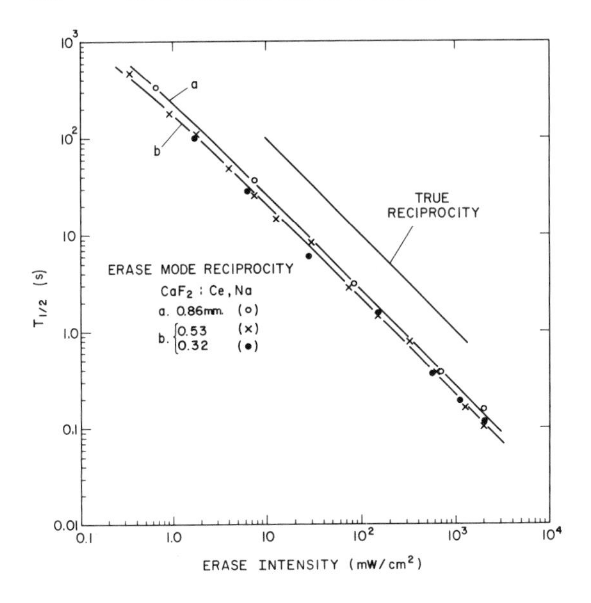


Fig. 5.14. Erase time $T_{1/2}$ (to $1/2\,\Delta D_{max}$) versus erase beam intensity for three wafers of $CaF_2$ : Ce, Na

Table 5.2. The time-of-erasure used is $T_{1/2}$, the time required for the readout optical density to be reduced to half its initial (saturated) value. The range of erase intensities spans nearly four decades or more. "True reciprocity" is indicated by straight lines with slopes of $-1.0$.

For erase beam intensities greater than about 10 mW/cm², almost the perfect inverse relationship required for true reciprocity is exhibited. Reciprocity begins to fail very gradually for smaller intensities, occurring earlier (i.e., at higher intensities) for $CaF_2$ : La, Na than for $CaF_2$ : Ce, Na, and slightly earlier for the thinner wafers than for thick ones. The values of $T_{1/2}$ corresponding to the first signs of reciprocity failure are $T_{1/2} \approx 10\,(\pm \times 3)$ s for $CaF_2$ : La, Na wafers and $T_{1/2} \approx 20\,(\pm \times 2)$ s for $CaF_2$ : Ce, Na. In the intensity range between 1 and 10 mW/cm², i.e., for approximately the first decade of reciprocity failure, $T_{1/2}$ varies as about the inverse 0.9 power of erase intensity.

The values of $T_{1/2,\text{thermal}}$ corresponding to room temperature thermal decay alone are about $2.6 \times 10^4$ s for $CaF_2$ : La, Na, about $4 \times 10^4$ s for $CaF_2$ : La, and about $4 \times 10^5$ s for $CaF_2$ : Ce. Thus, gradual reciprocity failure first begins while the ratio $T_{1/2}/T_{1/2,\text{thermal}}$ is still only about $10^{-3}$ or smaller.

## $SrTiO_3$ : Ni, Mo, Al and $CaTiO_3$ : Ni, Mo

Sensitometric characteristics for photochromic titanates are shown in Figs. 5.15 and 5.16. The solid curves indicate the erasures induced in two titanate wafers by exposure for varying times to erasure beams of selected *fixed intensities*. The abscissae are the corresponding erase exposure energy densities. Thus, selected sensitivity curves of Figs. 5.8 and 5.9 respectively, corresponding to beam intensities spanning more than three decades (more than four in Fig. 5.16), are

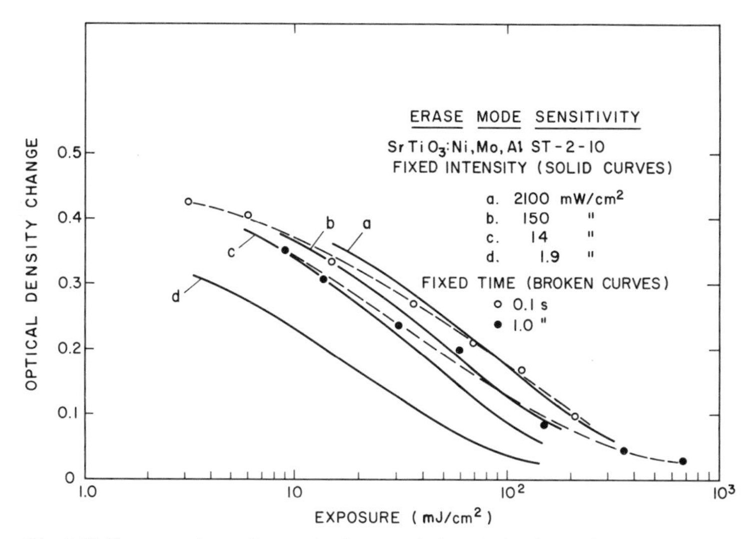


Fig. 5.15. Erase-mode sensitometric characteristics of a $SrTiO_3$ : Ni, Mo, Al wafer, showing failure of time-intensity reciprocity

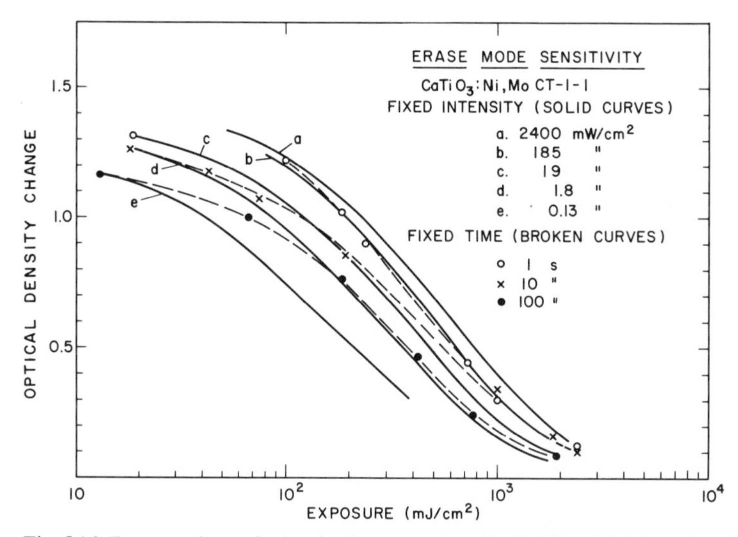


Fig. 5.16. Erase-mode sensitometric characteristics of a $CaTiO_3$ : Ni, Mo wafer, showing failure of time-intensity reciprocity

simply replotted with a new abscissa scale. Unlike the similarly derived curves for the $CaF_2$ materials, these curves have not coalesced to single sensitometric characteristics. The closest pairs of curves differ by more than 30% in exposure for the $SrTiO_3$, by about 20% for the $CaTiO_3$. Clearly, for the erase-mode operation of these titanate materials, time-intensity reciprocity does not hold over any extended region in this range of erase intensities. As a result, the

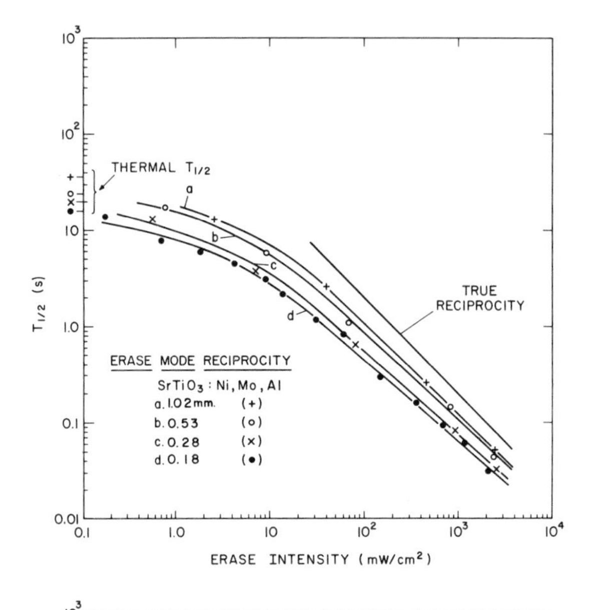


Fig. 5.17. Erase time $T_{1/2}$ (to $1/2\Delta D_{max}$) versus erase beam intensity for four wafers of $SrTiO_3$ : Ni, Mo, Al

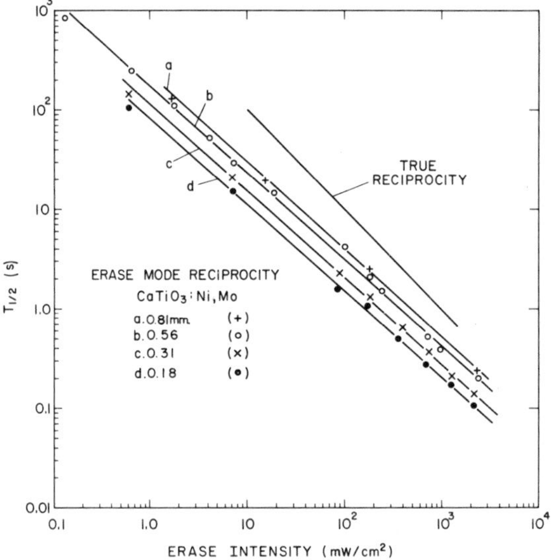


Fig. 5.18. Erase time $T_{1/2}$ (to $1/2\Delta D_{max}$) versus erase beam intensity for four wafers of $CaTiO_3$ : Ni, Mo

sensitivity data contained in the families of curves in Figs. 5.8 and 5.9 cannot be condensed into single characteristic curves.

The broken curves of Figs. 5.15 and 5.16 further support this conclusion. These curves indicate the erasures induced in selected *fixed times* by erase beams of various (six or more) intensities. These curves too are distinctly separate.

Figures 5.17 and 5.18 show the "reciprocity characteristics" for each of the $SrTiO_3$ : Ni, Mo, Al and $CaTiO_3$ : Ni, Mo wafers listed in Table 5.2. Even at the highest erase intensities used, about 2 W/cm², these materials do not show the true time-intensity reciprocity exhibited by the $CaF_2$ wafers. Instead, $T_{1/2}$ varies as about the inverse 0.9 power of the erase intensity in this range.

Since the titanate-based photochromics exhibit much more rapid thermal decay than do the $CaF_2$-based materials, it is not surprising that they suffer a more severe reciprocity failure as well. For the $CaF_2$ materials, an empirical criterion for true reciprocity was $T_{1/2}/T_{1/2,\text{thermal}} \lesssim 10^{-3}$ to $10^{-4}$.

For $SrTiO_3$:Ni, Mo, Al and $CaTiO_3$:Ni, Mo the values of $T_{1/2,\text{thermal}}$ are about $25 \pm 10$ s and about $2 \times 10^3$ s, respectively. For $SrTiO_3$, the shortest observed optical erasure times are $T_{1/2} \approx 0.04$ s, which corresponds to $T_{1/2}/T_{1/2,\text{thermal}} \approx 2 \times 10^{-3}$. For $CaTiO_3$, the smallest observed values of $T_{1/2}$ are about 0.2 s, or $T_{1/2}/T_{1/2,\text{thermal}} \approx 10^{-4}$.

Reciprocity fails very badly for the $SrTiO_3$ wafers at erase intensities smaller than about 40 mW/cm$^2$. In this range, $T_{1/2} > 1.5$ s and lies within about a factor of ten of its limiting value, the thermal half-life $T_{1/2,\text{thermal}}$. These limiting values are indicated for each wafer in Fig. 5.17. For the $CaTiO_3$ wafers, on the other hand, the inverse 0.9 power dependence of $T_{1/2}$ on erase intensity continues to somewhat below 10 mW/cm$^2$ and values of $T_{1/2} \approx 20 \pm 10$ s. For smaller intensities, the curves of Fig. 5.18 also begin to flatten.

## 5.3 Hologram Storage

Holograms are most commonly recorded in photochromic materials by erase-mode recording as described in Section 5.2.3, i.e., by selective optical erasing or bleaching of material previously darkened by uniform exposure to switching radiation. When the sample is exposed to an interference pattern of erasing light, the result is an absorption grating, maximum absorption where the incident light was weakest, minimum where it was strongest. For hologram readout, a single coherent beam is used. It is diffracted by the grating to reconstruct the stored information. To erase the hologram, the switching light recolors the crystal uniformly, returning it to its original state. This consitutes an extremely simple reusable storage medium that requires neither processing steps nor special recording techniques.

An added advantage of these materials is their volume storage capability. The storage process occurs within the bulk of the sample. Thus, in principle, one can record a multiplicity of thick holograms in one crystal by taking advantage of their angular selectivity. To prevent distortion of the reconstructed beams, the same wavelength must then be used for readout as was used for storage. This means that continued readout of any one of the holograms will eventually bleach the entire crystal. Some alkali halide photochromics have the advantage that readout is nondestructive if carried out at a temperature lower than that used for storage [5.31, 32].

Although the basic storage process in photochromic materials is relatively simple, the holographic behavior of these materials is quite complex, both in theory and in practice. The reasons have been outlined by *Collier* et al. [5.33],

*Tomlinson* [5.34], and *Ashcheulov* and *Sukhanov* [5.35]. The basic problem is that absorption processes are involved; a crystal must absorb light to record the hologram, and it must also absorb light to allow readout via diffraction. The absorption, however, also attenuates the storage, readout, and diffracted beams as they traverse the crystal. This leads to highly nonuniform gratings and maximum efficiencies much less than 100%. The situation is even more complex for storage of many holograms in the same crystal. *Tomlinson* [5.36] has pointed out that the situation can be improved, at least for the efficiency, by using the refractive index changes due to anomalous dispersion on the sides of the absorption band.

In the following, we treat some simple theoretical cases to give an overall picture of the sensitivities, diffraction efficiencies, and storage capacities expected for photochromic materials. Experimental results, where appropriate, are also discussed.

### 5.3.1 Sensitivity

We first calculate the maximum possible sensitivity. For simplicity, a simple sinusoidal grating, uniform from the front to the back of the crystal, is assumed. Although there exist more complicated theories that include the effect of beam coupling and sample bleaching during storage [5.34], our results should be a good approximation to the theoretical limits of these materials. The grating is defined as [see (1.38)]

$$a_0 = a + a_1 \cos Kx , \tag{5.1}$$

where $a$ is the average (amplitude) absorption coefficient, $a_1$ is the modulation due to the bleaching process during storage, and $K$ is the spatial frequency of the recorded grating. $K$ is given by

$$K = (4\pi/\lambda) \sin\theta_0 , \tag{5.2}$$

where $\lambda$ is the optical wavelength and $\theta_0$ is one-half of the angle between the storage beams. From *Kogelnik* [5.37], the diffraction efficiency (diffracted intensity/incident intensity during readout) of this grating in a crystal of thickness $d$ is given by

$$\eta = \exp(-2ad/\cos\theta_0) \sinh^2(a_1 d/2\cos\theta_0) . \tag{5.3}$$

The sensitivity is a measure of the rate at which $\eta$ increases during a given exposure. We must make some assumptions, however, because this increase is quite complicated. For example, $a_1$ begins at zero and increases linearly with exposure until the sample is overexposed, while $a$ begins at the initial value of the crystal absorption and decreases as the storage beams bleach the crystal. To calculate a sensitivity, we then consider only the early stages of storage where $a_1 \ll a$ and $a_1 d/2\cos\theta_0 \ll 1$. This gives

$$\sqrt{\eta} = e^{-L}(a_1 d/2\cos\theta_0), \tag{5.4}$$

where $L = ad/\cos\theta_0$.

We now define a bleaching coefficient, $C$, that gives the absorption decrease due to a given energy of absorbed light. It is given by

$$C = a_1 d/[\varepsilon(1 - e^{-2L})], \tag{5.5}$$

where $\varepsilon$ is the average energy density of the *incident* light, and $1 - \exp(-2L)$ is the fraction of the light that is absorbed. It is clear that this coefficient is valid only during the initial stages of the recording, the region of interest here. After that, the recording becomes nonlinear, and the concept of a $\gamma$ becomes important. The effect of this on the holographic storage behavior will not be discussed here.

Now defining the sensitivity as

$$S = \sqrt{\eta}/\varepsilon \tag{5.6}$$

we find

$$S = e^{-L}(1 - e^{-2L})C/2\cos\theta_0 \tag{5.7}$$

which has, at

$$L = \ln\sqrt{3}, \tag{5.8}$$

a maximum value of

$$S_{max} = [1/(3\sqrt{3})]\frac{C}{\cos\theta_0}. \tag{5.9}$$

The theoretical maximum of $C$ can be found by assuming a quantum efficiency of unity, i.e., each absorbed photon bleaches one color center. Using Smakula's equation for the center absorption [5.5], one then finds (for a center with unity oscillator strength and ~0.5 eV bandwidth) a $C$ of ~500 cm$^2$/J. Assuming $\cos\theta_0 \approx 1$, the theoretical $S_{max}$ is ~100 cm$^2$/J. Thus, a 1% hologram would require, ignoring saturation effects, only 1 mJ/cm$^2$ of incident power.

Before comparing this to measured values, let us consider the effect of anomalous dispersion. From *Tomlinson* [5.34], we see that the $S_{max}$ of (5.9) should be multiplied by a factor $F$ given by

$$F = [\sigma(\lambda)/\sigma_0][1 + \varrho^2(\lambda)]^{1/2}, \tag{5.10}$$

where $\varrho$ is a measure of the diffraction due to the anomalous dispersion, relative to that due to absorption, and $\sigma(\lambda)/\sigma_0$ is the normalized absorption of the band.

At the peak of the band, $\varrho=0$; away from the peak, it increases monotonically with wavelength, e.g., $\varrho \cong 1.14$ at $\sigma(\lambda)/\sigma_0 \cong 0.53$ and $\varrho \cong 3.66$ at $\sigma(\lambda)\sigma_0 \cong 0.14$. The increase of $\varrho$, nevertheless, is compensated in (5.10) by the decrease of $\sigma(\lambda)/\sigma_0$ away from the bandpeak. This is seen by an inspection of Tomlinson's [5.34] data for a Gaussian absorption band; e.g., $F$ goes from unity at the bandpeak to 0.8 at $\sigma(\lambda)/\sigma_0 \cong 0.53$ to 0.53 at $\sigma(\lambda)/\sigma_0 \cong 0.14$. The result is that anomalous dispersion cannot increase the sensitivity; the best wavelength for highest sensitivity is still at the bandpeak.

Measured values of the storage energy range from 10 to 500 mJ/cm$^2$ [5.38], much more than the best value calculated above. This is due, most likely, to low oscillator strengths and poor quantum efficiencies for photostimulated charge transfer, in addition to nonuniformities in the recorded gratings. There is still room for improvement, but the prospects are dim.

### 5.3.2 Maximum Efficiency

The strongest grating that can be stored in a photochromic material is one which uses up its entire recording range ($\Delta a_0$). The maximum diffraction efficiency possible in this case can be easily calculated. We let $a_1 = a = \Delta a_0/2$ in (5.3), and then, using $d$ as an independent parameter, maximize the efficiency. This gives (for $\Delta a_0 d/\cos\theta_0 = 2\ln 3$) $\eta_{\max} = 1/27 = 3.7\%$. Indeed, efficiencies of this order have been observed experimentally [5.33].

*Tomlinson* [5.34] points out that larger efficiencies are possible. One simply uses a wavelength away from the peak of the absorption band. This introduces an additional diffraction due to anomolous dispersion. The relative importance of this effect is given by $\varrho$ (see Sect. 5.3.1). For $\varrho > 10$, the maximum efficiency approaches 100% [5.34], the value for a purely phase grating. The problem is that the sensitivity decreases because of the drop in absorption away from the bandpeak. From the data of *Tomlinson* [5.34], one can show that for $\varrho \cong 10$ (the case where $\eta_{\max} \cong 50\%$), the sensitivity becomes only 0.40 of the value at the bandpeak. The situation becomes worse even further from the peak; at $\varrho \cong 37$, $\eta_{\max} \cong 80\%$, but the sensitivity is down to 0.04 of its value at the peak. Thus, there is a clear tradeoff between maximum efficiency and sensitivity. The optimum use of this tradeoff in a given application would require a selectivity in wavelength that is not generally available. The rapidly improving dye laser technology could prove useful here.

*Scrivener* and *Tubbs* [5.39] used anomolous dispersion to record phase holograms in KCl. They produced diffraction efficiencies of up to 4.2%, larger than that allowed by purely absorptive gratings. The wavelength for storage (514.5 nm) was close to the half power point of the F-band, a point where the refractive index variation has a large effect compared to the absorption. Indeed, in previous experiments with KBr [5.40] the same authors had found a maximum efficiency of only 2.3%; there the storage wavelength (632.8 nm) was close to the peak of the F-band, and the anomalous dispersion had little effect.

### 5.3.3 Storage Capacity

Since the resolution of these materials is not limited by grain size, the storage capacity of a thick crystal is determined primarily by the number of holograms which it can contain. This, in turn, depends on the diffraction efficiency needed for a suitable signal-to-noise ratio in readout; there is usually a tradeoff between the number of holograms stored and their diffraction efficiency. To calculate this tradeoff, we consider a set of $N$ purely absorption holograms, each one with an absorption modulation of $a_1$. For this case, the average absorption coefficient is given by $a=Na_1$. Placing this in (5.3) and maximizing, we find that the maximum efficiency of each hologram is

$$\eta_{\max}=[(2N+1)/(2N-1)]^{-2N}(4N^2-1)^{-1}. \tag{5.11}$$

For $N \gg 1$, (5.11) becomes

$$\eta_{\max}=\mathrm{e}^{-2}/4N^2=(1/N^2)\times 3.4\,\%. \tag{5.12}$$

Each hologram has an "inherent" diffraction efficiency of $1/4N^2$ that is attenuated by the crystal's transmission of $\exp(-2)$. Clearly one pays a large price for increasing the number of holograms. For $N=100$, the $\eta$ of each is only $3.4\times 10^{-4}\,\%$.

This can, in principle, be improved by the anomolous dispersion effect discussed by *Tomlinson* [5.34]. Following his theory and considering only efficiencies much lower than 100%, we find

$$\eta=(\mathrm{e}^{-2}/4N^2)(1+\varrho^2). \tag{5.13}$$

The absorption components maximize as before [e.g., the transmission of the crystal is still $\exp(-2)$], but the efficiency is enhanced by the factor $(1+\varrho^2)$. At $\varrho=10$, $\eta_{\max}\cong 3.4\,\%(10/N)^2$. The efficiency for a given number of holograms is increased by a factor of 100 over that allowed by absorption effects alone. As discussed in Section 5.3.1, such an improvement is at the expense of sensitivity.

In the above discussion, it was assumed that the material had sufficient recording range (maximum absorption change) to store the holograms. The required range can be calculated from the maximum possible absorption introduced by all of the holograms. This occurs where they all add up in phase. This summation added to the average absorption gives

$$\Delta a_0=2Na_1\,, \tag{5.14}$$

where $\Delta a_0$ is the required recorded range. From the value of $\Delta a_0 d$ required for the maximum efficiency derived above, one can show that

$$\Delta a_0=(2\cos\theta_0)/d\,. \tag{5.15}$$

For a 1 cm thick crystal, this becomes $\Delta a_0 \cong 2\ \mathrm{cm}^{-1}$, a value that can be easily obtained in most photochromic materials. A larger recording range is required if the efficiency enhancement via anomalous dispersion is to be used. This is because the wavelength required for a large effect is at the side of the absorption band. For $\varrho \cong 10$, the absorption at this wavelength is down to 0.039 of the value at the bandpeak [5.34]. Thus, the maximum recording range (at the peak) must increase by a factor of $\sim 27$.

These recording range arguments hold true only for uniform gratings. In practice, the gratings are not uniform. The ones recorded first are stronger at the front than at the back of the crystal. This is due simply to the attentuation of the storage beams by the sample's absorption. Thus, successive storage steps in the same crystal will eventually overexpose the front portion of the crystal. This will eventually destroy the holograms stored there. The onset of this loss is a practical limit to the number of holograms that can be superimposed by a sequential storage process. The best reported result is storage of 100 holograms in photochromic glass by *Friesem* and *Walker* [5.41].

## 5.4 Photodichroics

Photodichroic materials store information through selective alignment of anisotropic absorption centers. This storage process was first proposed by *Schneider* [5.42]. It was later demonstrated experimentally by *Schneider* et al. for bit storage [5.43] and by *Lanzl* et al. for hologram storage [5.44]. The alignment of centers is induced by exposure to linearly polarized light. It is observed as an induced dichroism in the optical absorption of the host crystal, i.e., the exposed crystal selectively absorbs light only of a certain polarization. Since this is a storage process that involves localized color centers, it has many of the features (and disadvantages) of the photochromic process discussed above. There are, however, some important improvements. The main one is that the material is, in effect, bidirectional, i.e., a sample's preferred orientation for absorption can be switched between two orthogonal directions. Thus, a single laser can be used for all steps (storage, readout, and erasure); only the direction of linear polarization is changed. In addition, there are new techniques, not available in the photochromic materials, for nondestructive readout. In the following, we present the storage model for these materials and then discuss their application for holographic storage.

### 5.4.1 Model

A good example of photodichromism is the center commonly referred to as the $F_A$ center, an alkali halide $F$ center perturbed by a nearest neighbor substitutional cation impurity [5.45]. As shown in Fig. 5.19, this center is anisotropic, i.e., it has an axis of orientation. The storage process is the realignment of the centers into a new direction.

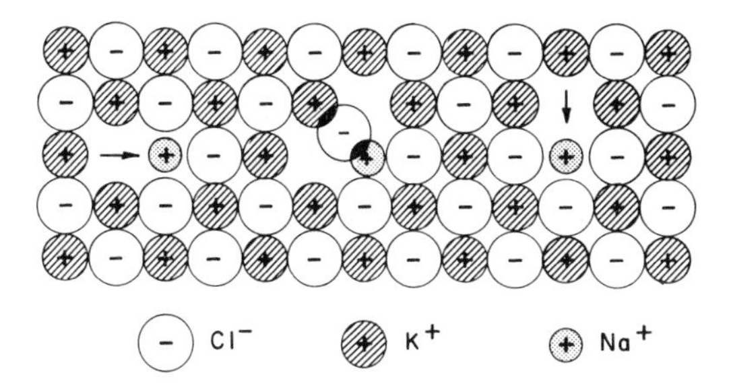


Fig. 5.19. Schematic diagram of $F_A$ centers in Na-doped KCl crystal, showing two possible orientations of an $F_A$ center. The arrows show the polarization direction of light absorbed by band 1 in Fig. 5.20. The middle configuration shows the midpoint of the reorientation process

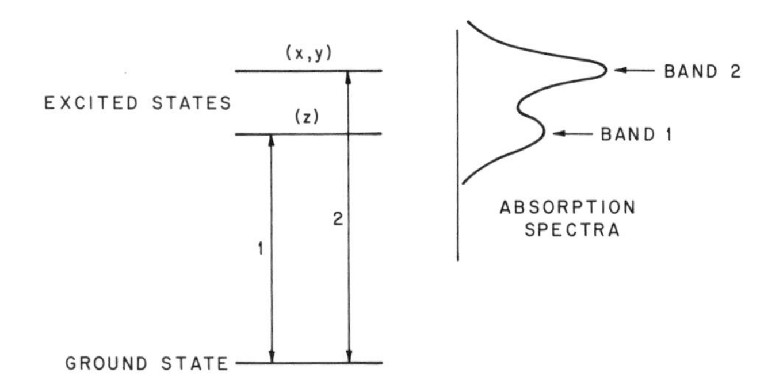


Fig. 5.20. Schematic illustration of the model for photodichroic behavior, showing optical transitions and corresponding absorption bands for an $F_A$ center. The bands are broadened by thermal vibrations

To understand this process, and how it is read out, one must first consider the optical absorption properties of the $F_A$ center. Its absorption arises from allowed transitions of the trapped electron from a single 1*s*-like state to three possible 2*p*-like states, each one aligned along one of three mutually orthogonal directions. In the unperturbed center these excited states are degenerate (all have the same energy) because of the symmetry of the center; the nearest neighbor cations are at corners of an octahedron. In the $F_A$ center, however, the electron is perturbed by an impurity at one of the corners of the octahedron (e.g., a Na ion replacing a K ion in KCl). Thus, its absorption is anisotropic, i.e., it exhibits linear dichroism. The reason is pictured in Fig. 5.20. The perturbation splits the excited state energies; the *z* excited state aligned toward the impurity is lowered relative to the other two states, *x* and *y*. This splits the *F* band into two separate highly anisotropic absorption bands (1 and 2). Band 1 absorbs light polarized parallel to the perturbation direction; band 2 absorbs light perpendicular to it.

The presence of anisotropic $F_A$ centers in a cubic crystal such as KCl does not necessarily lead to linear dichroism. A given center can be aligned along any one of three equivalent directions. Normally, the directions are equally populated. This averages out the microscopic anisotropy, giving a net absorption that is perfectly isotropic.

A crystal exhibits linear dichroism only if the orientation of the centers is changed. In photodichroics, this is done by exposure to linearly polarized light. Only centers aligned in the proper direction absorb the light. Upon absorption, their activation energy for reorientation ($\sim$0.1 eV) becomes much lower than normal ($\sim$1 eV for centers in the ground state). They can thus freely reorient into another direction at temperatures greater than $\sim$170 K. Upon relaxation to the ground state, the realigned centers are frozen in. The result is a spot of linear dichroism that can remain for weeks at room temperatures.

*Lüty* [5.45] has shown that the center reorients by hopping the vacancy around the impurity. This process is pictured in Fig. 5.19. A horizontal center (on the left) is exposed to horizontally polarized light that is absorbed by band 1 (Fig. 5.20). Upon absorption, the center rotates into a vertical orientation in which it no longer absorbs the light. The net result of such an exposure is to place most of the centers in the vertical direction. The crystal then exhibits linear dichroism.

Information stored in this manner is read out as a change in the absorption of light of a given polarization, either directly or via diffraction. Refractive index changes (i.e., birefringence) due to anomalous dispersion can also be used [5.46].

Other anisotropic centers can be used in this manner. They include the $M$ center [5.43] (two adjacent $F$ centers) and the $M_A$ center [5.47] (an $M$ center perturbed by an alkali-ion impurity). The photochromic center in rare-earth-doped $CaF_2$ shows similar reorientation effects [5.20], but it has not been evaluated for photodichroic storage applications.

### 5.4.2 Hologram Storage

The $F_A$ center in Na-doped KCl is by far the most popular photodichroic material for holographic storage applications. *Lanzl* et al. [5.44] found a sensitivity for the photodichroic process three orders of magnitude higher than that for the photochromic process in similar materials. Only 10 mJ/cm$^2$ of incident light (at 632.8 nm) was needed to produce an optical density change of 0.4. The high sensitivity was ascribed to a near unity quantum efficiency for reorientation (centers rotated per absorbed photon) at room temperature. Holographic images of excellent quality were recorded with 15 mJ/cm$^2$ (at 632.8 nm) and then read out at the same wavelength. Since readout was done at room temperature, an attenuated beam was needed to prevent hologram degradation. They pointed out, however, that nondestructive readout is possible at temperatures lower than 150 K.

*Roder* [5.48] later studied the volume storage capabilities of the same material. He found that the angular sensitivity of a hologram recorded in a 1.6 mm thick sample was $\sim$25% smaller than expected. He suggested that this was due to the attenuation of the storage beams as they traverse the crystal. The hologram was stored primarily toward the front part of the crystal, thus reducing its effective thickness. The maximum diffraction efficiency was only 0.15%, but the sensitivity was high; it required only 5 mJ/cm$^2$ at 632.8 nm. He also recorded up to eight holograms at different angles in the same crystal and found a severe degradation problem. The first recorded hologram is bleached as the subsequent recordings begin to overexpose the front part of the crystal. Eight such recordings decreased its efficiency by nearly a factor of 20.

*Blume* et al. [5.49] described a system for document storage and retrieval using $F_A$ centers in KCl. Holograms were stored at different angles, 14 min apart, with the 514.5 nm line of an argon laser. There was negligible crosstalk between the holograms upon readout with the same wavelength. Readout at 55 K was practically nondestructive. The results demonstrated that superposition of 100 holograms seems to be feasible, each one requiring an average exposure of 1 $mJ/cm^2$. A unique feature of their system was the use of orthogonally polarized object and reference beams. In this method, the optical interference pattern is composed of periodic patterns of alternating polarizations. As a result, the storage process does not lead to a bleaching process analogous to that in photochromic materials. Thus, the overall optical density of the crystal does not change during storage. This may be particularly helpful during multiple storage (see Sect. 5.3.3) but cannot eliminate erasure during storage. The beams used to record a new hologram will erase holograms previously recorded at different angles. This method also allows suppression of scattered noise during readout. A linear polarizer is positioned to pass the reconstructed image but block the light scattered from the orthogonally polarized reference beam.

Dichroic absorption by $M$ centers and $M_A$ centers has also been used for information storage, although most of this work has involved direct storage, i.e., bit storage [5.43, 50–52]. Nevertheless, holographic applications of these have been proposed [5.46, 52] with particular emphasis on the possibility of phase holograms due to anomalous dispersion [5.46]. Indeed, phase holograms have been recorded using the $M$ center in NaF, but with efficiencies of only up to 0.8 % [5.39]. $M$-center storage has the disadvantage of a low reorientation efficiency [5.43], and this leads to storage sensitivities lower than that for $F_A$ centers [5.44].

An important advantage of $M$ and $M_A$ centers is the possible use of suppressive writing effects [5.47, 52] for nondestructive readout. Here, the reorientation process is suppressed by a long wavelength (e.g., 820 μm) light beam. The beam can be used for storage (by simultaneous exposure with a broadband light beam that reorients the centers) and also for readout. Since by itself it cannot lead to any reorientation, the readout is nondestructive, even at room temperature. The suppressive effect occurs because the $M$ and $M_A$ centers reorient only after they have been ionized by the broadband light. The released electron is trapped at an available $F$ center, converting it to an F′ center. After this phototransfer process, the ionized $M$ and $M_A$ centers can absorb the broadband light and reorient. The long wavelength light suppresses this process by immediately exciting the extra electron out of the F′ center, returning it to the ionized center before it has a chance to reorient.

## References


5.1 G.H.Brown, W.G.Shaw: Rev. Pure Appl. Chem. **11**, 2 (1961)
5.2 R.Exelby, R.Grinter: Chem. Rev. **65**, 247 (1965)
5.3 J.J.Amodei: Direct Optical Storage Media. In *Handbook of Lasers*, ed. by R.J.Pressley (Chemical Rubber Co., Cleveland 1971) Chapt. 20
5.4 Z.J.Kiss: IEEE J. Quant. Electron. QE-**5**, 12 (1969); —Phys. Today **23**, 42 (1970)

5.5 B.W.Faughnan, D.L.Staebler, Z.J.Kiss: Inorganic Photochromic Materials. In *Applied Solid State Science*, Vol. 2, ed. by R.Wolfe (Academic Press, New York 1971)
5.6 M.Y.Hirshberg: Compt. Rend. **231**, 903 (1950);
Y.Hirshberg, E.Fischer: J. Chem. Soc. (London), Part I, p. 629 (1953); —J. Chem. Phys. **21**, 1619 (1953)
5.7 T.J.Pearsall: J. Royal Institution **1**, 77, 267 (1830)
5.8 P.Curie, M.Curie: C. R. Acad. Sci. (Paris) **129**, 823 (1899)
5.9 K.Przibram: *Irradiation Colours and Luminescence*, translated and revised by J.E.Caffyn (Pergamon Press, London 1956)
5.10 E.Ter Meer: Ann. **181**, 1 (1876)
5.11 D.Chen, J.D.Zook: Proc. IEEE **63**, 1207 (1975)
5.12 M.Chomat, M.Miller, I.Gregora: Opt. Commun. **4**, 243 (1971)
5.13 P.J.van Heerden: Appl. Opt. **2**, 393 (1963)
5.14 A.N. Carson: USAF Contract Report, unpublished (1965)
5.15 E.N.Leith, A.Kozma, J.Upatnieks, J.Marks, N.Massey: Appl. Opt. **5**, 1303 (1966)
5.16 D.R.Bosomworth, H.J.Gerritson: Appl. Opt. **7**, 95 (1968)
5.17 D.L.Staebler, Z.J.Kiss: Appl. Phys. Lett. **14**, 93 (1969)
5.18 D.L.Staebler, S.E.Schnatterly, W.Zernik: IEEE J. Quant. Electron. QE-**4**, 575 (1968)
5.19 W.Phillips, R.C.Duncan, Jr.: Metallurg. Trans. **2**, 769 (1971)
5.20 D.L.Staebler, S.E.Schnatterly: Phys. Rev. B **3**, 516 (1971)
5.21 C.H.Anderson, E.S.Sabisky: Phys. Rev. B **3**, 527 (1971)
5.22 R.C.Alig: Phys. Rev. B **3**, 536 (1971)
5.23 R.C.Duncan, Jr., B.W.Faughnan, W.Phillips: Appl. Opt. **9**, 2236 (1970)
5.24 R.C.Duncan, Jr.: R C A Rev. **33**, 248 (1972)
5.25 B.W.Faughnan, Z.J.Kiss: Phys. Rev. Lett. **21**, 1331 (1968)
5.26 B.W.Faughnan, Z.J.Kiss: IEEE J. Quant. Electron. QE-**5**, 17 (1969)
5.27 J.J.Amodei: Ph.D. Thesis, Univ. of Pennsylvania (1968)
5.28 B.W.Faughnan: Phys. Rev. B **4**, 3623 (1971)
5.29 L.Merker: J. Amer. Ceram. Soc. **45**, 366 (1962), also private communication
5.30 H.C.Kessler, Jr.: J. Phys. Chem. **71**, 2736 (1967)
5.31 D.Huhn, W.Martienssen: Opto-Electron **2**, 47 (1970)
5.32 B.Stadnik, Z.Tronner: Opt. Commun. **6**, 199 (1972)
5.33 R.J.Collier, C.B.Burkhardt, L.H.Lin: *Optical Holography* (Academic Press, New York 1971)
5.34 W.J.Tomlinson: Appl. Opt. **14**, 2456 (1975)
5.35 Y.V.Ashcheulov, V.I.Sukhanov: Opt. Spectr. **34**, 201 (1973); **34**, 324 (1973)
5.36 W.J.Tomlinson: Appl. Opt. **11**, 823 (1972)
5.37 H.Kogelnik: Bell System. Tech. J. **48**, 2909 (1969)
5.38 K.S.Pennington: Holographic Parameters and Recording Materials. In *Handbook of Lasers*, ed. by R.J.Pressley (Chemical Rubber Co., Cleveland 1971) Chapt. 21
5.39 G.E.Scrivener, M.R.Tubbs: Opt. Commun. **10**, 32 (1974)
5.40 G.E.Scrivener, M.R.Tubbs: Opt. Commun. **6**, 242 (1972)
5.41 A.A.Friesem, J.L.Walker: Appl. Opt. **9**, 201 (1970)
5.42 I.Schneider: Appl. Opt. **6**, 2197 (1967)
5.43 I.Schneider, M.Marrone, M.N.Kabler: Appl. Opt. **9**, 1163 (1970)
5.44 F.Lanzl, U.Roder, W.Waidelich: Appl. Phys. Lett. **18**, 56 (1971)
5.45 F.Lüty: $F_A$ Centers in Alkali Halide Crystals. In *Physics of Color Centers*, ed. by W.B.Fowler (Academic Press, New York 1968) Chapt. 3
5.46 I.Schneider: Phys. Rev. Lett. **32**, 412 (1974)
5.47 I.Schneider: Appl. Opt. **11**, 1426 (1972)
5.48 U.Roder: Opt. Commun. **6**, 270 (1972)
5.49 H.Blume, T.Bader, F.Lüty: Opt. Commun. **12**, 147 (1974)
5.50 I.Schneider, M.Lehman, R.Barker: Appl. Phys. Lett. **25**, 77 (1974)
5.51 J.Burt, H.Knoebel, V.Krone, B.Kirkwood: Appl. Opt. **12**, 1213 (1973)
5.52 D. Casasent, F.Caime: Appl. Opt. **15**, 215 (1976)

# 6. Thermoplastic Hologram Recording

J. C. Urbach

With 14 Figures

The creation of surface relief holograms by thermoplastic deformation has proven to be an unusually satisfactory recording technique for many types of holography. These holograms, which can be used in either transmission or reflection, are rather pure examples of thin phase holograms, and can be treated accordingly, as described by *Kogelnik* [6.1]. In this chapter we shall briefly discuss the historical background of the method, describe several of its more important device and process variants, and consider some of the experimental techniques involved in device fabrication and operation. Then we shall discuss both the optical and physical properties important in hologram recording by means of thermoplastic deformation. The former include such characteristics as resolution and bandwidth, sensitivity, diffraction efficiency, nonlinear distortion effects, and noise properties. The latter include cyclic reusability, mechanical durability, and the shelf life of the unused device, the latent image and the completed hologram.

It can be said that thermoplastic hologram recording [6.2] is descended directly from the recording of screened conventional images ("carrier frequency photography") [6.3, 4]. This in turn evolved from the use of thermoplastics for in-vacuo electron beam recording [6.5], which in its turn is a descendant of the Eidophor television system [6.6].

All of these processes have in common the creation of localized electrostatic forces which deform a fluid surface (usually a free surface adjacent to vacuum or gas) into a pattern of surface relief capable of altering the phase of a transmitted or reflected optical readout wave. This phase alteration, if on a scale large compared with the wavelength of the readout light, can be regarded as deflecting that light according to the laws of reflection and refraction. The resulting geometric optics description of the readout process, while simple, is inappropriate for most holographic cases in which the scale of the deformation is often on the order of a few wavelengths. Hence, the more general diffraction theory description [6.1] of readout is normally used.

## 6.1 Process Description

There are two basically different ways of making holograms or holographic spatial filters. One is the photographic approach, in which the hologram is formed by using a photosensitive process to record a naturally occurring interference pattern. The other is the synthetic approach, in which a hologram is computed and then created artificially. In the latter approach, it may be possible to create the hologram without the use of any light-sensitive process. This distinction is particularly important when considering thermoplastic holograms, since the direct recording on thermoplastics of patterns created by a scanning, modulated electron beam, already noted [6.5] for historical reasons, can be a powerful tool for the creation of synthetic holograms without the need for optical and photographic intermediate steps. Interest in the electron beam recording process has flourished intermittently since the 1960's at the University of Michigan, and has recently been active at its Environmental Research Institute [6.7].

The different forms of thermoplastic recording technology which are appropriate to these two types of holography all depend, of course, on the use of electrostatic forces to induce plastic flow and thus create the relief pattern which constitutes the phase image or hologram. They differ rather drastically in practical embodiments, and are best described separately. Although most of this chapter will deal with the light-sensitive version used for natural holography, it seems appropriate for both historical and tutorial reasons to begin with a simple description of the electron-beam form of the process.

### 6.1.1 Thermoplastic Electron Beam Recording

If a layer of insulating thermoplastic is deposited on a conductive substrate and placed in a vacuum with the substrate grounded, it can be charged by a scanning electron beam. Such a beam deposits negative charge on the surface of the thermoplastic. Electrical neutrality is maintained by attraction of an equal and opposite positive charge onto the side of the thermoplastic layer in contact with the grounded substrate. The attraction between charges produces a certain electrostatic pressure, which macroscopically is proportional to the square of the surface charge density, and which is exerted on the thermoplastic layer. If the thermoplastic is heated above its softening temperature (a temperature which is actually not very well defined in most such materials), it becomes in principle capable of fluid deformation. In order for deformation to occur, two major conditions must be satisfied. First, the charge density must be nonuniform. Uniform charge, hence uniform electrostatic pressure, will not significantly deform a substantially incompressible fluid such as molten thermoplastic. (An exception to this condition should be noted immediately: If the surface charge density is high enough, the fluid will break up into random deformation patterns, commonly known as "frost" because of their visual appearance. This phenomenon is of considerable significance, and will be discussed subsequently.) Second, the conductivity of the molten thermoplastic must be low enough to permit the fluid flow of deformation to proceed significantly prior to electrical discharge of the thermoplastic. This includes both conductivity normal to and parallel to the surface, since the former collapses the driving field causing electrostatic pressure, while the latter can adversely affect resolution.

All deformation of this type occurs in opposition to the force of surface tension, which tends to restore the free surface to the minimum area of its flat state. Continued heating, which maintains the thermoplastic's conductivity at an elevated level, will eventually discharge the surface, and eliminate the forces inducing deformation. Since the thermoplastic remains in the fluid state while heated, the surface tension takes over and restores the surface to its flat state. For this reason, thermoplastic development does not proceed to a steady state deformation pattern, but rather if prolonged or resumed eventually returns the surface to an undeformed state. Thus the process is inherently reversible, sometimes a significant advantage. Linked with this advantage, however, is the need to terminate the development process at or near an optimum state, and by relatively rapid cooling to harden the thermoplastic while it is still deformed, and thus to "fix" the deformation pattern. This need is sometimes a rather difficult one to satisfy properly, and can give rise to more or less severe processing problems.

In the actual use of this process for electron beam recording, beam current, dither or focus modulation can be used to modulate the surface charge density and thus the latent electrostatic image. Uniform heating then initiates deformation, and uniform cooling interrupts it to fix the image. Both removable film versions [6.5], using a flexible web substrate, and fixed faceplate versions [6.7, 8] of the process have been operated successfully. Extensive materials work was done at General Electric Co., IBM Corp., and elsewhere to develop thermoplastics compatible with the requirements of vacuum operation and electron beam bombardment.

When used in conjunction with high resolution electron optical systems and high optical quality substrates, this form of the process may still turn out to be one of the most suitable means of recording synthetic holograms, either from nonoptical signals, as in synthetic aperture radar, or in the form of computer-generated holograms. When the hologram information is contained in an electrical signal prior to actual hologram recording, some form of direct electron beam recording is often the most desirable way to transform this signal into an optical hologram. In that case, the inherent advantages of thermoplastic electron beam recording over other types of electron beam recording may prove significant. There are also advantages of this form of the process over the photosensitive version discussed subsequently. These include, of course, simplicity of fabrication, absence of a thermoplastic/photoconductor interface with its attendant chemical and electrical problems, and absence of the light absorbing photoconductor which can reduce the efficiency of optical transmission readout.

### 6.1.2 Photosensitive Recording

The remainder of this chapter will be devoted to the photosensitive forms of thermoplastic recording. It is these forms that have seen widespread use in optical hologram recording, and it is these that have been investigated most extensively.

## Photoplastic Devices

There are two basic versions of the photosensitive devices. The first, shown in Fig. 6.1a, has been termed "photoplastic" by *Gaynor* and *Aftergut* [6.3], and is the simpler in terms of structure and process. In this version, the thermoplastic of the electron beam recording process is replaced by a photoconductive thermoplastic. Otherwise, the structure remains unchanged from the elementary form. Now the process steps are as follows:

1) Uniform charging (conductive substrate grounded).
2) Exposure to light.
3) Heating to develop the deformation.
4) Cooling to fix the deformation.
5) Heating, if desired, to erase the deformation.

The uniform charging step is generally accomplished by corona charging, but other charging methods, discussed subsequently, are also applicable here. It creates a uniform internal electric field in the material. The exposure step now replaces the former process of electron beam imaging as the means for inducing the nonuniform surface charge density needed for creating a developable latent image. The remaining process steps are the same as in the electron beam recording version.

The simplest form of charging is of course the use of a corona discharge in air, typically a DC discharge requiring several thousand volts applied to one or more fine wires or an array of small-radius points. The radius of the wires or points must be small enough to assure that local fields exceed those required for air breakdown, so that ionization can occur. The grounded substrate of the device together with the choice of corona potential assures that a relatively weak average electric field exists to move ions of the desired sign toward the device surface. Usually it is necessary to charge that surface to a potential of several hundred volts, corresponding, at typical device thicknesses, to several tens of volts per micron of internal electrical field. For the single-layer form of device, it is sufficient to assure that the surface charge density is uniform. This requirement imposes some restrictions on charging geometry. Either a relatively dense fixed array of wires or points must be used, somewhat larger than the device to reduce edge effects, or a one-dimensional array or single wire must be scanned at a uniform velocity in the orthogonal direction to assure uniform surface charge. Since all theoretical predictions, e.g., *Gaynor* [6.9], suggest that the device sensitivity depends at least quadratically on the electric field or charge density, it is clear that uniformity must be good. Estimates of tolerable deviations from uniformity vary widely and are of course dependent on applications, but in general it is advisable to hold these variations under 5%.

The operation of the exposure step is in principle relatively simple. Since the thermoplastic has been made photoconductive, incident light results in the photogeneration of charge carriers throughout the sensitive layer. These are separated by the action of the internal electric field, and, if both holes and electron are mobile, each migrates to the oppositely charged surface, and

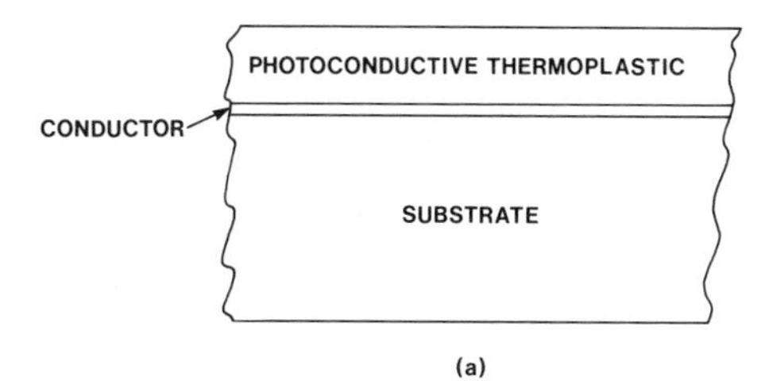


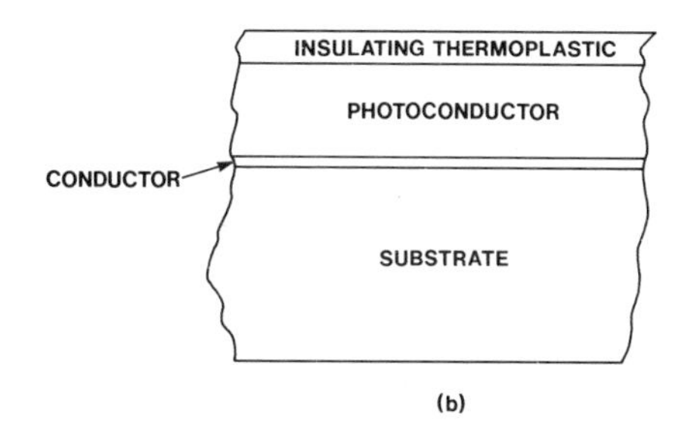


Fig. 6.1a and b. Device cross section (not to scale). (a) "Photoplastic" device configuration: Photoconductive thermoplastic on a conductive substrate. (b) Thermoplastic-overcoated photoconductor device configuration: Nondeformable photoconductor on a conductive substrate

partially or entirely neutralizes the original charges deposited there during the charging step. If only one sign of charge is mobile, bulk exposure is less effective in neutralizing the surface charge density, but the operating principle is similar. In any case the effect of exposure-induced reduction in the surface charge density reduces the electrostatic pressure in regions with high exposure relative to that in regions with low exposure. The result is an imagewise variation of electrostatic pressure, hence the developable latent image.

Uniformly illuminated areas have uniform surface charge density, so that their latent image are not developable. In terms of spatial frequency response, this means that the process has no spatial DC response, a characteristic requiring special compensation techniques in ordinary photographic use, but naturally suited to offset reference beam holography, matters which will be discussed in a subsequent section.

Development is usually carried out by uniform heating as in the electron beam version. Several ways of achieving this heating have been used. These include radiant heating with non-actinic (to prevent immediate photodischarge of the charge pattern) radiation, ohmic heating by passing a current through the conductive substrate coating, ohmic heating by use of RF fields capacitively coupled to the conductive coating and inducing RF alternating currents in it, and heating by means of a hot air stream directed at the thermoplastic surface. In addition to thermal development, use of solvent vapors to soften the thermoplastic is also possible [6.10], although generally more difficult to control than the thermal methods. The rate of development by any of these means can vary from tenths of milliseconds to hours, but is typically on the order of one second in most embodiments of the process [6.47]. The achievement of uniformity in development is rarely easy, and is of course easier in slow thermal development where the device is always close to thermal equilibrium and internal as well as external heat transfer processes have enough time to help reduce heating nonuniformities.

### Thermoplastic-Overcoated Photoconductor Devices

The other version of the device [6.11] is more complex in structure but has certain advantages that have led to its use in most holographic experiments and applications. It is shown in Fig. 6.1b. It differs from the photoplastic device insofar as the thermoplastic and photoconductor functions have been separated, and are now performed by two different layers, the upper transparent thermoplastic layer having no photoconductivity in this version. Terminology here was somewhat confused by the tendency of certain authors, e.g., *Colburn* and *Dubow* [6.12], to use the term "photoplastic" to denote these two-layer devices. For clarity, we use it only in its original meaning of single-layer device. The process steps required for operation of the two-layer devices are slightly more complex. The requirement for this more complex process can be readily, albeit somewhat simplistically and not quite accurately, understood from the following considerations: Exposure and consequent photogeneration in the photoconductor cause opposite charges to migrate to its two interfaces, with the thermoplastic and substrate, respectively. While this reduces the surface potential of the device, it does not reduce the surface charge density, thus creating no imagewise variation of that density on the thermoplastic, hence no developable latent image. This problem is solved by the introduction of another process step, the recharging of the surface to a constant potential. This step deposits additional charges in exposed regions, roughly in proportion to the decrease in surface potential, hence to exposure. Thus the exposure-induced surface potential variation is converted by recharging into a corresponding variation in surface charge density, and therefore into a developable latent electrostatic image. From here on, development, fixing and erasure proceed as in the first version.

The requirement for recharging to a constant surface potential can be met by using a screened corona charging device commonly known as a "scorotron" [6.13, 14]. In such a device, a perforated sheet or wire mesh screen is inserted between the corona wires and the device surface and is held at a voltage above (or below) ground approximately equal to the desired surface potential. Flow of ions from the corona to the thermoplastic proceeds through the screen, albeit with slightly reduced efficiency, until the surface potential approaches the screen potential. After the electric field from the screen to thermoplastic approaches zero, further ion flow is to the screen rather than to the thermoplastic. Thus the thermoplastic surface asymptotically approaches the screen potential. Such an arrangement has the additional advantage of helping to assure uniform charging since the potential of the screen itself is uniform because it is made of a highly conductive material. Therefore it forces all areas of the thermoplastic to approach the same uniform potential.

Although these two advantages generally justify the use of screened coronas, it should be noted that even recharge with an unscreened corona will render the latent image developable. This is presumably a result of preferential deposition of charge in those regions of the surface which are at a lower potential after exposure, leading to the desired exposurewise variation of charge density after recharge.

A useful feature of this two-layer version of the device and the corresponding process is that the recharge step largely stabilizes the electrostatic latent image, permitting subsequent exposure to actinic light without significant degradation of the latent image. This occurs for the same reason that recharge is required in the first place: Exposure of the two-layer device alters only the potential, not the surface charge density. Therefore such exposure after the recharge step does not alter the developable latent image, although it does of course affect the undevelopable potential distribution. This holds true only if the thermoplastic is indeed a good insulator, a condition easily satisfied by most thermoplastics. If this exposure is not accompanied or followed by additional charging, it neither erases nor alters the developable latent image. The practical consequence of this is often quite desirable: it permits inspection or even automatic feedback control of the developing image (or hologram) with actinic light. In the case of many panchromatic photoreceptors, all visible light is actinic, hence this feature makes possible inspection or closed loop development control in a manner not possible with the single layer device, or for that matter with most other photographic processes.

It should be noted here that the qualitative explanations given thus far for all electrostatic imaging effects, including this last one, are based on a semi-macroscopic view. A detailed analysis of the microscopic aspects of the electrostatic fields and pressure distribution does predict a more complex situation. In particular, a weakly developable latent image is formed in the two-layer device even without the recharge step. This prediction has been confirmed experimentally. Likewise, post-recharge exposure does have some effect on the developable latent image. In most practical situations, however, these effects represent a relatively minor correction on those expected from the semi-macroscopic view, hence the much simpler qualitative explanation provided by that theory is usually a satisfactory guide for experimental work.

### Device and Process Variations

As might be expected in a device and process having a number of possible degrees of freedom, many interesting variants have been proposed and tried. We shall omit those that have not, to our knowledge, been used experimentally in the holographic applications of thermoplastic electrophotography. Perhaps the most obvious device variation is the addition of a reflective overcoat to the thermoplastic. This of course greatly increases the efficiency of image reconstruction by reflected light. If the reflective surface layer is so compliant or fragile that it has little effect on the rheological or electrical properties of the thermoplastic, and if the exposure of the device is through a transparent substrate, this variant can make possible a significant sensitivity increase, as will be discussed subsequently. Under certain conditions, the reflective layer can also be made conductive, as in the Gamma Ruticons of *Sheridon* [6.15]. This permits operation of the device in an electroded configuration, eliminating the need for corona charging and resulting in a more convenient process.

Another form of device/process configuration which eliminates some of the drawbacks of corona charging has been described by *Colburn* and *Tompkins* [6.16]. This uses a transparent charging electrode closely spaced above the thermoplastic layer. A voltage is applied between this electrode and the electrode underlying the photoconductor; this voltage is capacitively divided among the device layers and the gap. When the voltage is high enough to prooduce an electric field in the gap exceeding the breakdown field for air (or other gas) in the gap, a discharge forms across the gap and the thermoplastic surface is charged. This configuration makes possible exceptionally rapid charging of the device.

Both this configuration and conventional corona charging have been used in two variants of the process sequence previously described for the thermoplastic-overcoated photoconductor device. One is to combine the charging and exposure steps (1 and 2) into a single step of simultaneous charge and exposure. This is usually compatible only with exposure through the substrate if corona charging is used, since most corona charging devices, with their wire discharge and screen arrangements, block exposure from the thermoplastic side. The aforementioned configuration of *Colburn* and *Tompkins* [6.16], on the other hand, permits this process variant to function with exposure from either side, and in fact performs better (in terms of resulting diffraction efficiency) if used with this variant.

A more radical variant is the combination of Steps 1–3. That is, charging, exposure and development are concurrent, as a result of heating either just prior to or during a combined charge and expose step. This process has been described by various authors [6.10, 17]. It tends to produce especially strong deformations under a given set of conditions, since the charge density continuous to increase in thinner regions as deformation proceeds, thus creating a kind of positive feedback which amplifies the deformation excursion. Some of the strongest deformations reported in [6.17] were obtained by using this type of process.

### 6.1.3 Process Models and Theories

Unlike many other image recording systems, whose operation involves highly complex chemical processes, thermoplastic deformation imaging is an essentially simple physical process, involving as it does only basic electrostatics and fluid flow. It thus promises to be more amenable to detailed theoretical analysis than are most of the other imaging systems. It has been a tempting subject for analyses, many of which were undertaken during the decade from 1963 to 1973.

Many of the published theories, for example References [6.18–21] concerned themselves only with the random, or frost, mode of deformation, where there was no initially nonuniform pattern of charge density. A few, e.g., [6.5, 22, 23] did consider the variations in charge density which correspond to image formation. In this section, we shall not discuss these theoretical investigations in detail, but shall note some of their important features. Considerable difficulty is entailed in using the results of these theories as a guide to practical device and process parameters. All of them model the actual experimental situation in a simplified way, while in fact it is often hard to know what model of that situation is actually justified. For example, parameters of importance in the theories, such as surface tension. viscosity, electrical conductivity, and dielectric strength may vary in depth and in time in a manner almost impossible to measure experimentally and difficult to predict theoretically. Even when such a prediction is made, e.g., *Mathies* et al. [6.21], p. 107, its inclusion would complicate the theory to an unmanageable degree. Thus there still exists no complete and definitive theory of thermoplastic deformation imaging. It now seems that the apparent simplicity of this imaging system was somewhat deceptive. The theoretical results obtained apply to an oversimplified model of the real systems, and have thus far provided only an incomplete guide to process and device optimization.

One of the most complete attempts to analyze the imaging system formed by the two-layer class of devices was carried out by *Schmidlin* in 1967 and is documented in a series of unpublished internal technical memoranda and reports. One of his results [6.24] is given below in condensed form and without derivation only to illustrate their general characteristics.

*Schmidlin* assumed a sinusoidal charge density, created by exposure, at the thermoplastic-photoconductor interface, with the recharge step producing a charge pattern on the top surface of the insulating thermoplastic according to conventional electrostatic theory. He then computed the relative deformation $\delta D/D$, where $D$ is the thermoplastic thickness. His result was:

$$\frac{\delta D}{D}=\frac{K_2 \Delta E V(\mathrm{e}^{\omega t}-1)}{\beta f \cosh(2\pi f D)[g+\tanh(2\pi f D)](\phi Q-2\pi f D)}, \tag{6.1}$$

where $t$ is time, $\Delta E$ is exposure excursion, $V$ is voltage, and $K_2$ is a constant, and where, unfortunately, $\omega$, $g$, $\beta$, $\phi$, and $Q$ are all functions of spatial frequency $f$, themselves complicated enough to effectively prevent any sense of a simple relationship between $\delta D$ and $f$.

This theory takes into account the possibility of an effective skin tension on the thermoplastic which may differ from the normal surface tension of that material. It also considers the effect of charge leakage during development, and of viscosity. It does assume, as do other theories, that temperature is uniform and constant during development, a generally invalid assumption necessary to avoid further complication in the already complex analysis. The results of the theory, plotted numerically, are in qualitative agreement with some experiments, predicting the rather sharply peaked spatial frequency response curve often observed.

It does not seem fruitful at this point to develop a more complete theory. Among other things, it would be hard to test experimentally, since, as noted before, it would be extremely difficult to obtain accurate measurements of the key parameters.

### Theories of Random Deformation

The theories of random deformation were usually intended to provide models for frost imaging, in which relatively high frequency random deformations are modulated by image information. Because of its inherently noisy nature, this mode of imaging is of little or no interest in hologram recording. Nevertheless, since random deformation of this type is the chief source of noise even when it is not being used as an information carrier, it is worthwhile to predict conditions under which such noise can occur. Therefore the theories of frost are of primary interest here as theories of noise in hologram recording. Of the many discussions of frost deformation, perhaps the most extensive is the joint theoretical and experimental report of *Matthies* et al. [6.21]. The theoretical treatment is based on that of *Budd* [6.19] and extends his work to include threshold as well as spatial frequency effects. Earlier treatments of the subject are special cases of this more recent theory.

In common with other theories, this one does not treat the problem as a true two-dimensional random deformation, but instead models the real situation by considering the growth of a sinusoidal perturbation of the surface. Both equipotential surface and uniform charge density cases are considered. The underlying physical effects here are that a charged fluid surface tends to increase its surface area as a result of mutual repulsion of the charges, while this increase is opposed by both surface tension and viscous drag. While deformation is under way, the finite conductivity of the fluid tends to discharge it, thus dissipating the charge which drives the deformation.

*Budd* [6.19] predicted that there is a wavelength dependence of the growth rate of the surface perturbation. In general, the wavelengths of most rapid frost growth are well within the range of the most rapid response to images, so that the noise spectrum of the deformation largely overlaps the useful imaging spatial frequency range. *Matthies* et al. [6.21] show that in all the cases which they consider, including equipotential surface and uniform charge density, and both sequential and simultaneous charging and development, there is a threshold value of initial surface potential. Below this value frost does not occur at the "dominant", i.e., fastest growing, wavelength predicted by Budd's theory. Such thresholds are in fact observed experimentally, and, under the controlled experimental conditions of [6.21], good agreement was found between theoretically predicted and experimentally observed values of threshold voltages.

The existence of this threshold is significant in holographic applications insofar as it suggests that there may be conditions under which imaging, i.e., charge-excursion-driven, deformations can grow at a significant rate, while frost deformation cannot develop. This is an area calling for further study, but one which could prove important by permitting dramatic improvements in signal-to-noise ratios. *Matthies* et al. (Ref. [6.21], p. 93) suggest that in general frost-free deformations will require the use of relatively thin thermoplastic layers, which will run into other limitations arising from finite dielectric breakdown strength. Application and perhaps extension of existing "driven-case" imaging theory could help define these optimum performance regions.

One other result of this theory, significant mainly in a negative sense, is a proof that gravitational forces are in fact negligible for thermoplastic layers of interest in holography. Expected on intuitive grounds, this result confirms that the gravitational field can safely be ignored in analysis and experiment (Ref. [6.21], p. 104).

### 6.1.4 Materials

The choice of materials and fabrication procedures for thermoplastic imaging devices is quite broad. The description of the device structure itself suggests that a wide range of substrates, photoconductors and thermoplastics can in principle be used in such devices.

#### Substrates and Conductive Coatings

Considering the device structure one layer at a time, we see that the substrate needs to be transparent and of high optical quality if transmission holography with front exposure is contemplated. If exposure is through the substrate, and optically compensating conjugate image reconstruction is used, requirements on optical homogeneity can of course be relaxed. More generally, it is advisable to select substrate quality according to the usual principle that a hologram is an optical element and that its optical quality must be appropriate to its intended placement in the imaging path. The requirement for transparency can of course be removed if reflected light reconstruction and front exposure are planned, but this is in fact a rare case, since it usually implies such undesirable practices as exposure through a reflective layer, or reconstruction by Fresnel reflection from the thermoplastic surface with a coherent (interfering) image reconstructed after substrate reflection and two-fold transmission, or post-development reflective coating of the finished hologram. It may be most practical if the hologram is to become a master for mechanical replication and thus not used optically. Mechanical requirements on the substrate are generally dictated by the optical ones. Usually, although not always, rigidity is demanded; this requirement can often be relaxed somewhat in the case of transmission holograms. The net consequence of the substrate requirements usually leads to the choice of glass, but occasionally transparent plastics have been used [6.17].

The optical requirements on the conducting coating used for the ground electrode are similar to those on the substrate itself. Usually they are easy to satisfy because the conductive coating can generally be quite thin, so that its homogeneity is not a problem from the optical standpoint. The electrical requirements may be more stringent, depending on the development mode chosen. If direct or RF coupled ohmic heating of this conductive layer is used for development, than it becomes necessary to assure a well-selected (relatively low, typically tens of ohms per square) and highly uniform resistivity of this layer. If its only purpose is as an electrostatic ground plane, on the other hand, it is quite adequate if the surface resistance is thousands of ohms per square, and is relatively nonuniform. This is a consequence of the extremely low currents that flow during the relatively slow charging steps, resulting in near-equilibrium operation of this layer.

## Photoconductors

If the device is of the single operational layer (photoplastic) variety [6.3, 25, 26, 46], then some rather special constraints are imposed by the simultaneous requirements of photoconductivity and thermoplastic deformation. Probably the most limited materials range is that required by the photovoltaic version of these single-layer devices, described by *Gaynor* and *Sewell* [6.27], but we shall omit discussion of those device here because of their generally low sensitivity. Even the photoconductive version has serious restrictions, because the thermoplastic polymer matrix must support conduction. This has usually been achieved in two ways, either by incorporating a photosensitive dye in the thermoplastic, or by producing a dispersion of photoconductive particles capable of injecting the charge carriers they create into the polymer, which in either case must be capable of transporting these charges. Both these approaches have been used [6.3, 25] in versions of photoplastic recording. In the former case, the materials used included TCNE-Pyrene and leuco-malachite green with trace malachite green while in the latter, the dispersed photoconductor was either copper phthalocyanine or copper-doped cadmium sulfide in the form of small (generally submicron) particles. The thermoplastic polymer was usually plasticized polystrene of unspecified molecular weight.

In addition to the perhaps limited number of possible combinations of sensitizing dyes or particles with suitable thermoplastic matrixes, there is a fundamental tradeoff in the photoplastic devices which has no direct counterpart in two-layer devices. This involves the choice of thickness of the layer. For a given material, one can expect some upper limit to the concentration of photoconductive material, whether it be in dye or particulate form. This limit will probably be imposed by either excessive dark decay or by interference with the desired rheological properties. In practice this has meant that the material must be many microns thick to absorb a reasonably large fraction, say 50%, of the incident light. (Absorption much above 50% is not likely to be desirable if image reconstruction is by transmitted light, since this light will have to pass through the same layer, and is likely to be of a wavelength similar to the exposing light.) Typical thicknesses of photoplastic films tend to be in the 20 to 25 μm range [6.25, 26]. Now one will encounter the thickness-imposed resolution or bandwidth limitation of all thermoplastics. The frequency of maximum deformation response and generally the useful spatial frequency bandwidth are inversely proportional to thermoplastic thickness (see pp. 186, 189). If this thickness must be increased to assure more absorption, hence higher photosensitivity, it will be at the expense of bandwith. This is not a linear tradeoff since absorption is exponential (although actual sensitivity sometimes is not [6.25]), while bandwidth is inversely proportional to thickness. While different quantitatively, this tradeoff is qualitatively somewhat reminiscent of the well-known bandwidth-sensitivity tradeoffs in silver halide emulsions. The consequences of this tradeoff appear to be quite serious in practice, because it prevents attainment of reasonably high sensitivity in the ultrathin (less than 1 μm) layers which both theory and experiment have shown to be the most promising in terms of bandwith and signal-to-noise ratio.

The two-layer system has attained favor in holography both because it circumvents this particular tradeoff and because of the greater ease of independently selecting favorable photoconductor and thermoplastic properties. The thickness of the thermoplastic in this case is independent of that of the photoconductor, although macroscopic electrostatic considerations based on capacitive field division suggest that it should be substantially thinner [6.11]. This means that a thin thermoplastic selected for maximum bandwidth can be used over a photoconductor chosen for maximum light absorption, which property in turn can be achieved by either a high absorption coefficient or considerable thickness. Although there are further constraints and tradeoffs, having to do with development gain and dielectric strength, the extra design freedom of the two-layer device has helped make possible all the published examples of high performance attained by thermoplastics in holographic applications.

The other advantage, that of combining almost any photoconductor type with any thermoplastic, has also helped contribute to the superior performance of two-layer devices in holographic applications. This too is not without some restrictions. In particular, the interface between the two must have satisfactory electrical properties. This means that carrier injection from the photoconductor into the thermoplastic must be negligible, as must lateral migration of carriers at the interface. Moreover, when erasability is important, excessively deep trapping sites must not exist at the interface (or in either layer, for that matter) if residual (ghost) image retention is to be avoided. Another restriction on arbitrary pairing of thermoplastic with photoconductor is the one of chemical compatibility, both during fabrication (including solvents involved in the fabrication process) and during the subsequent life of the device. An example of the latter problem might be undesired crystallization of a vitreous photoconductor because of contact with an unsuitable thermoplastic.

Assuming compatibility has been achieved, there are few restrictions on photoconductor properties other than that of adequate transparency if transmitted light reconstruction is planned. If the sequential mode of process operation is to be used, then the dark decay of the photoconductor must be low enough to assure that random buildup of dark decay charge fluctuations at the thermoplastic-photoconductor interface is an insignificant source of noise. In rapid process cycles, and especially in the simultaneous charge-expose-develop process, dark decay is rarely significant. Other than this, the large range of photoconductor choices has allowed broad experimentation, which has led to considerable success.

One of the most widely used photoconductors in two-layer devices is poly-n-vinyl carbazole (PVK), sensitized either with Brillant Green dye or with 2, 4, 7 trinitro-9-fluorenone (TNF). The former sensitizer yields lower sensitivity, and has a transmission window in the green region of the spectrum, further reducing its sensitivity there. The latter achieves boths high and nearly pachromatic photosensitivity. The use of TNF, however, requires special precautions against inhalation or ingestion, because this compound has been found to be highly carcinogenic in animal experiments. The proportion of TNF can be increased from relatively low levels up to 1:1 molar, with progessively increasing sensitivity in most instances. The actual composition of the material consists of a solid solution of PVK, free TNF and a PVK-TNF charge transfer complex. The photogeneration efficiency and mobility of electrons and holes in this material are field dependent, but mobility is different for electrons and holes, being higher for the holes, especially when TNF concentrations are low.

The theory and practice of dye sensitization of such photoconductors is discussed in detail by *Colburn* et al. [6.28] (in their Sects. 3 and 4). That discussion covers problems and techniques of both visible and infrared sensitization.

One of the most serious problems observed in the cyclic use of two-layer devices, especially those having PVK photoconductors, is the so-called "ghost image" effect. This effect is the reappearance of a previously erased deformation image after a subsequent charge, recharge and development sequence. Ghost images appear both in the absence of a second exposure and when such an exposure is present, in which case they can be a severe source of background "noise", because their intensity is often close to that of the desired image. Although it is in principle possible that ghost images could arise from trapping effects in the thermoplastic or at the thermoplastic-photoconductor interface, in practice it has been shown that they are normally the result of photoconductor problems. In addition to the indirect theoretical and experimental evidence pointing toward this conclusion, a set of direct experiments conducted several years ago in the Xerox Research Laboratories demonstrated that certainly the thermoplastic itself and probably the interface as well were not involved in the formation of the ghost image. In these experiments, the normal charge-expose-recharge-erase sequence with device and process parameters known to produce ghost images was followed by chemical dissolution of the entire thermoplastic layer in a solvent which did not attack the photoconductor. The devices were then recoated with a new layer of thermoplastic. Charging and development then resulted in reappearance of the original image on the surface of the new thermoplastic.

In an analysis based on previously published studies of charge generation and mobility, *Colburn* and *Dubow* (Ref. [6.12], p. 29 et seq.), conclude that the low electron mobility of PVK/TNF is the primary cause of ghost imaging. A residual exposure-wise pattern of electrons remains behind in the bulk of the photoconductor even after the more mobile holes have been swept through the photoconductor and neutralized at the ground electrode. These relatively immobile electrons are left behind after photodischarge and erasure have largely collapsed the originally strong electric field in the photoconductor. The variation in thermoplastic surface potential caused by the presence of this residual pattern of electrons in the photoconductor results in a corresponding uneven buildup of surface charge density upon the next charging step. It is this buildup that produces the driving force for the ghost image.

There are several ways to minimize the ghost image effect. All are based on promoting migration of the residual electrons to the thermoplastic-photoconductor interface, where they can be neutralized by positive ions moving through the thermoplastic to that interface during erasure. At the basic level of material composition, selection of photoconductors to assure comparable hole and electron mobilities can minimize the problem. Of course one must also avoid materials that have deep traps for either sign of carrier. In the particular case of PVK/TNF, *Colburn* and *Dubow* [6.12] predicted theoretically and confirmed by experiment that higher TNF concentrations, in addition to increasing absorption and sensitivity, greatly increased electron mobility, resulting in sharply reduced ghost imaging. For example increasing TNF/PVK ratio from 1/10 to 1/4 by weight decreased ghost image effects by as much as an order of magnitude, with the improvement slowly falling off for long erasure times.

For a given photoconductor composition, *Colburn* and *Dubow* [6.12] suggest several other ways of minimizing the ghost image problem. These include use of longer exposure times for a given total exposure energy and an increased elapsed time between recharge and development. Both these techniques simply allow more time for electrons to be swept to the interface by the residual fields, thus becoming available for later neutralization at that interface. It would also appear that additional delay between development and erasure should have a similar effect. Higher development and erase temperatures can also contribute to effective erasure, because electron (and hole) mobilities increase with temperature. The former methods, of course, are constrained by process and application limitations which may dictate available times for partial and total cycle operation. The latter is often constrained by the danger of thermoplastic or photoconductor degradation at elevated temperatures.

Perhaps more satisfactory in practice than these methods is one investigated experimentally by *Colburn* and *Dubow* [6.12]. This is the use of a strong (generally white light) flood exposure prior to or during erasure. These authors attribute the effectiveness of this method to "increased drift" of the electrons through the bulk of the photoconductor. It might also be the result of massive but spatially uniform photogeneration of holes and electrons in the photoconductor, with a resultant high probability of recombination of some of these holes with most of the ghost electrons, leaving behind a substantially more uniform distribution of electrons in the bulk, while remaining holes are, as usual, swept through the photoconductor to neutralization at the ground electrode. Since uniform surface charge distributions are not developable, the subsequent recharge step does not produce a developable ghost pattern. Whatever the explanation, this method was found to reduce the relative ghost image intensity for given erasure conditions by more than an order of magnitude.

## Special Photoconductors

Although the great bulk of reported experimental results is based on the use of PVK/TNF photoconductors, a variety of other photoconductors has in fact been used for special purposes. As noted earlier, PVK sensitized with Brillant Green is usable, although typically about an order of magnitude less sensitive than PVK/TNF, and still more subject to ghost images because of low electron mobility in the absence of TNF. A typical concentration of brilliant green is 1 part in 200 by weight relative to the PVK. This gives an absorption peak at 640 nm, closely matched to the 633 nm wavelength of the He–Ne laser [6.29].

More significant in terms of extending the spectral range beyond the visible is the use of infrared sensitizers. *Colburn* et al. [6.30] found that by replacing TNF with 2, 4, 5, 7-tetranitrofluorenone or (2, 4, 7-trinitrofluorenylidene)-malononitrile, sensitivity could be extended as far as the 1.15 μm line of the He–Ne laser. Diffraction efficiency was generally on the order of a few percent, and the usual high spatial frequency response was attainable. Sensitivity was approximately 400 mJ/cm$^2$ for 1% diffraction efficiency, far less than in the visible, but still at a usable level. A later advance in infrared sensitization was made by the same authors [6.28] when they turned to the use of a laser *Q*-switching dye, Eastman *Q*-Switch Solution A9740, in conjunction with PVK, giving 1% diffraction efficiency with about 5 mJ/cm$^2$ and reaching 7% with about 50 mJ/cm$^2$ exposure, again at 1.15 μm. Although specific data are unavailable, occasional trade press reports suggest that research in the USSR has also led to successful infrared-sensitive photoconductors for thermoplastic recording.

In general, it should be expected that devices of this type could be unusually successful in the infrared. Attempts to extend photoconductor sensitivity into the infrared tend to result in more rapid dark decay at or near room temperature because thermal energy begins to be comparable with the available photon energies. In most electrophotographic processes, there is usually a significant delay between charging and exposure and between exposure and development. During these delays, dark discharge competes with the exposure-induced photodischarge, and noise in the dark decay process may become limiting to the performance of the system. An unusual advantage of thermoplastic deformation as an electrophotographic development process in this context is the possibility of a very rapid sequence of the key events of charging, exposure, recharging and development. In its ultimate version, simultaneous charging, exposure and development (especially if development is brought about by rapid heating of the thermoplastic from its surface inward to minimize excess dark decay produced by heating the photoconductor) might permit use of infrared sensitive photoconductors otherwise unsuitable for electrophotography.

Another direction in which photoconductors have sometimes been pushed is greater sensitivity. Even PVK/TNF is far from being a highly sensitive photoconductor. The author has explored the use of $As_2Se_3$, a fast panchromatic chalcogenide photoconductor, in an effort to increase sensitivity [6.31]. This photoconductor is nearly opaque in normal device thickness throughout the visible portion of the spectrum, with some transmission in the longer wavelength part of the red region. Therefore readout must either be by transmission at extremely low efficiency, or by reflection from the thermoplastic surface, or by a combination of transmission through the thermoplastic, reflection from the thermoplastic-photoconductor interface, and a second transmission through the thermoplastic. In normal reflection readout, the latter two readout effects coexist simultaneously. Because interference among the three succesively diffracted waves obtained in this way can complicate data interpretation, the holograms produced in these experiments were subsequently aluminized to produce pure first surface reflection readout and unambiguous deformation measurement. Using this method, it was found that sensitivities were at least two orders of magnitude greater than those obtained with PVK/TNF. Of this increase about one order of magnitude is attributable to the more sensitive photoconductor, the rest to the readout method (see pp. 193, 196). As with infrared-sensitive photoconductors, this high sensitivity material has relatively high dark decay and benefits from a rapid sequential or simultaneous type of process. The use of this type of photoconductor has thus far been only exploratory, but its ability to achieve these sensitivities at a normal (400 cyc./mm) spatial frequency promises unusually high informational sensitivity, which in some uses may offset the more difficult readout requirements.

## Thermoplastic Materials

The essential requirement for any thermoplastic is the ability to retain charge as it passes through its glass transition temperature and becomes soft enough for reasonably rapid deformation. If the device is to be used repeatedly, of course, this basic requirement must be supplemented by an ability to retain its electrical and rheological properties through many successive cycles of reuse.

Many thermoplastics have been investigated for use in electron beam recording devices as well as in both single- and two-layer photosensitive devices. Thermoplastic-like materials of the type useful in information recording have all been organic polymers. In principle they could be either an amorphous or a crystalline type of polymer. The crystalline type has the potential advantage of a sharply defined melting temperature, which might make process control easier to achieve. Crystalline thermoplastics generally consist of many small crystals embedded in an amorphous matrix having a somewhat different refractive index. Consequently substantial light scattering occurs in such polymers, rendering them cloudy and entirely useless for transmitted light readout. Although they might be of some use in the type of reflected light readout appropriate to opaque photoconductors such as those noted in the preceding section, they have in practice been neglected because almost all investigations have concentrated on transmitted light readout.

The amorphous polymers which have universally been used in thermoplastic recording change from a substantially solid phase (in which deformation, if it occurs at all, is of the elastic type) to a substantially fluid (generally non-Newtonian) state over a finite range of temperatures. The change is known as the glass transition, and the midpoint of the temperature range over which it occurs is called the glass transition temperature, usually denoted $T_g$. In general, this range tends to be narrower if the polymer has a narrow distribution of molecular weights, or if the average molecular weight is low. The value of $T_g$ and the width of the transitional range do not appear to correlate in any simple way with the suitablity of a polymer for thermoplastic recording. However, for image retention it is necessary to be sure that $T_g$ and the transition width are so chosen that the lower end of the plastic flow range is well above room temperature; if any significant amount of "creep" is possible, gradual erasure of the deformation pattern will occur under the influence of surface tension.

It should be noted here that elastic deformation of the polymers ordinarily used for thermoplastic imaging is entirely negligible, so that no measurable deformation ordinarily occurs in the absence of the heating step. On the other hand, a deliberate choice of polymers capable of extraordinarily high elastic deformation and having substantially no fluid flow at room temperature has made possible a different type of cyclic imaging device, the Ruticon, described by *Sheridon* [6.15], operating entirely on the principle of elastic rather than flow deformation.

In this section we shall not attempt a detailed description of the selection, design or synthesis of thermoplastics especially suited for deformation recording, but shall survey some of the salient features of this aspect of the technology. Many different polymers have been tried with varying degrees of success since the early research on thermoplastic electron beam recording. Polystyrene of various molecular weights was an early and moderately successful material in the former application. Polystyrene was also used with suspended or dissolved photoconductors in single-layer photosensitive devices [6.9, 26]. Although polystyrene has also been used in two-layer photosensitive devices [6.32], much of the early work on both frost mode image recording and on holographic recording was done with Staybelite Ester 10 (available from Hercules, Inc., Wilmington, Del.). This material, a derivative of a natural tree resin, has never been fully characterized. Although it has excellent resistivity and can lead to high deformation holograms, Staybelite has poor cycling properties, usually being limited to the range of 10 to 100 cycles before its properties degrade significantly. It can be recommended for the casual experimenter or for low recycling applications because of its ready availability and ease of preparation. However, because of its natural origin and uncharacterized structure, significant batch-to-batch variations can occur, sometimes frustrating orderly experimental progress. Another thermoplastic which gives comparable, but generally more consistent results is Piccopale H-2 [6.29] (available from Pennsylvania Industrial Chemicals Corp., Clairton, Pa.).

Some significant advances in thermoplastic materials were made in the course of work aimed primarily at electron beam recording. The research of *Anderson* et al. [6.33] was intended to develop materials capable of repeated cycling in that mode of operation. Although some of the work reported by these authors was done in order to minimize the deleterious effects of outgassing and radiation damage unique to electron beam recording in a vacuum, many of their experiments were carried out with corona charging in air, and some of their main concepts are transferable to cyclic devices of the photosensitive type.

In particular, they emphasize the need for using a nonvolatile plasticizer to avoid hardening of the polymer (and consequent change of $T_g$ and development requirements) with time and repeated heating cycles. They also proposed a key principle for long-term cycling by balancing the two opposing effects that usually degrade a polymer subjected to radiation (or, in our case, the similar damaging effects of heating, ozone and ion bombardment). These effects are, first, cross linking of polymer chains which tend to harden the material, increasing viscosity and $T_g$; and second, scission of polymer chains, which tends to lower viscosity and $T_g$. Obviously it is best to minimize both these effects, but because they cannot be entirely eliminated, the optimum design of polymers requires that the two effects both occur to an equal and therefore offsetting degree. Such self-compensating polymers can have remarkably good cycling characteristics. Because the two effects may have different types of dependence on the degrading influences, it is clear that optimization of a thermoplastic may need to be specific to a particular set of process parameters. Thus it has always been necessary to test thermoplastics under more or less realistic use conditions to assure that this design principle is effectively utilized.

*Anderson* et al. [6.33] explored a variety of styrene-methacrylate copolymers, compensating the cross-linking tendencies of the styrene with the scission characteristics of the methacrylates. Of the many different methacrylates tried, particularly good results were obtained with octyl-decyl methacrylates.

The best cycling results reported in [6.33] under electron beam recording conditions were 50000 cycles using a copolymer of 85% styrene with 15% octyl-decyl methacrylate. No significant changes in required process parameters (charging current density, development temperature, development time) and deformation were observed during this test sequence.

The same investigators (Ref. [6.33], p. 149) extended their tests to single-layer photoconductive devices, adding *n*-vinyl carbazole to the copolymers used for electron beam recording. Partial success was reported, with a moderate increase in development and erase temperatures during 11500 cycles. It appears simpler, as well as more effective in terms of recording properties, to use the copolymers in two-layer devices, for the reasons of relatively independent constituent optimization noted previously. This was done by several workers [6.12, 16] with considerable success. For example, *Colburn* and *Tompkins* [6.16] reported the use in two-layer devices of the styrene-octyl-decyl-methacrylate copolymer (which they refer to as a terpolymer in their paper), with a mole ratio of 85%, 9%, and 6% for the respective constituents. Used with a corona discharge in air, this material showed little degradation for about 100 cycles, then degraded slowly in holographic performance to about 1000 cycles, and finally fell off rapidly above 1000 cycles. Little change in $T_g$ occurred throughout a 5000 cycle test. When used with the parallel plane charging apparatus mentioned previously, performance degraded very rapidly, becoming unacceptable in 30 to 40 cycles. Attributing this degradation to ozone attack on the thermoplastic in the confined charging structure, *Colburn* and *Tompkins* [6.16] replaced air with argon gas in a sealed-cell version of this apparatus, and found substantially stable holographic performance for 5000 cycles, although in this case a moderate increase in $T_g$ was observed. The latter fact is further confirmation of our earlier remark regarding the need to optimize polymer formulation for a given set of process conditions.

Once ozone degradation is eliminated as a major cause of thermoplastic deterioration, the main remaining source of degradation in parallel plane charging devices appears to be sputtering of the thermoplastic in the high fields characteristic of this configuration. Such sputtering removes material in a random manner from the thermoplastic surface, resulting in a roughened surface, whose light scattering degrades signal-to-noise ratio. Further sputtering can remove so much material that diffraction efficiency also is degraded. The lower fields typical of corona charging in air should substantially reduce the amount of sputtering to be expected. *Colburn* and *Tompkins* [6.16] nonetheless reported some surface damage to thermoplastics in this mode of use as well.

An alternative to operation in an atmosphere of an appropriate inert gas is the development of an ozone-resistant thermoplastic. One such material reported in [6.16] is a vinyl-toluene copolymer of unspecified constituent proportions. Although incapable of achieving diffraction efficiencies in excess of 1%, this material, operated so as to give efficiencies of a few tenths of a percent, was able to cycle well over 5000 times using corona charging in air, thus demonstrating a very high resistance to ozone attack. Although it had a high and stable signal-to-noise ratio, this thermoplastic showed poor response to spatial frequencies above 500 cyc./mm (Ref. 1 [6.12], p. 54). The combination of low diffraction efficiency and limited spatial frequency response may often prove to be an excessive price for good cycling in air.

A hydrogenated version of this vinyl-toluene thermoplastic was reported by *Colburn* and *Dubow* [6.12]. This material gave diffraction efficiencies as high as 2%, but its efficiency was unstable, rising by over an order of magnitude from start to about 1000 cycles, then falling rapidly.

Still another thermoplastic discussed in [6.12] is a "highly stabilized ester resin". Whether this is a refined or improved version of the Staybelite Ester 10 mentioned earlier is unclear from the description given in that report. This material, like the styrene-methacrylate type, showed fairly rapid decline of diffraction efficiency when corona charging is used in air, but had constant response up to 3000 cycles in an argon-filled parallel plane charging device. However, its softening temperature rose significantly more during those 3000 cycles. This rise in $T_g$ is also shared by the two vinyl-toluene materials. Another thermoplastic used with considerable success [6.34] is Foral 105 (from Hercules, Inc.).

Overall, the styrene-methacrylate materials developed by *Anderson* et al. [6.33] still appear to be the best choices for most applications of thermoplastic hologram recording. Capable of several thousand cycles with minor degradation when protected from ozone, they enhance process stability while retaining the traditional thermoplastic virtues of high spatial frequency response and diffraction efficiency.

Even the advanced work on thermoplastics, noted above, has not given us a deformable material free from all problems. Further optimization of these materials in the context of specific devices and applications will be needed if better cyclic performance is required.

### 6.1.5 Fabrication Techniques

Thus far we have discussed the materials themselves without discussing their actual fabrication into working devices. While it is not the intent of this section to provide a detailed instruction manual on the fabrication of these devices, some major aspects of fabrication will be considered here. This is appropriate because of the crucial role of fabrication techniques in the making of any complex optoelectronic device, above all if that device must perform the multiple functions of thermoplastic imaging repeatably and uniformly.

#### Substrates and Conductive Coatings

The substrate which forms the foundation for these devices need not be further discussed here. Suitable substrates, either glass or plastic, are generally obtainable commercially, and are no different from any other optical recording medium substrate. Although glass has been the most widely used substrate material, mylar, cronar, and other plastics have also been used with excellent results [see, e.g., Ref. [6.17], p. 210).

Likewise, fabrication of the conductive layer poses few special problems. Several commercially coated glass substrates are available, for example, indium oxide coated glass from Pittsburgh Plate Glass Co. In addition to these, many others have been custom fabricated. Thin conductive layers of metal have been coated by evaporation on plastic or glass, with just enough thickness to achieve electrical continuity while still remaining sufficiently transparent for satisfactory transmitted light reconstruction. Metals used include aluminum, gold, chromium-gold alloys, and silver. Of these, chromium-gold is probably most satisfactory. Conductive compounds often give better results when low resistivity is desired. Aside from the indium oxide already mentioned, satisfactory coatings have been made from tin oxide and antimony oxide, or mixtures of these oxides. Coating techniques for these materials often involve evaporation or reactive sputtering of these materials in an oxygen atmosphere, or in moist air. Some conductive coating fabrication techniques which require oxidation reactions upon contact of a liquid with a hot glass substrate in air result in severe residual stresses of the suddenly cooled glass, which distort it and produce substantial stress birefringence. Since, as noted above, a hologram must be regarded as an optical element, it is evident that substrates with these optical defects are not usable in certain applications.

It has already been stated that the resistivity of conductive coatings used for direct ohmic heating should generally be in the range of 10 to 100 ohms per square. If uniform development is required, the resistivity must be controlled rather tightly. It has been suggested (Ref. [6.35], p. 110) that the extreme sensitivity of development rates to thermal power input requires that resistivity be kept uniform to within 1 or 2 percent. This degree of uniformity cannot be obtained from any commercial coatings. The authors of [6.35] found that custom reprocessing of commercially available conductive substrates was required to achieve satisfactory uniformity in resistivity.

### Photoconductor Fabrication

Fabrication of the photoconductive coating falls into two radically different categories. The rare examples of devices utilizing a vitreous photoconductor such as $As_2Se_3$ require vacuum evaporation of the photoconductor onto a (typically) glass substrate. Because such devices have, to our knowledge, only been used for hologram recording in a few experiments, further discussion of this rather difficult coating technique does not seem warranted here. We shall therefore restrict this section to the fabrication techniques for organic photoconductors and specifically for the various forms of poly-n-vinyl-carbazole (PVK) almost universally used in the two-layer devices of most interest here.

The object of photoconductor coating is of course to produce uniform optical, electrical and thermal properties. This requires that the material being coated be homogeneous and that the coating thickness be uniform. While quantitative tolerances on these parameters have never been definitely determined, it is clear that serious response variation within a sample, or among samples, can follow from such photoconductor variations, because they will affect field strengths within and field division between the photoconductor and the thermoplastic, hence local device sensitivity. They can also affect sensitivity by way of altering light absorption. Thickness or thermal conductivity variation can also affect the uniformity of development, because the photoconductor is in the path of heat conduction to the thermoplastic if ohmic heating is used, and in the path of heat conduction from the thermoplastic if direct surface heating is used.

Coating of PVK has generally been done from a solution. The various techniques of coating all have thickness uniformity as their object, hence are naturally unable to compensate in any way for solution inhomogeneities. It is not particularly easy to obtain entirely uniform solutions of PVK/Brillant Green or PVK/TNF. Various procedures have been described in the literature. For example, the solid PVK and TNF in the desired proportions were dissolved in a 1:1 mixture of 1, 4 dioxane and dichloromethane to obtain about a 6% concentration of solute [6.17]. This solvent requires coating in a relatively dry (less than 25% relative humidity) atmosphere to avoid fogging of the coating which can result when the cooling of the photoconductor by rapid evaporation of the solvent causes condensation of atmospheric moisture on the surface. Other solvents which create a less severe fogging problem include 1,1,2-trichloroethane and tetrahydrofuron. A relatively detailed prescription for effective use of the latter is given in (Ref. [6.3.5], p. 111). PVK is first dissolved in the proportion of 67 grams per liter of this solvent. This mixture is stirred for 8 to 12 h to ensure complete dissolution of the PVK. It is then filtered, and the desired amount of TNF is added, followed by another four hours of stirring. Although not explicitly mentioned in this prescription, a second filtering after adding the TNF would seem to be desirable. Because of solvent loss during mixing, additional tetrahydrofuran is added back to restore proportions to the desired level for subsequent coating. It is noted in [6.35] that even this relatively elaborate procedure does not assure complete homogeneity of the photoconductor coatings, which remain subject to small defects attributable to the solution rather than the coating procedure.

This procedure itself can take several forms. Dip coating, or the equivalent method of drain coating, results in deposition of a thin layer of solution on the substrate as it is being withdrawn vertically from the free surface of the solution. Various forms of doctor-blade or roller coatings have also been used, with positive mechanical control of coating thickness. Spin coating is a highly developed technique in other fields (such as photoresist coating) which could be adapted to this operation. Thus far, however, dip coating remains the most common technique, and, when properly carried out, can give highly uniform results.

In dip coating, the thickness is determined primarily by the density, viscosity, surface tension and vapor pressure of the solution and by the rate of withdrawal of the substrate from it. The solution (which of course must be able to wet the substrate) tends to run back into the container from the substrate surface. If the solvent is sufficiently volatile, its loss raises the viscosity of the coated surface to the point where further draining back of the coating into the solution tank no longer occurs, and the coating thickness is stabilized. Over some range of withdrawal speeds, the coating thickness increases with this speed. This range is usually in the low tens of cm per min, (for example, 5 to 24 cm/min for the solution described in [6.35] and summarized above) but depends on the solvent and viscosity selected. In practice, control of thickness in the coating is usually obtained by selecting solute concentration to adjust viscosity, and then adjusting withdrawal speed to obtain the exact thickness desired.

Uniformity of the coating requires careful construction and control of the apparatus. Naturally, withdrawal must be smooth in order to avoid local variations in instantaneous velocity, which would result in corresponding variation in coating thickness. Likewise, vibration must be carefully avoided because it can easily result in agitation of the free surface of the solution, again producing localized variations in instantaneous withdrawal speed. Excessively high solvent vapor pressure, in addition to producing fogged ("blushed") coatings in high relative humidity ambients, can also result in significant variation of local viscosity near the solution surface during the course of a coating. This produces a taper in coating thickness, which can prove objectionable when using large substrates and low withdrawal speeds. In principle, this effect could be compensated by using a variable withdrawal speed, but in practice this has generally not proven necessary if reasonable choices of solvents were made.

Despite all the difficulties of obtaining high quality dip coatings, the method has generally been more satisfactory than doctor-blade coating, perhaps since the former depends on the choice of macroscopic parameters rather than on mechanical precision to assure uniformity.

Once the photoconductor coating has been applied, it must be cured by careful drying at an appropriate temperature and in a contamination-free ambient. If there are no problems of solvent incompatibility, complete drying of the photoconductor may not be required prior to the next stage of thermoplastic coating, since the PVK solvent can usually diffuse and evaporate through the thermoplastic layer.

## Thermoplastic Fabrication

The same choices of coating techniques are available for thermoplastics as for organic photoconductors. Once again, dip coating has generally been the preferred method in experimental practice. There is of course an additional constraint on the thermoplastic solvent, namely, that it not attack or otherwise undesirably interact with the underlying photoconductor coating.

For each thermoplastic used with a given photoconductor, it has generally been possible to find a solvent which meets this requirement. The particular solvents used also depend on the thermoplastic. If the photoconductor is PVK/TNF and Staybelite Ester 10 is used as the thermoplastic, suitable solvents include petroleum ether and hexane, and typical thermoplastic concentrations are around 20% solid. If the same photoconductor is to be used with styrene/methacrylate copolymers, the latter can be dissolved in naphtha, typically at fairly similar concentrations, around 20% to 25%. If the thermoplastic chosen is Piccopole H-2, a suitable solvent is iso-octanes, with solid again at similar concentrations.

After coating to the desired thickness, the relatively low volatility thermoplastic solvent must be removed, usually by oven baking at temperatures in the range of 50 °C to 75 °C for times of one to several hours.

The typical completed device will have a thermoplastic thickness in the range of 0.2 to 2 μm. Macroscopic capacitance arguments suggest that a high ratio of photoconductor to thermoplastic thickness will lead to higher fields across the thermoplastic, hence higher sensitivity when all else is equal. Analysis of microscopic driving forces at high spatial frequencies predicts rapid (exponential) fall-off of high spatial frequency response with thermoplastic thickness. Both of these factors, together with the near-reciprocal dependence of the frequency response peak location and width upon thermoplastic thickness, have resulted in an emphasis on thin thermoplastic coatings, with the result that coatings over 1 μm in thickness are not widely used, and most devices are made with thermoplastic coatings in the vicinity of 0.5 μm.

### Overcoating and Matrixing

The fabrication techniques described thus far are all that are needed in making conventional two-layer devices. If reflective or conductive metallic overcoatings are desired, vapor deposition in vacuum of metallic indium or gold-indium alloys can provide such coatings. These must generally be quite thin, usually under a few tens of nanometers in average thickness, if their mechanical stiffness is not to inhibit deformation of the thermoplastic, reducing absolute response and moving the peak of the frequency response curve toward lower spatial frequencies. The overcoating must also not inject charge into the thermoplastic, and must not provide nucleation sites for the initiation of random frost deformations. If a reflective coating is intended to serve as an electrode in order to permit direct electrical charging of the device, it must have enough integrity to maintain electrical continuity. If in addition the device is to be reused, then the overcoat must not break up or agglomerate during the perturbation of development and erase cycles. This last requirement has never been adequately satisfied in practice; hence overcoated reflective devices can rarely exceed a few, if any, cycles of reuse.

Post-deformation metallization of devices not intended for reuse is a much simpler matter, and can be done by vapor deposition of metals such as aluminum, silver or gold in relatively arbitrary thicknesses. The main concern here is to avoid heating of the thermoplastic above $T_g$ by radiation from the evaporation source. Substrate cooling is often required to prevent erasure of the developed image when metallizing thermoplastic holograms in this way for high efficiency reflective readout.

The most elaborate types of device fabrication attempted to date are those aimed at matrix-addressed control of ohmic development of individual small holograms in an array of holograms intended for use in an optical memory. Following the first description of such an array by *Lin* and *Beauchamp* [6.36], extensive further work was done at RCA and Harris. A description of the required array fabrication techniques is given in Appendix B (p. 84) of *Colburn* and *Dubow* [6.12]. Because of the previously noted sensitivity of device performance to development parameters, considerable care must be taken in the fabrication of such addressing arrays to assure uniform performance in all locations. Whether this type of array fabrication can become a practical process for large-scale production is still uncertain.

## 6.2 Holographic Properties

### 6.2.1 Bandwidth and Resolution

The theoretically expected bandpass response of thermoplastic imaging devices has been observed in almost all experiments. Because of the difficulty of accurately modeling the complete experimental situation, close agreement between theory and experiment has never been observed. Therefore we remain with a semi-empirical description of the dependence of spatial frequency response upon the main device and operating parameters. The most important of these parameters is thermoplastic thickness. The frequency $\nu_p$ of the peak of the frequency response depends more or less inversely upon the thickness $D$ of the thermoplastic. A typical empirical relationship is

$$1/3D \lesssim \nu_p \lesssim 1/2D$$

with the value of $\nu_p$ increasing more or less logarithmically with voltage for a given thickness.

The bandwidth of usable frequency response around this peak is quite variable, depending upon material properties, voltage (field strength) and manner of development. Under some circumstances, relatively broad peaks in frequency response are observed, with resultant high usable bandwidths. With other modes of operation, the peaks are sharper, and the bandwidth much lower. Since, for low deformations, the diffraction efficiency depends quadratically upon the deformation, it is often reasonable to define the bandwidth in deformation response as the bandwidth measured to the spatial frequency values at which deformation has fallen to $1/\sqrt{2}$ of its peak value, which corresponds to diffraction efficiency at 50% of its peak value. For some applications, however, especially those involving Fourier holograms, a more uniform frequency dependence of diffraction efficiency is required, resulting in a smaller usable bandwidth.

Using the foregoing definition of bandwidth, values of the full bandwidth (note that the frequency response curve is generally not symmetric about its peak) have been found to range from as much as $3/4D$ to as little as $1/6D$. The higher values have been observed in our laboratories under rather atypical process conditions, including very slow development taking minutes rather than

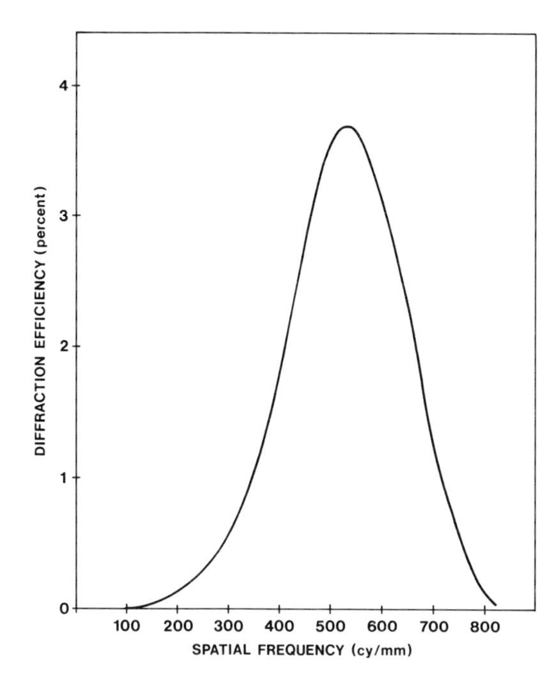


Fig. 6.2. Spatial frequency response of thermoplastic-overcoated photoconductor device, thermoplastic thickness 1.1 μm, beam ratio 1:1 (from *Urbach* and *Meier* [6.38])

tens or hundreds of milliseconds. Even more atypical results have occasionally been observed. One of these is a double-peaked frequency response, with a peak at very low spatial frequencies, and another near the $1/2D$ frequency. This result is sometimes time dependent in the following way: the low frequency peak occurs early in a relatively slow development process, then erases as the high frequency peak appears with further development. Under some circumstances, not fully understood, the low frequency peak does not erase entirely before the high frequency one develops, resulting in the retention of the double peak in the final recording as reported by *Lee* [6.37]. *Lo* et al. [6.32] have reported similar results, noting that the low frequency peak was observed at relatively low surface charge densities.

Figures 6.2 through 6.5 show some fairly representative spatial frequency response curves reported in the literature. These should make it abundantly clear that there is no such thing as a single frequency response curve. Moreover, since these curves are always dependent on development time and temperature, even setting the basic process parameters of thermoplastic thickness and voltage does not uniquely determine the frequency response of the device. Figure 6.4 illustrates this by showing, for the aforementioned slow development and high bandwidth process, how the frequency response curve evolves with development time, with its peak moving to higher spatial frequencies as development proceeds.

All of the foregoing remarks apply specifically to the sequential process mode. The highest spatial frequency response in this mode was reported by

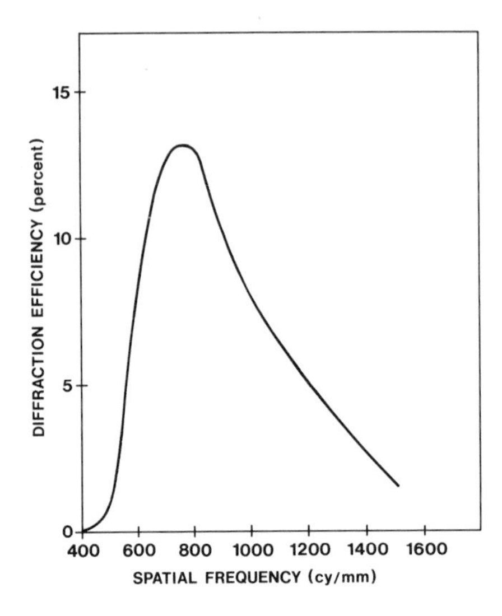


Fig. 6.3. Spatial frequency response of thermoplastic-overcoated photoconductor device. Thermoplastic thickness approximately 0.4 μm (after *Colburn* and *Dubow* [6.12])

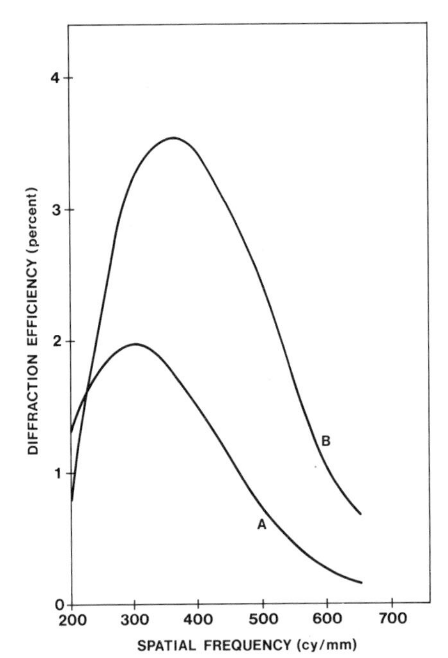


Fig. 6.4. Spatial frequency response of slowly developed thermoplastic-overcoated photoconductor at two development times. Thermoplastic thickness approximately 1 μm. Beam ratio 1:1. Development times: *A* 7.5 min; *B* 12.5 min. Development temperature near ambient (~22 °C) (after *Urbach* [6.39])

*Credelle* and *Spong* [6.17], who observed an 1850 cyc./mm response in devices with 0.25 μm thermoplastic thicknesses, albeit with relatively low (2 to 5%) values of diffraction efficiency, and with a requirement for very short (2 to 3 ms) heating pulses. A more typical result reported in [6.17] is a $\nu_p$ value of 1170 cyc./mm with $D = 0.45$ μm and diffraction efficiencies as high as 22%.

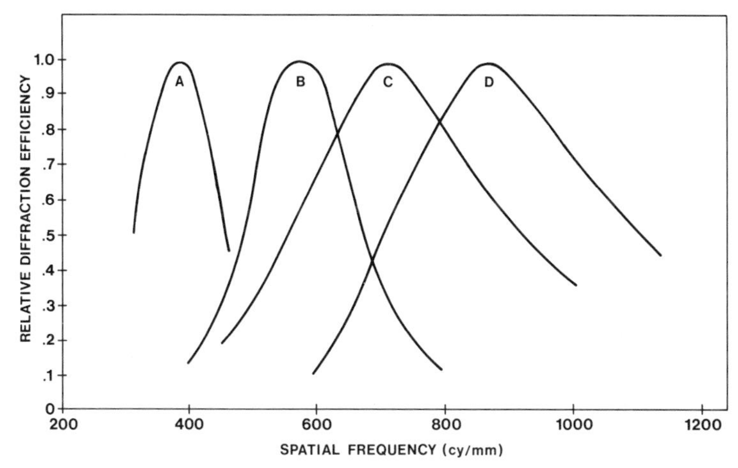


Fig. 6.5. Normalized spatial frequency response of thermoplastic-overcoated photoconductor devices for four thicknesses: *A* 1.1 μm, *B* 0.8 μm, *C* 0.55 μm, *D* 0.50 μm (after *Lo* et al. [6.32])

The same authors also reported radically different results when the simultaneous charge-expose-develop version of the process is used. In particular they found that the usual relationships between thermoplastic thickness and frequency response do not hold for this process, and that much thinner, for example 0.1 to 0.2 μm, thermoplastics can be used to give center frequencies and bandwidths in the 1200 and 500 cyc./mm range, respectively. This frequency response was associated with very high diffraction efficiency (sometimes in excess of the 33.9% predicted as the theoretical limit for sinusoidal gratings), evidently the result of the "positive feedback" type of surface charge contrast amplification typical of the simultaneous process. At the upper limit of frequency response, diffraction efficiencies around 1% were reported at spatial frequencies of 4100 cyc./mm using the same values of $D$, and at the intermediate frequency of 1850 cyc./mm, the efficiency was still over 10%.

The great range of frequency response characteristics discussed thus far and their strong dependence on several process parameters complicate attempts to characterize the process, but give the resourceful experimenter much leeway in tailoring process response to his requirements. Considering frequency response as well as diffraction efficiency and noise (to be discussed subsequently), it is clear that the simultaneous process mode will give superior performance if the application in question will allow its use.

The implication of all the foregoing is clearly that it is the thermoplastic layer and the parameters affecting it, such as the electric field across it, together with the process sequence that primarily determine the frequency response of these devices. This is largely true of many devices described in the literature, but is far from universally true. Frequency response can be limited by the photoconductor in several ways. It happens that, partly by coincidence and partly by more or less empirical optimization of practical devices, the choice of materials and fabrication parameters has usually led to a thermoplastic-dominated frequency response; under other conditions, and even in some of the experiments reported, photoconductors can limit high spatial frequency response. If one is operating a device in such a regime that photoconductor effects dominate at high spatial frequencies, the overall response will still be of the familiar bandpass type, because the thermoplastic will continue to have poor response at low spatial frequencies, while the photoconductor will limit the high frequency response before the thermoplastic does. One clue that device operation is in such a regime is the occurrence of the peak in response to sinusoidal exposure patterns at a spatial frequency well below that of the peak in random frost deformation of uniformly exposed areas. For example, the behavior of thin thermoplastics as described above on the basis of the *Credelle* and *Spong* [6.17] research on the simultaneous mode process could be an example of photoconductor-limited spatial frequency response. The reported occurrence of the imaging response peak at spatial frequencies a factor of two to four below those expected on the basis of frost deformation suggests that this may well be the case (see also Sect. 6.2.3 below).

There are several possible mechanisms by which the photoconductor can limit resolution. The most obvious is lateral diffusion of charge carriers, first discussed by *Gaynor* and *Aftergut* [6.3] for homogeneous single-layer devices. It was later analyzed by *Schmidlin* and *Stark* [6.40] for two-layer devices, where trapping and release of carriers at the photoconductor-thermoplastic interface can become the main determinant of resolution. Of course in the case of particulate dispersions of photoconductors, typical of many later single-layer devices, the dimensions and separations of the photoconductor particles can also limit resolution.

Even if charge diffusion effects are negligible, resolution is limited by basic electrostatics. It has been shown by *Schaffert* [6.41] that the electric field above a planar sinusoidal charge distribution falls off exponentially both with distance from the plane and with spatial frequency. Depending on the particular kind of charge trapping in the photoconductor, and also on the optical absorption coefficient of the photoconductor, the effective charge pattern that drives development will be distributed in some manner through the depth of the photoconductor. If enough is known about the physics of the particular device and its operation, appropriate integration of Schaffert's equations could provide a measure of the driving field. In two-layer systems, with their almost perfectly insulating thermoplastics, the recharge step results in formation of a charge image on the thermoplastic surface. The separation of this surface from the original image in the photoconductor results in an unavoidable fall-off of charge density at high spatial frequencies.

Because many of the PVK-based photoconductors seem to exhibit very low dark decay and effective charge trapping at the interface, there is some reason to believe that the geometric fall-off effect noted above is the predominant one in devices based on these photoconductors. In some of the high sensitivity vitreous inorganic photoconductors, this may not be the case. At least some of these appear to have resolution limits primarily dominated by the success or failure of interfacial trapping, since the charge image formed by exposure often migrates rapidly to the interface, and there tends to diffuse laterally. In view of the results obtained by *Credelle* and *Spong* [6.17], it seems reasonable to speculate that the high resolution limits that they found with ultrathin thermoplastics and relatively thin photoconductors in the simultaneous process mode may well have been dominated by the electrostatic limitation imposed by a charge image trapped in the bulk of their thin but nevertheless finite photoconductor.

### 6.2.2 Sensitivity and Diffraction Efficiency

As has been remarked before, an unusually desirable feature of the two-layer device is the somewhat separable control of resolution by the thermoplastic and of sensitivity by the photoconductor. The limitations on resolution imposed by the photoconductor show that, as one might expect from fundamental informational arguments, this separation is ultimately limited. For example, attempts to fix thermoplastic thickness and increase sensitivity by altering the photoconductor will eventually result in photoconductor resolution limitations dominating over those originally imposed by the thermoplastic. Both of the chief types of limitations mentioned in the foregoing discussions can come to the fore in such attempts to increase sensitivity.

If the photoconductor is weakly absorbing, as is typical of PVK-Brillant Green, sensitivity can be increased by two means: Increase the thickness of the photoconductor, or increase the concentration of sensitizing dye. In the first case, the electrostatic limitations will become significant even in those rare cases where resolution is not lost simply as a result of the fact that the holographic interference pattern does not have its isophotes everywhere perpendicular to the photoconductor layer, thus leading to an integration of a tilted fringe pattern and a reduction of average contrast. Attempts to increase sensitivity by increasing dye concentration are successful up to the point where increased dark discharge, i.e., conductivity, results from higher dye concentrations. Then the limitations based on that effect take over, preventing an indefinite increase of sensitivity by such means. It should be noted that the effect of this limitation increases with the time between exposure and recharge required in the process sequence, and thus will be least in the simultaneous process mode.

Attempts to increase sensitivity by outright change to a more sensitive photoconductor also run into the correlation between higher sensitivity and higher dark conductivity. In practice, as noted above, this approach may instead be dominated by difficulties with interfacial trapping of migrated charge images.

Despite these restrictions, the sensitivity of early devices was sufficiently far removed from fundamental limitations on the ultimate informational sensitivity of photoconductors that considerable independence was found between resolution and sensitivity, a rare occurrence with any photographic system.

Just as it is impossible to obtain a unique spatial frequency response curve for any given device, so is it impossible to uniquely define sensitivity. Because there is no large area response, there can be no equivalent to the traditional large area characteristic curves for silver halides such as the amplitude transmittance vs. exposure curve often used in holography. Each exposure response curve is applicable only to the particular spatial frequency for which it is measured. A useful sensitivity-defining convention might be to plot the diffraction efficiency as a function of exposure at the peak of the spatial frequency response curve. Because, however, the location and magnitude of that peak, as well as the shape of the curve, depend upon process conditions, even this definition has severe limitations on its generality.

## Diffraction Efficiency Considerations

As is well known from the scalar diffraction theory of thin sinusoidal phase gratings [6.1], the intensity $I$ diffracted into the first order is given by

$$I = J_1^2(\theta) I_0, \tag{6.2}$$

where $\theta$ is the optical phase excursion in radians and $I_0$ is the incident intensity (losses are neglected). Because for small values of $\theta$, $J_1(\theta) \approx \theta/2$ and because for transmission readout, the phase shift $\theta_t$ is

$$\theta_t = \frac{2\pi}{\lambda}(n-1)d, \tag{6.3}$$

where

$d$ = deformation amplitude;

$n$ = refractive index of thermoplastic,

it follows that for small deformations $I$ is proportional to $d^2$.

Because $d$ appears to be roughly proportional to exposure $E$ for many devices, it is clear that the initial slope of the $\log I$ vs. $\log E$ curve at a given spatial frequency will be 2. In this limited sense, the devices can be said to have a gamma of 2, and to have the generally desired linear dependence of diffracted amplitude upon exposure. Over a larger range, of course, this dependence becomes sublinear. The consequences of this will be discussed in the next section. Some other results of the elementary diffraction efficiency relations are worth mentioning here.

We have already noted that it is possible to consider reconstruct phase holograms either by transmission or reflection. In the latter case, the phase shift $\theta_r$ is given by

$$\theta_r = \frac{4\pi}{\lambda} d. \tag{6.4}$$

Thus the small-excursion first-order diffraction efficiency $\eta_r$ is

$$\eta_r = \frac{4\pi^2 R d^2}{\lambda^2} \tag{6.5}$$

for reflective readout, where $R$ is the intensity reflection coefficient. For the more common transmissive readout, the equivalent efficiency $\eta_t$ is

$$\eta_t = \frac{\pi^2 (n-1)^2 T d^2}{\lambda^2}, \tag{6.6}$$

where $T$ is the intensity transmission coefficient. Therefore, one can directly compare the reflected and transmitted light diffraction efficiencies for a given deformation excursion $d$ by taking the ratio

$$\frac{\eta_r}{\eta_t} = \frac{4R}{(n-1)^2 T}. \tag{6.7}$$

For readout light normally incident on the dielectric thermoplastic, one can assume ordinary Fresnel reflection, so that $R=(n-1)^2/(n+1)^2$ and (6.7) becomes

$$\frac{\eta_r}{\eta_t} = \frac{4}{(n+1)^2 T}. \tag{6.8}$$

This in turn has some interesting consequences. With a typical organic photoconductor, $T$ is usually near 0.5; $n$ is around 1.5 for most thermoplastics, so that in this case $\eta_r = 1.28\ \eta_t$. This shows that, despite the low value of $R$ for dielectric reflection, the reconstruction efficiency of these holograms is about the same for either reflection or transmission.

In the event that maximum efficiency is desired from a given small deformation $d$, it is interesting to compare the idealized limits achieveable if $T=1$ (which could be approached by replicating the surface relief onto a clear plastic provided with antireflection coatings on both sides) and if $R=1$ (e.g., by coating with a perfectly reflecting layer). Then (6.7) becomes

$$\frac{\eta_r}{\eta_t} = \frac{4}{(n-1)^2} \tag{6.9}$$

which for $n=1.5$ implies that $\eta_r = 16\,\eta_t$, suggesting that much higher maximum efficiency can be obtained by reflective readout. Once more, it should be noted that all these results hold only for deformations so small that $J_1(\theta_r) \approx \theta_r/2$ (the more stringent requirement because $\theta_r > \theta_t$).

For completeness, we might also note that readout from the dielectric side is a possibility, if aberrations and other constraints permit it.

Then,

$$\theta_r = \frac{4\pi nd}{\lambda}$$

and

$$\eta_r = \frac{4\pi^2 n^2 R T^2 d^2}{\lambda^2} \tag{6.10}$$

because the readout light must pass through the device twice, hence the factor of $T^2$ (we neglect reflection losses from the readout beam). Thus

$$\frac{\eta_r}{\eta_t} = \frac{4n^2 RT}{(n-1)^2} \tag{6.11}$$

and for the previous assumption of normal Fresnel reflection giving the value of $R$,

$$\frac{\eta_r}{\eta_t} = \frac{4n^2 T}{(n+1)^2} \tag{6.12}$$

so that if $n = 1.5$ and $T = 0.5$ as before, $\eta_r = 0.72\,\eta_t$. It is interesting to observe that the larger phase shift caused by reflective readout from the dielectric side comes fairly close to compensating for the strong attenuation of two passes through the photoconductor.

## Experimental Exposure Response Results

Turning from these elementary relationships to experimentally observed results, we now consider the actual sensitivity of these devices. This can be described in terms of either deformation excursion $d$ or diffraction efficiency $\eta$ as a function of exposure. The latter has been used almost universally, but the former, readily deducible from it by use of (6.2) and (6.6) or (6.5) if deformations are sinusoidal, relates more directly to material response. Some typical exposure response curves are shown in Figs. 6.6, 6.7, 6.9, and 6.10.

We have already commented on the absence of a single well-defined exposure response characteristic for these devices. Not only are exposure response curves dependent on spatial frequency, but like the frequency response curve itself, exposure response is also affected by development time [6.29, 39]. This dependence is illustrated in Fig. 6.8. Curve $C$ of Fig. 6.8 shows an actual reversal of diffraction efficiency with increasing development, an effect typical of strong

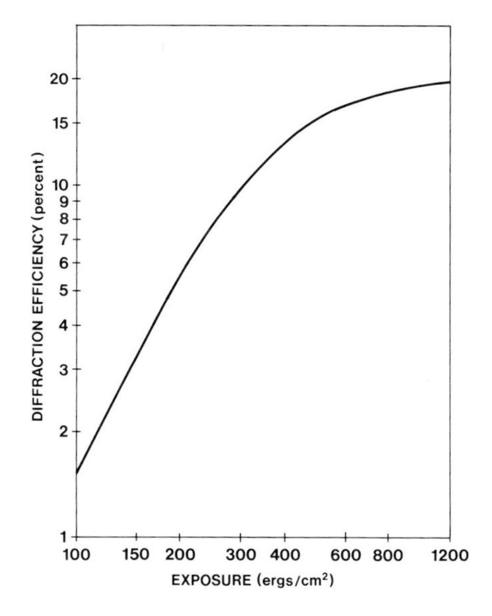


Fig. 6.6. Exposure response (diffraction efficiency vs. exposure) of device with PVK-Brillant Green photoconductor. Thermoplastic thickness 1.1 μm, spatial frequency 400 cyc./mm, beam ratio 1:1. Device and process similar to those used in obtaining data of Fig. 6.2 (after *Urbach* [6.39])

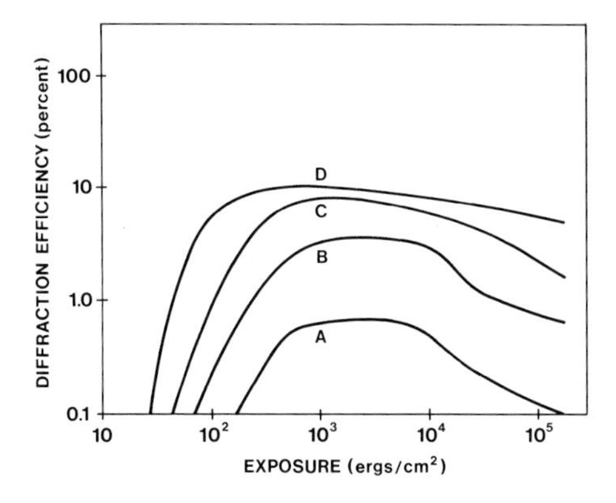


Fig. 6.7. Exposure response of device with PVK/TNF photoconductor at reference-to-object beam ratios of *A* 40:1, *B* 10:1, *C* 4:1, *D* 1:1 (after *Colburn* and *Tompkins* [6.16])

deformations and predicted by (6.2) for sinusoidal deformations. The absence of an actual zero in Curve *C* suggests that deformation in this case was not sinusoidal.

In addition to these effects, the response, being driven in part by the electric field excursion, is of course also dependent on the fringe modulation, hence on the beam ratio [6.14, 29] Fig. 6.7. All of these relationships complicate attempts at universally applicable device characterization. They suggest that tests under the specific conditions of a proposed application are even more important with these devices than with more conventional recording materials, because complete performance prediction from a few characterization curves is usually not possible.

The response curves of Figs. 6.6 through 6.8 were all obtained with relatively insensitive PVK-Brilliant Green or PVK/TNF photoconductors. As noted

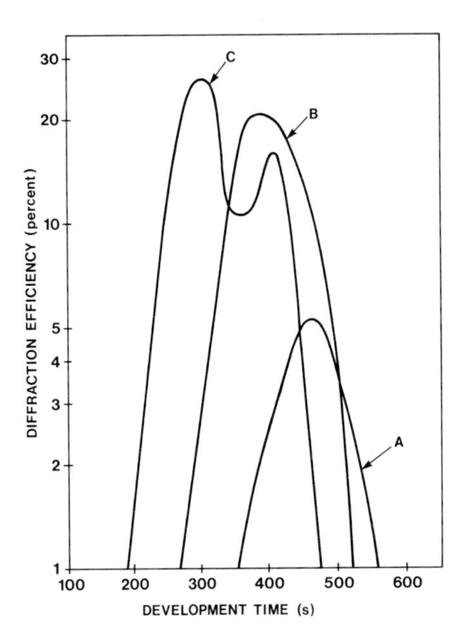


Fig. 6.8. Development response at three levels of exposure: *A* 62 ergs/cm$^2$, *B* 250 ergs/cm$^2$, *C* 1000 ergs/cm$^2$. Thermoplastic thickness 1 μm, spatial frequency 350 cyc./mm, beam ratio 1:1 (after *Bergen* [6.29])

previously (p. 177), greater sensitivity can be achieved by employing the more sensitive vitreous photoconductors such as selenium or its alloys. The response of a device using $As_2Se_3$ as a photoconductor is shown in Fig. 6.9. Because this photoconductor is nearly opaque throughout the visible portion of the spectrum, $T$ is very low in (6.6), and reasonable values of diffraction efficiency can be obtained only by reflective readout. The data for Fig. 6.9, therefore, were obtained in reflection after post-development aluminizing. This means that (6.5) and (6.7) are applicable, so that in comparing Fig. 6.9 with Figs. 6.6 through 6.8, where transmission readout was used, one must ascribe at least an order of magnitude of sensitivity increase to the effect of reflective readout alone. Of the approximately two orders of magnitude increase in sensitivity shown by $As_2Se_3$ compared with PVK/TNF, therefore, only a factor of 5 to 10 is attributable to the more sensitive photoconductor.

The theoretical limit on first-order diffraction by a thin, normally illuminated sinusoidal phase grating is 33.9%, and this value is in fact rarely exceeded in practice. Implicit in the foregoing statement, however, are three possible ways of exceeding this upper limit on $\eta$. First, the grating could be thick, second it could be nonsinusoidal and third it could be illuminated at an angle different from 90°. The nature of thermoplastic relief deformation precludes the first of these possibilities, but the second and third have in fact been used to increase the diffraction efficiency of thermoplastic holograms beyond the value predicted by elementary theory.

Both these methods and their results were described by *Credelle* and *Spong* [6.17]. When deformations, such as those obtained with the simultaneous process, are extremely strong, the relief profile can be distinctly nonsinusoidal,

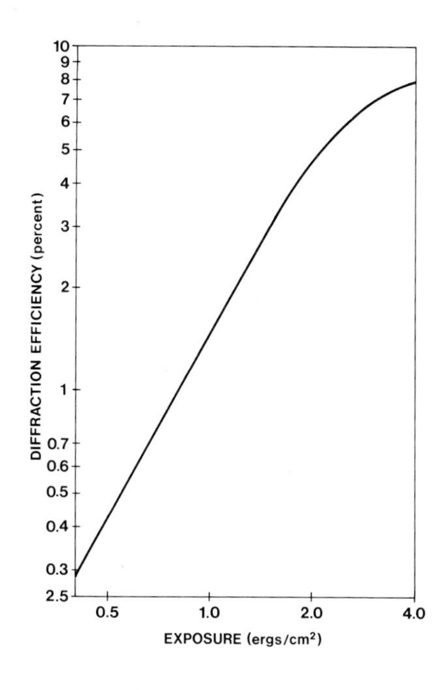


Fig. 6.9. Exposure response of device with $As_2Se_3$ photoconductor. Thermoplastic thickness 1.1 μm, spatial frequency 400 cyc./mm, beam ratio 1:1. Thermoplastic surface aluminized after development; reconstruction by reflected light

and the diffraction efficiency may increase beyond the predicted sinusoidal limit, to values as high as 40%. When the holograms are illuminated off-normal, a primitive "blazing" effect can occur in which geometric refraction (or reflection) of light coincides with one of the first diffracted orders, and efficiencies as high as 60% were reported.

Using the sequential version of the process usually leads to lower values of $\eta$, with the upper limit of 22% reported in [6.17] fairly typical of most results. In most laboratory situations, values in the 1% to 10% range are more common, but it is not difficult to attain higher efficiencies if necessary.

One method for increasing diffraction efficiency beyond that normally obtained is to repeat the charging and development portions of the process sequence [6.42]. The principles behind this method are similar to those underlying the simultaneous mode of the process: a kind of positive feedback tends to amplify existing relief deformations because the higher capacitance of thinner regions results in higher charge densities there when charging to a uniform potential. This kind of post-process amplification is likely to amplify random noise as effectively as it amplifies signal, so it is most useful only in low-noise situations.

### 6.2.3 Noise

Turning now to the subject of noise itself, we should distinguish two types, the true random noise caused by random surface relief, and the so-called intermodulation "noise", a form of image distortion which arises from the effects of nonlinearities. The latter, in turn, can result from either the "intrinsic" nonlinearity inherent in any phase hologram or from the material nonlinearities caused by the particular characteristics of device response to exposure [6.38]. Since the former nonlinearity is always present in thermoplastic holograms, it is usually not necessary to distinguish between the two types of nonlinearity.

It is on the other hand important to distinguish between the random and intermodulation types of noise. The former is somewhat akin to the familiar grain noise of silver halide emulsions but with important differences. "Frost" deformations in thermoplastics are random in orientation, but, like systematic deformations, tend toward a preferred periodicity or spatial frequency [6.4, 11, 18–21]. The angular spectrum of scattered light from such deformations is typically ring-shaped, peaking at an angle determined primarily by the average spatial frequency. Usually this spatial frequency is fairly close to the frequency of peak response to sinusoidal exposures, thus suggesting that the noise and signal are both maximized at or near the preferred spatial frequency.

This characteristic does not mean, however, that the signal-to-noise ratio is largely frequency independent. Thermoplastic random noise differs from that of silver halides insofar as there is often a direct competition between frost noise and exposure-driven deformations [6.4]. If the latter are sufficiently strong, they will tend to suppress the former, resulting in absolute reduction of random noise, and hence in a better signal-to-noise ratio than one might expect on the basis of independent signal and noise measurements. While this result has frequently been observed, the magnitude of the noise suppression effect varies widely, and can change from highly significant to negligible over a fairly narrow range of device and process parameters. The author has in fact sometimes observed a sizable variation of noise suppression within the confines of a single hologram approximately $100 \times 150$ mm in size. Such variations resulted from small and incompletely characterized nonuniformities in fabrication and development.

Although there is still no universal agreement on theoretical explanations for thermoplastic device noise, most theories (see p. 170) predict a threshold voltage below which frost does not form, a result largely confirmed by experiment. Since there should be no such threshold voltage for exposure-driven relief deformations, it might be expected that operation of devices below frost threshold voltages should lead to high signal-to-noise ratios. In practice, this is not ordinarily a usable manner of operation in sequential recording, because the relief mode deformation does not develop rapidly enough at low charge densities. A different situation appears to prevail in simultaneous mode recording. Here the "positive feedback" amplification of exposure-dependent deformation proceeds so rapidly at charge densities below frost threshold that these systematic deformations can become large enough to compete succesfully with frost deformation before the frost threshold is reached. Thus the desired signal deformations will dominate over noise in this mode of recording. The experimental results reported by *Credelle* and *Spong* [6.17] bear out this expectation, since they were able to make virtually noise-free holograms in the simultaneous mode. These holograms were also made with very thin thermoplastic films, having extremely high natural peak response frequency, so that the signal deformations occurred at frequencies far below the expected frost frequencies. Although this may be coincidental, it agrees with the prediction of *Matthies* et al. [6.21] that best signal-to-noise performance might be expected with thin layers.

Intermodulation "noise" or distortion would normally be expected to disturb performance even in the absence of frost noise. The intrinsic nonlinearity of phase holograms, unless accidentally or intentionally compensated by material nonlinearities, will always make thermoplastic devices subject to this kind of distortion [6.38]. However, there is an important effect of frequency response upon intermodulation distortion. As noted by *Credelle* and *Spong* [6, 17], the intermodulation distortion, being the result of beats between the object spectrum and low spatial frequency autocorrelation terms, is sharply diminished if low spatial frequency response is reduced. Because there is very little low frequency response in thermoplastics, there is correspondingly little intermodulation distortion. This was illustrated in [6.17] by results of recording a hologram of a test pattern. Despite an extremely high (35%) reconstruction efficiency, effects of intermodulation distortion were surprisingly low.

It is interesting to note in this context that the intermodulation reduction scheme of *Lamberts* and *Kurtz* [6.43] used an ingenious process for minimizing the low frequency response of bleached silver halide phase holograms with the specific purpose of minimizing intermodulation distortion. Their relatively complex processing method achieved artificially with silver halides the desirable type of bandpass response naturally inherent in thermoplastic devices.

### Experimental Noise Measurements

The exposure dependence of signal-to-noise ratio is compared with that of diffraction efficiency in Fig. 6.10, showing rather typical results reported by *Colburn* and *Tompkins* [6.16]. As with other measurements of signal-to-noise ratio, these were specific to a particular set of experimental conditions. The "signal" was a diffuse object consisting of a trans-illuminated ground glass square with the central one-fourth of its area left opaque to provide a nominally zero intensity object region in which noise could be measured. The signal-to-noise ratio is defined here as the ratio of measured intensity in the illuminated region of the reconstructed image of the diffuser to intensity in the originally dark region of that image. Although specialized, this definition and experimental arrangement are similar to those often used in signal-to-noise measurements for other recording materials, thus facilitating comparison with those measurements.

The particular measurements shown in Fig. 6.10 were made with a reference-to-signal beam ratio of 4:1. As is common with phase holograms, the peak of the signal-to-noise ratio occurs at a lower value of exposure than the peak of the diffraction efficiency, in this case, about an order of magnitude lower. The former peak is somewhat sharper than the latter.

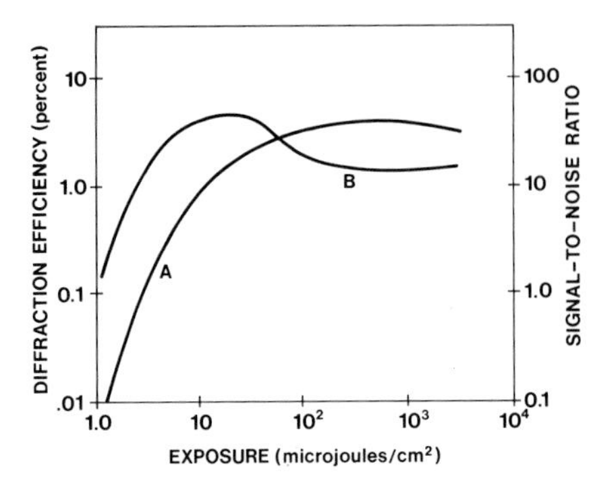


Fig. 6.10. *A* Diffraction efficiency and *B* signal-to-noise ratio vs. exposure for diffuse object with reference-to-object beam ratio of 4:1 (after *Colburn* and *Tompkins* [6.16])

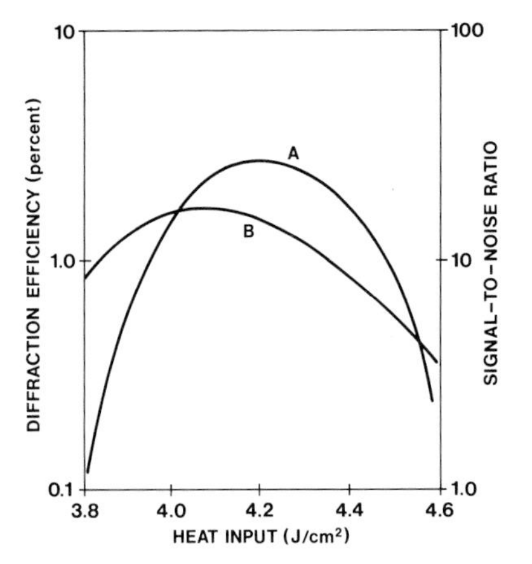


Fig. 6.11. *A* Diffraction efficiency and *B* signal-to-noise ratio vs. heat input to conductive layer in 130 ms pulse (after *Colburn* and *Tompkins* [6.16])

It is interesting to contrast this result with another from [6.16], shown in Fig. 6.11. Here diffraction efficiency and signal-to-noise ratio are plotted as a function of development heat input, when ohmic heating is produced by 130 ms pulses of direct current through the conductive layer between substrate and photoconductor. Here the diffraction efficiency is substantially more sensitive than signal-to-noise ratio to changes in development energy, and both are much more rapidly affected by such changes than by corresponding variations in exposure.

Once more we should emphasize that all of these results are dependent on many other experimental parameters, such as time of development and surface charge density, once more precluding simple device characterization. The particular temperature history produced by the 130 ms ohmic heating pulses used in this set of experiments is illustrated in Fig. 6.12, giving the thermoplastic temperature, measured with a low-mass thermistor probe, as a function of time after the initiation of such a heating pulse. The slow decline of temperature shown here is typical of all such heating arrangements. It implies a strong history dependence in process response. If insufficient time has elapsed after an erasure pulse (whose decay is similar [6.16]), the temperature starting point will be

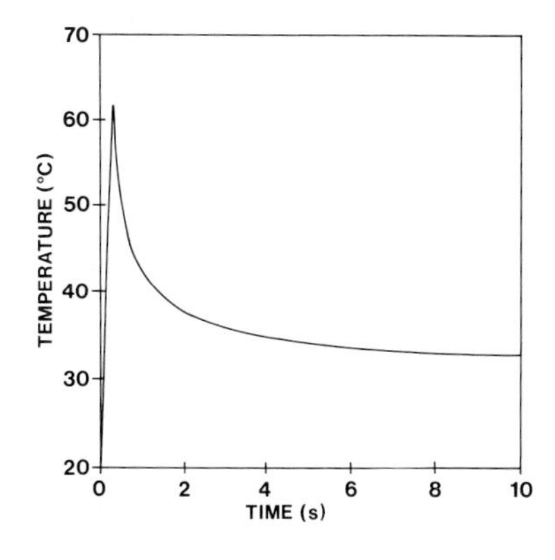


Fig. 6.12. Thermoplastic temperature vs. time after initiation of a 130 ms development pulse (from *Colburn* and *Dubow* [6.12])

higher than ambient, and relative overdevelopment will occur with a development pulse optimized for an ambient starting point. This problem can be overcome, at the expense of complexity, by process control based on recording the thermal history or instantaneous temperature of the thermoplastic, and adjusting development energy accordingly.

## 6.2.4 Cyclic Operation

One of the primary attractions of thermoplastic imaging devices is their ability to be erased and reused. In early devices, this ability was severely limited by thermoplastic degradation. Extensive refinements have led to devices capable of thousands of cycles of reuse. These refinements have been discussed at some length on pp. 179–181. Here we shall only summarize some of the experimental results obtained with the more recent materials, illustrating that earlier discussion of thermoplastics. Figure 6.13 shows the diffraction efficiency and signal-to-noise ratio of a copolymer of styrene with octyl and decyl methacrylate as a function of imaging cycles in air.

Since, as noted on p. 180, operation in the presence of oxygen severely degrades the performance of this type of thermoplastic, it is also interesting to study its properties in an inert gas. This is illustrated by Fig. 6.14, showing cyclic operation of a similar copolymer device in an argon atmosphere. Such operation clearly extends the several hundred cycles of substantially undegraded operation to several thousand. Particularly noteworthy in Fig. 6.14 is the stability of the signal-to-noise ratio up to about 5000 cycles of reuse. Its lower initial value is simply the incidental result of the use of a parallel-plane charging device, whose glass surfaces produced reflections which lowered the measured SNR. The stability of the diffraction efficiency is less significant in this experiment because it was maintained by increasing development energy gradually to compensate an increasing softening temperature.

Most recently *Lo* et al. [6.44] have reported up to 80 000 cycles of operation in an inert atmosphere with compensated development. It is clear that the several dozen reuse cycles of early devices using Staybelite thermoplastic have been

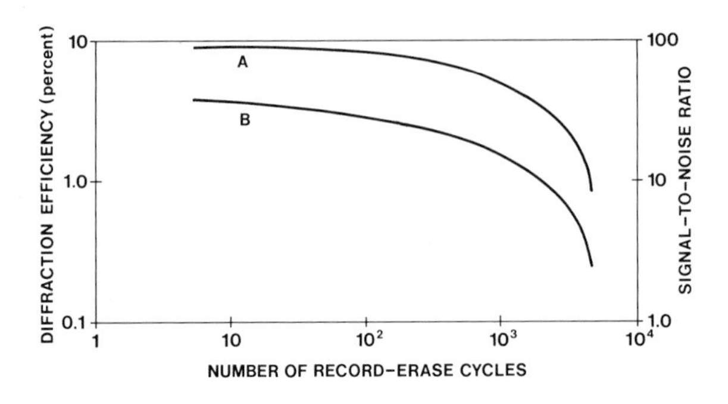


Fig. 6.13. *A* Diffraction efficiency and *B* signal-to-noise ratio vs. cycles of reuse for copolymer thermoplastic in air (after *Colburn* and *Tompkins* [6.16])

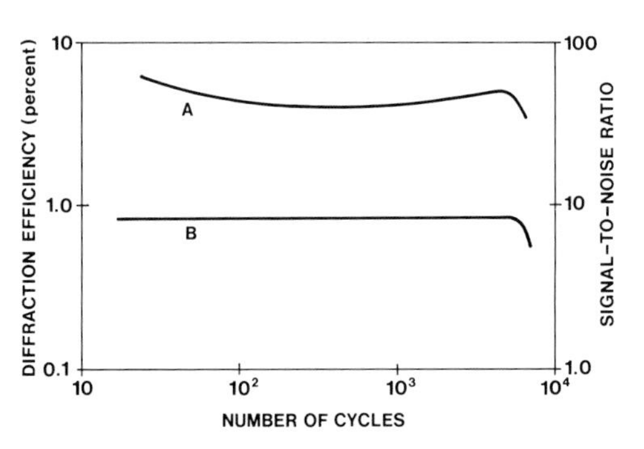


Fig. 6.14. *A* Diffraction efficiency and *B* signal-to-noise ratio vs. cycles of reuse for copolymer thermoplastic in argon atmosphere (after *Colburn* and *Tompkins* [6.16])

significantly improved, but it is also evident that thermoplastics are still far removed from unlimited cycling. The nature of their degradation, together with the ghost images arising from charge trapping in the photoconductor and at the interface, suggests that there will be no early or simple solution to the problems preventing indefinite reuse.

### 6.2.5 Other Physical Properties

Although little discussion of these characteristics has appeared in the literature, potential applications of thermoplastics may be strongly affected by such properties as shelf life, archival stability and mechanical durability. These characteristics have not been the subject of much quantitative measurement, as they are with well established, commercially available materials. Nevertheless, some information on them has been acquired, often as a more or less incidental byproduct of investigations into the optical properties discussed previously.

#### Shelf Life

There appears to be every reason to expect excellent shelf life of the unused devices. Like all electrophotographic devices, they are optically inert when not charged, and suffer no degradation from exposure to moderate levels of heat, actinic radiation, or even some types of ionizing radiation. The charge trapping effects in organic photoconductors, discussed previously, have been known to affect the behavior of uncharged devices given strong exposure to actinic light shortly before actual use. Sensitivity can be degraded this way, and can be restored by dark storage, especially at moderately elevated temperatures, as would be expected. Although not observed in actual experiments (since none were attempted), it is likely that nonuniform exposure of uncharged devices could produce weak ghost image patterns by the same mechanism.

Aside from the adverse effects of such pre-use exposure in some photoconductors, the shelf life should be very long. This was not true of devices using some of the early thermoplastics, such as Staybelite, because oxidation and perhaps loss of residual solvents altered $T_g$ within a few months. The newer thermoplastics, able to withstand the rigors of thousands of cycles of reuse, can be expected to have much better shelf life, quite probably on the order of several years at room temperature.

## Latent Image Life

In the single-layer devices, the latent electrostatic image is subject to whatever degree of dark decay exists in the photoplastic. Because charge generation, injection and transport can be expected to occur at a low rate in the dark, and because fields exist across the still-charged portions of the device, a fairly rapid decay of the undeveloped latent image can be expected. Development within some minutes of exposure is likely to be advisable with many of these devices.

Two-layer devices (used in the sequential process), on the other hand, have a different type of latent image, insofar as most of it exists as fields across a nonphotoconductive thermoplastic. Any residual fields in the photoconductor soon decay in the manner described above for photoplastics. The latent image on the thermoplastic can be extremely long lived. The low conductivity of the thermoplastic is further aided by charge trapping at its air and photoconductor interfaces, leading to remarkably stable latent images. Development of latent images into good holograms has been observed in our laboratories after delays of several months following exposure and recharge. These particular devices used Staybelite thermoplastics, and appeared to be ultimately limited more by its hardening than by latent image decay. Development of electron beam recordings (presumably on polystyrene) has been observed as much as one year after image formation. Although no results are available on latent image life with the styrene-methacrylate copolymers, similar longevity might be expected.

This latent image stability, although often unimportant, does suggest that these devices are suitable for uses in which it is either impractical or undesirable to develop the latent image shortly after exposure.

## Archival Life

The stability of the developed image is normally excellent. If $T_g$ is well above ambient temperature, there is no mechanism causing any alteration in the surface deformation of a thermoplastic, or reducing its signal-to-noise ratio. All changes in the thermoplastic which tend to raise $T_g$ (such as cross linking) only enhance this stability. The thermoplastics used to date show little or no tendency to lower $T_g$ with time, so gradual softening and spontaneous erasure have not been observed. Images made with Staybelite Ester 10 have remained in excellent condition for over a decade.

The main threat to developed image stability of course comes from elevated temperatures in either storage or readout. There is ordinarily no protection available against this threat, since it is a consequence of erasability. If it is known that a device will not be reused, and if protection of the recorded image is important, then considerable protection can be obtained by inducing an irreversible increase in $T_g$. This is most readily done by exposure to ultraviolet radiation in the presence of oxygen, a process which in most thermoplastics tends to promote cross linking more rapidly than scission, thus elevating $T_g$. With enough increase in $T_g$, it is usually possible to stabilize the finished image against a reasonable range of ambient and readout temperatures.

### Mechanical Durability and Contamination

Even when $T_g$ is well above ambient temperature, thermoplastics tend to have a soft surface, easily marred or scratched by mechanical contact and abrasion. This can be a serious problem. While no quantitative measurements of surface hardness have, to our knowledge, been made, there is no doubt that thermoplastics usable in these devices are far softer than the hardened gelatin of a properly fixed silver halide emulsion. This means in practice that care must be taken to avoid all forms of mechanical contact with the device surface.

This requirement can usually be satisfied fairly readily when glass substrates are used and few devices are needed. It becomes more problematic when a continuous web of flexible plastic is used as substrate for many devices. The problems arise from the normal practice of storing the web as a roll before and after use. Contact with the adjacent substrate backing often causes significant damage to the thermoplastic both before and after holograms are recorded on it. One way of preventing such damage is the use of a "dog biscuit" substrate cross section with narrow raised edges which prevent contact between the central regions of successive layers on a roll. While increasing the bulk of a rolled web substantially, this approach can solve the contact problem.

Partly related to the softness characteristics is the high sensitivity of thermoplastics to dust and dirt. The soft surface makes it easy for contaminants to become imbedded in it. Moreover, it makes surface cleaning by mechanical means difficult if not impossible. This problem is further aggravated in two ways. First, like any charged surface, the thermoplastic during the charged portion of the process sequence attracts oppositely charged dust particles, and in regions of field gradient, can attract polarizable neutral particles as well.

Second, if a foreign particle is on the thermoplastic surface prior to development, it can have a drastic effect on the local behavior of the device. It can serve as a nucleation site for frost deformation which might otherwise be below threshold. It can also nucleate a more local perturbation, or can penetrate the thermoplastic and promote temporary arcing through it, especially in the simultaneous process or upon reuse. In all these cases, it can create a surface perturbation whose optical cross section far exceeds that of the particle itself.

This last effect, then, can in itself be an extraneous but very serious source of noise. Depending as it does on incidental environmental conditions, it was not an appropriate topic for discussion in Section 6.2.3. It is, nevertheless, often the dominant source of noise in practical operation. Moreover, because of the combination of softness, charge, and tackiness of the device, dust and dirt accumulation is irreversible, and is often an effective bar to cyclic operation in unprotected ambients.

## 6.3 Conclusions

The advantages and drawbacks of thermoplastic overcoated photoconductor devices have become more sharply defined during the past decade of research and device refinement. The primary advantages are:

1) Relatively high diffraction efficiency typical of thin phase holograms.

2) Very high limiting resolution (up to 4000 cy/mm), and, more important, high useful bandwidth, the latter in excess of 1000 cycles/mm on some devices.

3) Suppression of intermodulation distortion as a result of this bandpass response, permitting full use of the high diffraction efficiency without excessive intermodulation effects on the reconstructed images.

4) High sensitivity: Even with relatively insensitive organic photoconductors, these are the only high resolution recording devices or materials which are comparable in sensitivity to silver halide emulsions.

5) Rapid dry development, including ease of *in situ* development, useful in optical storage and in holographic interferometry.

6) Reusability, with many thousands of cycles of reuse now attainable.

7) Replicability by mechanical means (like other surface relief holograms).

8) Insensitivity of device both prior to charging and after recharging, permitting handling in ambient light before charging as well as after formation of the stabilized latent image.

The main drawbacks are perhaps more subtle, but sometimes severe:

1) High sensitivity of device performance to details of fabrication, charging and development; lack of "development to completion".

2) In part deriving from 1), the relatively complex apparatus needed for adequately controlled charging and development.

3) Lack of unique specification of major response characteristics resulting from the complex interdependence of diffraction efficiency and frequency response, together with their sensitivity to a variety of device and process parameters.

4) Relatively complex and precisely controlled fabrication techniques required for high quality devices.

5) Sensitivity of the thermoplastic to abrasion and dust, the latter aggravated by electrostatic attraction of dust during device operation.

6) Relative ease of accidental erasure by overheating.

7) Tendency toward ghost image formation by charge trapping, requiring special precautions in cyclic processes.

These devices have been used in many experiments by a variety of technical groups in several countries. Their use in optical storage experiments has been especially widespread, e.g., [6.14, 34, 36]. The general lack of utility of early holographic memory systems may have, by association, cast doubt upon the usefulness of the thermoplastic devices which were once believed to be an essential ingredient in such systems, but this is an incorrect association, because these storage systems would have been impractical with any recording material. Finally, it is evident that the characteristics of thermoplastic devices are ideally suited to many forms of holographic interferometry [6.45] and they should find increasing application in that field.

## References


6.1 H. Kogelnik: Reconstructing Response and Efficiency of Holograms Gratings. In Proc. of Symposium on Modern Optics, ed. by J. Fox (Polytechnic Press, Brooklyn 1967) pp. 606–609

6.2 J.C. Urbach, R.W. Meier: Appl. Opt. **5**, 666 (1966)

6.3 J. Gaynor, S. Aftergut: Phot. Sci. Eng. **7**, 209 (1963)

6.4 J.C. Urbach: Phot. Sci. Eng. **10**, 287 (1966)

6.5 W.E. Glenn: J. Appl. Phys. **30**, 1870 (1959)

6.6 E.I. Sponable: J. Soc. Motion Picture Television Engrs. **60**, 337 (1953)

6.7 I. Cindrich, G.D. Currie, W.D. Hall, C. Leonard: Real-Time Optical Data Modulator, Report AFAL-TR-75-77, Radar and Optics Division, Environmental Research Institute of Michigan (1975)
I. Cindrich, G.D. Currie, C. Leonard: Report AFAL-TR-75-77, Radar and Optics Division, Environmental Research Institute of Michigan (1975)

6.8 R.J. Doyle, W.E. Glenn: IEEE Trans. Electron Devices, ED-**18**, 739 (1971)

6.9 J. Gaynor: IEEE Trans. Electron Devices, ED-**19**, 512 (1972)

6.10 R.G. Cross: Deformation Image Processing, IBM Technical Disclosure Bulletin **4**, 35 (1961)

6.11 R.W. Gundlach, C.J. Claus: Phot. Sci. Eng. **7**, 14 (1963)

6.12 W.S. Colburn, J.B. Dubow: Photoplastic Recording Materials, Technical Report AFAL-TR-73-255, Air Force Avionics Laboratory, Air Force Systems Command (1973)

6.13 J.H. Dessauer, H.E. Clark: *Xerography and Related Processes* (Focal Press, London 1965) p. 205

6.14 W.C. Stewart, R.S. Mezrich, L.S. Cosentino, E.M. Nagle, F.S. Wendt, R.D. Lohman: RCA Rev. **34**, 3 (1973)

6.15 N.K. Sheridon: IEEE Trans. Electron Devices, ED-**19**, 1003 (1972)

6.16 W.S. Colburn, E.N. Tompkins: Appl. Opt. **13**, 2934 (1974)

6.17 T.L. Credelle, F.W. Spong: RCA Rev. **33**, 206 (1972)

6.18 P.J. Cressman: J. Appl. Phys. **34**, 2327 (1963)

6.19 H.F. Budd: J. Appl. Phys. **36**, 1613 (1965)

6.20 J.M. Schneider, P.K. Watson: Phys. Fluids **13**, 1948 (1970)

6.21 D.L.Matthies, W.C.Johnson, M.A.Lampert: Experimental and Theoretical Study of the Frost Instability in Thermoplastic Recording Materials, Technical Report No. 25, Device Physics Laboratory, Dept. of Electrical Engineering, Princeton University (1973)
6.22 P.Gushcho, P.A.Ionkin: Zh. Nauch. i Prikl. Fotogr. i Kinematogr. **12**, No. 3, 166 (1967)
6.23 A.I.Babchin, M.C.Borolkina, F.Z.Dzhabarov, Ch.A.Maksimova, V.I.Uspiensky, V.I.Shiebierstov: Zh. Nauch. i Prikl. Fotogr. i Kinematogr. **12**, No. 3, 149 (1967)
6.24 F.W.Schmidlin: Private communication (1967)
6.25 J.J.Bartfai, V.Ozarow, J.Gaynor: Phot. Sci. Eng. **10**, 60 (1966)
6.26 S.Aftergut, J.J.Bartfai, B.C.Wagner: Appl. Optics (Suppl. 3: Electrophotography), 161 (1969)
6.27 J.Gaynor, G.J.Sewell: Photo. Sci. Eng. **11**, 204 (1967)
6.28 W.S.Colburn, J.C.Dwyer, L.M.Ralston: Materials for Holographic Optical Elements, Final Technical Report AFAL-TR-73-364 (1973)
6.29 R.Bergen: Phot. Sci. Eng. **17**, 473 (1973)
6.30 W.S.Colburn, L.M.Ralston, J.C.Dwyer: Appl. Phys. Lett. **23**, 145 (1973)
6.31 J.C.Urbach: Hologram Recording Properties of Thermoplastic Xerography, Paper Summaries, SPSE Annual Conference, p. 150 (1969)
6.32 D.S.Lo, L.H.Johnson, R.W.Honebrink: Appl. Opt. **14**, 820 (1975)
6.33 H.R.Anderson, Jr., E.A.Bartkus, J.A.Reynolds: IBM J. Res. Develop. **15**, 140 (1971)
6.34 Electro-Optics Operations, Harris Corporation, "Updated Optical Read/Write Memory System Components", Final Technical Report, NASA Contract NAS8-26672, Section IV (1974)
6.35 Holographic Recording Materials, Final Technical Report, Contract No. F30602-74-C 0030, Electro-Optics Operation, Harris Electronic Systems Division (1974)
6.36 L.H.Lin, H.L.Beauchamp: Appl. Opt. **9**, 2088 (1970)
6.37 T.C.Lee: Digest of Technical Papers, Topical Meeting on Optical Storage of Digital Data, Optical Society of America, paper WB-4 (1973)
6.38 J.C.Urbach, R.W.Meier: Appl. Opt. **8**, 2269 (1969)
6.39 J.C.Urbach: Advances in Holograms Recording Materials. In *Developments in Holography*, Proc. of SPIE Seminar, ed. by B.J.Thompson, J.B.deVelis, Vol. 25, pp. 31–32 (1971)
6.40 F.W.Schmidlin, H.M.Stark: Lateral Conduction Mechanisms in Xerography and Their Limitations on Resolution, Paper Summaries, SPSE Annual Conference, p. 152 (1969)
6.41 R.M.Schaffert: *Electrophotography*, Revised ed. (Focal Press, New York, London 1975) pp. 481, 483, 509
6.42 A.E.Jvirblis, J.C.Urbach: Deformation Imaging Method, U.S. Patent No. 3795514 (1974)
6.43 R.L.Lamberts, C.N.Kurtz: Appl. Opt. **10**, 1342 (1971)
6.44 D.S.Lo, L.H.Johnson, R.W.Honebrink: J. Opt. Soc. Am. **66**, 1084 (A) (1976)
6.45 J.C.Bellamy, D.B.Ostrowsky, M.Poindron, E.Spitz: Appl. Opt. **10**, 1458 (1971)
6.46 D.R.Terrell: Phot. Sci. Eng. **21**, 66 (1977)
6.47 W.T.Malone, R.L.Gravel: Appl. Opt. **13**, 2471 (1974)

# 7. Photoresists

R. A. Bartolini

With 11 Figures

For optical information storage and retrieval applications, holography offers certain advantages that have attracted the current interest of investigators. One of these advantages is mass replication [7.1, 2] offered by thin relief phase holograms that can be easily replicated by a simple two-step process: a) making a metallic master and b) embossing this master into a suitable material such as vinyl.

A material well suited for recording thin relief phase holograms is photoresist. Photoresists are light-sensitive organic materials which form imaged relief patterns upon exposure and development. There are two types of photoresists—negative and positive. In negative type photoresist, the exposed areas become insoluble as they absorb light so that upon development only the unexposed (soluble areas) are dissolved away. In positive type photoresists the opposite is true; the exposed areas become soluble and dissolve away during the development process.

The choice of photoresist for recording relief phase holograms is based on a tradeoff among sensitivity, resolution capability, and ease of application and processing. In evaluating the first of these it is evident that the spectral response characteristic of the photoresist must correlate with the output spectrum of the recording source, and the photoresist must be sensitive enough to permit reasonable exposure speed. With regard to resolution, the photoresist must be capable of recording the details of the interference pattern which constitutes the hologram. Typically, optical interference patterns have fringe spacings on the order of a micrometer. This implies that the photoresist must have resolution capability exceeding $1000\, l \cdot mm^{-1}$.

For ease of application and processing, the photoresist must, of course, adhere well to its substrate when first applied and also through development. In the case of negative photoresists the action of absorbing light is to harden the organic material by forming cross linkages among the molecules. The developer solution then fixes the material by washing away the unexposed molecules while having little or no effect on the exposed molecules. Thus, to ensure substrate adhesion with negative photoresists it is necessary to form many cross linkages through sufficient exposure [7.3]. This requirement causes difficulty in hologram recording because the exposure necessary for good adhesion may exceed that necessary for optimizing the holographic process. Thus, during the holographic process, the hologram consisting of very fine lines of photoresist

Table 7.1. Photoresists

| Material | Usable thickness | Preparation | Processing | Recording processes | Recording wavelength range | Recording sensitivity | Resolution | Ref. |
|---|---|---|---|---|---|---|---|---|
| Shipley AZ-1350 (pos) | μm's | Spin or spray coating | Wet chemical | Absorption and cross linking | UV to 500 nm | $10^{-2}$ J·cm$^{-2}$ at 441.6 nm | $>1500$ $l$·mm$^{-1}$ | [7.4–9, 11, 12, 17, 18] |
| Shipley AZ-111 (pos) | μm's | Spin or spray coating | Wet chemical | Absorption and cross linking | UV to 500 nm | $10^{-2}$ J·cm$^{-2}$ at 441.6 nm | $<1000$ $l$·mm$^{-1}$ | [7.4, 5, 9, 16, 17] |
| Kodak KPR (neg) | μm's | Spin or spray coating | Wet chemical | Absorption and cross linking | UV to 450 nm | $10^{-2}$ J·cm$^{-2}$ at 400 nm | $<1000$ $l$·mm$^{-1}$ | [7.4, 10, 11] |
| Kodak KTFR (neg) | μm's | Spin or spray coating | Wet chemical | Absorption and cross linking | UV to 450 nm | $10^{-2}$ J·cm$^{-2}$ at 400 nm | $<1000$ $l$·mm$^{-1}$ | [7.4, 10, 11] |
| Kodak KMER (neg) | μm's | Spin or spray coating | Wet chemical | Absorption and cross linking | UV to 450 nm | $10^{-2}$ J·cm$^{-2}$ at 400 nm | $<1000$ $l$·mm$^{-1}$ | [7.4, 10, 12] |
| Kodak KAR 3 (pos) | μm's | Spin or spray coating | Wet chemical | Absorption and cross linking | UV to 450 nm | $10^{-2}$ J·cm$^{-2}$ at 400 nm | $>1500$ $l$·mm$^{-1}$ | [7.9, 10] |
| GAF PR-102 (pos) | μm's | Spin or spray coating | Wet chemical | Absorption and cross linking | UV to 500 nm | $10^{-2}$ J·cm$^{-2}$ at 400 nm | $\sim 1000$ $l$·mm$^{-1}$ | [7.9, 13, 17] |
| Kodak KOR (neg) | μm's | Spin or spray coating | Wet chemical | Absorption and cross linking | UV to 550 nm | $10^{-2}$ J·cm$^{-2}$ at 488 nm | $\sim 1000$ $l$·mm$^{-1}$ | [7.4, 10–12] |
| Horizons LHS7 (neg) | μm's | Spin or spray coating | Heated air | Absorption and cross linking | UV to 550 nm | $5 \times 10^{-3}$ J·cm$^{-2}$ at 488 nm | $>500$ $l$·mm$^{-1}$ | [7.14] |
| Way Coat LSI395 (pos) | μm's | Spin or spray coating | Wet chemical | Absorption and cross linking | UV to 500 nm | $10^{-2}$ J·cm$^{-2}$ at 441.6 nm | $>1500$ $l$·mm$^{-1}$ | [7.15] |
| Kodak Micro Positive Resist 809 (pos) | μm's | Spin or spray coating | Wet chemical | Absorption and cross linking | UV to 450 nm | $10^{-1}$ J·cm$^{-2}$ at 400 nm | $>1000$ $l$·mm$^{-1}$ | [7.10] |

tends to detach from the substrate due to insufficient exposure. Most negative photoresists succumb to this exposure problem.

One can avoid the above-mentioned exposure problem by using positive photoresists, which have exposure and development properties opposite to those of negative photoresists; i.e., the absorbed light softens the organic material by destroying the cross linkages (polymerization) with the subsequent fixing (development) washing away the exposed (destroyed cross linkages) molecules.

## 7.1 Types of Photoresists

Table 7.1 lists some of the most widely used photoresists and some of their more useful properties. The most widely studied photoresist for holographic applications is Shipley AZ-1350 [7.6–9, 11, 12, 17–23] a commercially available photoresist manufactured by the Shipley Co., Inc., Newton, Massachusetts 02612. Reasons for this choice are given later in this chapter. Thus, the design procedure for recording holograms developed in a later section of this chapter while applicable to all types of photoresist will primarily be applied to Shipley AZ-1350 photoresist.

## 7.2 Theory of Mechanism

The characteristics of photoresist materials are described by a change in layer thickness $\Delta d$ caused by exposure $E$; i.e.,

$$\Delta d = f(E) . \tag{7.1}$$

To determine an explicit relationship between $\Delta d$ and $E$, we refer to the field of photochemistry. Light energy absorbed by a positive photoresist layer at a depth $x$(cm) can be described as [7.24]

$$I(x) = I_0 \exp(-\alpha N_1 x) , \tag{7.2}$$

where $I_0$ = irradiance on the photoresist surface ($\mathrm{mW \cdot cm^{-2}}$), $\alpha$ = absorption cross section of the molecule ($\mathrm{cm^2}$), $N_1$ = density of the absorbing molecules ($\mathrm{cm^{-3}}$).

If the photoresist layer is thin (on the order of a few micrometers) and the percent absorption is small, and the number of photons $N$ ($\mathrm{cm^{-2} \cdot s^{-1}}$) available for absorption throughout the material is

$$N = \frac{I_0}{h\nu} , \tag{7.3}$$

where $h$ = Planck's constant ($6.23 \cdot 10^{-31}\,\mathrm{mJ \cdot s}$); $\nu$ = frequency of light ($\mathrm{s^{-1}}$), then equating the time ($t$) rate of decrease of unabsorbed positive photoresist molecules, $-dN_1(t)/dt$, with the rate of creation of absorbed molecules, we have [7.24] for small absorption, i.e., $I(x) \sim I_0$,

$$-\frac{dN_1(t)}{dt} = \eta_q \frac{I_0}{h\nu} \alpha N_1(t), \tag{7.4}$$

Because not all molecules become excited molecules (removed during development) when they absorb a photon, we introduce $\eta_q$, the quantum efficiency, defined as the number of excited molecules per number of absorbed photons. Solving (7.4) for $N_1(t)$ yields

$$N_1(t) = N_0 \exp\left(-\frac{n_q \alpha}{h\nu} I_0 t\right) \tag{7.5}$$

or

$$D_u = \frac{N_1(t)}{N_0} = \exp(-\alpha_0 E), \tag{7.6}$$

where $D_u$ = fraction of unabsorbed photoresist molecules, $N_0$ = initial density of total photoresist molecules ($cm^{-3}$), $\alpha_0 = n_q \alpha / h\nu \triangleq$ exposure constant for positive photoresist ($cm^2 \cdot mJ^{-1}$), $E = I_0 t$ = photoresist exposure ($mJ \cdot cm^{-2}$).

If we define

$$D_a = \frac{N_2(t)}{N_0}, \tag{7.7}$$

where $D_a$ = fraction of absorbed photoresist molecules, $N_2(t)$ = density of positive molecules that have absorbed a photon ($cm^{-3}$), then with $D_a + D_u = 1$ we have

$$D_a = 1 - D_u = 1 - \exp(-\alpha_0 E). \tag{7.8}$$

If the mechanism for positive photoresist development is one in which exposed (absorbed) and unexposed (unabsorbed) portions are attacked by the developer but at different rates, then the following etch rate model for the amount of material, $\Delta d$(μm), removed at any one point in the photoresist is given as

$$\Delta d = (D_a r_1 + D_u r_2) T, \tag{7.9}$$

where $r_1$ = rate of removal (etching) of absorbed molecules during development ($\mu m \cdot s^{-1}$), $r_2$ = rate of removal (etching) of unabsorbed molecules during development ($\mu m \cdot s^{-1}$), $T$ = development time (s).

If we now substitute (7.6) and (7.8) into (7.9) with $\Delta r = (r_1 - r_2)$ we have

$$\Delta d = T[r_1 - \Delta r \exp(-\alpha_0 E)]. \tag{7.10}$$

If $E = 0$, then $\Delta d = r_2 T$; i.e., as anticipated, $\Delta d$ is determined solely by the unabsorbed etch rate constant. If $E$ is very large than $\Delta d \sim r_1 T$; i.e., $\Delta d$ is determined solely by the absorbed etch rate constant. If $\alpha_0 E \ll 1$ then

$$\Delta d \sim \Delta r T \alpha_0 E + r_2 T . \tag{7.11}$$

Thus, with this linearization $\Delta d$ is determined by a constant term $r_2 T$ and the exposure term $\Delta r T \alpha_0 E$. The importance of this linearization will be discussed next.

## 7.3 Hologram Nonlinearities

Images retrieved from thin phase holograms recorded in photoresist are deteriorated by two kinds of nonlinearity—the inherent or *intrinsic* nonlinearity of the phase holography recording process and the *material* nonlinearity caused by the exposure characteristics of the photoresist. For example, because the transmittance function $T_a(x, y)$ of a phase recording material is a nonlinear function of the recorded phase shifts, i.e.,

$$T_a(x, y) = \exp j\phi(x, y) , \tag{7.12}$$

phase recording materials are inherently nonlinear even if there is no material nonlinearity [7.25]. The material nonlinearity results from the fact that the phase shift $\phi(x, y)$ itself is a function of the relief depth $\Delta d$ for a relief phase hologram; i.e.,

$$\phi(x, y) = \frac{2\pi}{\lambda_R} (n-1)\Delta d , \tag{7.13}$$

where $\lambda_R$ = readout wavelength and $n$ = index of refraction of material. Phase is thus affected by material nonlinearity since the relief depth is a nonlinear function of exposure $E$ as shown in (7.10).

Equation (7.10) represents the material nonlinearity and (7.13) the intrinsic nonlinearity. These nonlinearities cause several noise effects in the images from the holograms; e.g., harmonic and intermodulation (IM) distortion. Noise effects from harmonic terms can be avoided by spatially separating the desired first-order images from the others by recording the hologram with the off-axis configuration.

The origin of intermodulation distortion can be determined by reference to the irradiance $I_0$ at the hologram plane:

$$\begin{aligned} I_0(x, y) = {} & 0_0^2(x, y) + R_0^2 + 0_0(x, y) R_0 \exp[j(\phi_R - \phi_0)] \\ & + 0_0(x, y) R_0 \exp[-j(\phi_R - \phi_0)] , \end{aligned} \tag{7.14}$$

where $0_0 = 0_0(x, y) \exp j\phi_0$ = complex field amplitude (CFA) of the object beam and $R = R_0 \exp j\phi_R$ = CFA of the reference beam. For a complex object (two or more object points), the term $0_0^2(x, y)$ is not always a constant. For holographic recording where the transmittance is approximately a linear function of irradiance, as with amplitude holograms, this fact is of no practical significance because the noise associated with this term can be avoided by making the term $(\phi_R - \phi_0)$ large enough to spatially separate the desired signal from the noise spectrum [7.25]. Unfortunately, spatial filtering is not a complete solution in the case of phase transmission functions. Consequently, the signal itself is a source of noise. This intermodulation noise appears as a ghost image in the background of the desired image, thus reducing desired image contrast. The question of controlling these nonlinearities and thus controlling IM distortion is considered next.

The controllable recording parameters that are of most interest in our analysis are hologram exposure, reference-to-object beam ratio, and development processing. As shown in the next section material nonlinearity can be controlled with proper development processing for some photoresists and thus $\Delta d$ in (7.10) can be approximated as suggested in (7.11) by

$$\Delta d = gE, \tag{7.15}$$

where $g$ = hologram exposure constant.

Control of the intrinsic nonlinearity requires a different approach. The irradiance $I_p$ of the primary hologram image is [7.26]:

$$I_p = R_0^2 J_1^2 \left[\frac{4\pi}{\lambda_R}(n-1)gtO_0R_0\right], \tag{7.16}$$

where $R_0$ = readout beam CFA and $J_1$ = Bessel function of first kind, order 1. The nonlinearity manifests itself here in the Bessel function. Control of this nonlinearity can be achieved if the following approximation is made:

$$J_1(a) \sim \frac{1}{2} a, \tag{7.17}$$

where $a$ is small and given by

$$a = 2\frac{2\pi}{\lambda_R}(n-1)gtO_0R_0 = \frac{4\pi}{\lambda_R}(n-1)\frac{gE}{I_0}O_0R_0. \tag{7.18}$$

In a typical situation, parameters in (7.18) are on the order of:

| | |
|---|---|
| $n = 1.7$ | photoresist index of refraction |
| $\lambda_R = 632.8$ nm | He–Ne laser readout |
| $\Delta d = gE = 0.1$ μm | typical exposure |
| $I_0 \sim 10 O_0^2$ | hologram irradiance |
| $R_0 \sim \sqrt{10}\, O_0$ | beam ratio of 10:1 used. |

Substitution in (7.18) yields

$$a=0.44 \tag{7.19}$$

and thus the linear approximation $J_1(a) \sim (1/2)a$ is valid.

## 7.4 Diffraction Efficiency and Signal-to-Noise Ratio

The efficiency of the hologram is the fraction of incident light flux diffracted into the first-order image so we have from (7.16),

$$\text{Efficiency} \triangleq \eta = \frac{I_p}{R_0^2} = J_1^2 \left[\frac{4\pi}{\lambda_R}(n-1) g t O_0 R_0\right]. \tag{7.20}$$

It follows that a tradeoff exists between intermodulation noise and hologram efficiency.

We thus define the tradeoff parameters as *hologram efficiency and hologram image* s/n (where the main source of noise is intermodulation distortion noise) and the recording parameters as exposure, reference-to-object beam ratio, and development processing. The relationship among these parameters can be derived as follows:

On p. 10 it is shown that:

$$\eta = J_1^2(a). \tag{7.21}$$

Substituting (7.18) into this relation gives

$$\eta = J_1^2 \left\{ \left[\frac{2\pi(n-1)}{\lambda_R}\right] (\Delta r T) \left(\frac{2\sqrt{K}}{1+K} \alpha_0 E_0\right) \right\}, \tag{7.22}$$

where $K = R_0^2/(O_0^2(x, y))$ = reference-to-object beam ratio, $g = \alpha_0 \Delta r T$, and $E_0$ = average exposure defined by (7.24). The maximum efficiency in this case is 33.9%. If we limit the efficiency such that the approximation of (7.17) can be used, then (7.22) becomes

$$\eta = \left[\frac{2\pi(n-1)}{\lambda_R}\right]^2 (\Delta r T)^2 \left(\frac{\sqrt{K}}{1+K} \alpha_0 E_0\right)^2. \tag{7.23}$$

The bracketed term in (7.23) contains the readout parameters; i.e., $n$ = the index of refraction of readout material and $\lambda_R$ = the readout wavelength. The second term in parentheses contains the development parameters: $\Delta r$ = difference between development etch rates and $T$ = development time. The third term contains the exposure parameters: $\alpha_0$ = the exposure constant, $K$ = the reference-to-object beam ratio, and $E_0$ the average exposure; i.e.,

$$E_0 = (O_0^2 + R_0^2)t. \tag{7.24}$$

*Lin* [7.27] determined the characteristics of ideal hologram-recording materials to be

$$\eta = \frac{4S^2 E_0^2 K}{(1+K)^2} \tag{7.25}$$

for a two-beam case, where $S$ is the material sensitivity. Using Lin's definitions and photoresist characteristics we find that

$$S = \frac{\pi(n-1)}{\lambda_R} \Delta r T \alpha_0 \, (\mathrm{cm}^2 \cdot \mathrm{mJ}^{-1}), \tag{7.26}$$

which makes (7.25) equal to (7.23).

*Dammann* [7.28] found that for a thin phase hologram of a diffuse object,

$$\eta = \frac{1}{2} \phi_1^2 \exp(-\phi_1^2), \tag{7.27}$$

where $\phi_1$ is the mean phase shift. In this case the maximum efficiency occurs at $\phi_1 = 1$ and $\eta_{max} = 18.4\%$. Using Dammann's definition and photoresist characteristics we have

$$\phi_1 = \sqrt{2} \frac{2\pi}{\lambda_R} (n-1) \frac{\Delta r T \alpha_0 \sqrt{K} E_0}{1+K}. \tag{7.28}$$

If the efficiency is limited as in the case of (7.22) then (7.28) becomes equal to (7.23), because $\exp(-\phi_1^2) \sim 1$. The question of limiting the efficiency depends on how much nonlinearity one tolerates. *Urbach* and *Meier* [7.25] calculate that for a 5% deviation from linearity [using (7.22)], the maximum efficiency is 9%. *Cathey* [7.29] suggested a limit of 1% on efficiency to produce negligible nonlinear effects. It is thus clear that we need a relationship between $\mathrm{s/n_{IM}}$ (where $\mathrm{n_{IM}}$ is intermodulation noise) and efficiency.

*Lee* and *Greer* [7.30] determined a relationship between $\mathrm{s/n_{IM}}$ and efficiency for photographic emulsions. Their theory is easily applied to the case of photoresists with the following results:

$$\frac{\mathrm{s}}{\mathrm{n_{IM}}} = \frac{K}{\eta}, \tag{7.29}$$

where $K$ is the reference-to-object beam ratio and $\eta$ is defined as in (7.23).

*Jenney* [7.31] found that the $\mathrm{s/n_{IM}}$ for a three-beam hologram (two object points and a reference beam) is

$$\frac{\mathrm{s}}{\mathrm{n_{IM}}} = \frac{4K}{4\eta + \eta^2 \frac{(1+K)^2}{K}}. \tag{7.30}$$

For limited efficiency values ($<10\%$) and $R<100$, (7.30) reduces to

$$\frac{\mathrm{s}}{\mathrm{n_{IM}}} \sim \frac{K}{\eta}, \tag{7.31}$$

which is the same as (7.29).

*Dammann* [7.28] has shown also that for a diffuse object the relationship between $\mathrm{s/n_{IM}}$ and $\eta$ for small modulation is

$$\frac{\mathrm{s}}{\mathrm{n_{IM}}} = \frac{2}{\eta^2}. \tag{7.32}$$

Thus, we find here disagreement within the literature concerning the effect of IM "noise" on the reconstructed wavefront from the hologram. In the next sections we empirically determine the $\mathrm{s/n_{IM}}$ vs. $\eta$ relationship for a particular photoresist and compare the results with these theoretical predictions.

These results have all assumed that there is no material nonlinearity; i.e., $\Delta d = gE$. In the next section we shall see that this approximation is valid with proper development processing for some photoresists.

## 7.5 Shipley AZ-1350 Photoresist Characteristics

As mentioned previously, the most widely used photoresist for holographic application is Shipley AZ-1350 positive photoresist. Its widespread use is due to the fact that the material nonlinearity described by (7.10) can be relieved for this photoresist by proper development processing as described below.

The properties of Shipley AZ-1350 photoresist are:

**Physical Properties**[1]

| | |
|---|---|
| Type of solution | solvent |
| Appearance—liquid | clear, amber red |
| Appearance—solid | pale yellow |
| Solids content | 19.5% |
| Viscosity | 5 centipoises (20 °C) |
| Specific gravity (liquid density) | 1.0–1.1 (20 °C) |
| Flash point | 46 °C |

**Optical Properties**

| | |
|---|---|
| Useful light sensitivity range[1] | 340–450 nm |
| Index of refraction over useful sensitivity range [7.32] | 1.66–1.70 |

[1] Shipley Company, Inc., Newton, Massachusetts 02612.

### 7.5.1 Shipley AZ-1350 Material Nonlinearity Considerations

The mechanism for positive photoresist development (when using AZ-1350 developer provided by the Shipley Company) is to dissolve away the exposed areas of the photoresist while leaving the unexposed areas essentially intact. This is essentially a binary process which is ideal for applications such as the fabrication of integrated circuits. However, for holographic recording, an analog process is desired. Moreover, to avoid noise caused by nonlinear distortion, as mentioned previously, it is desirable to have a depth vs. exposure characteristic which is approximately linear over a wide exposure range. When using AZ-1350 developer, the photoresist appeared to act nonlinearly, limiting the hologram diffraction efficiency for a given intermodulation noise to a greater degree by this material nonlinearity than by the intrinsic nonlinearity of thin phase holograms.

Accordingly, the development process needed is one that can lead to a net relief pattern that provides for higher hologram diffraction efficiency limited by the intrinsic nonlinearity. This can be accomplished with Shipley AZ-1350 photoresist if AZ-303 developer, an alkaline solution, is used in place of the AZ-1350 developer [7.6]. Since AZ-1350 photoresist is alkaline soluble, the resulting etching development process is one in which exposed portions are etched at a faster rate than unexposed portions.

In order to control this etch rate in practice, it was found necessary to dilute the AZ-303 developer with four parts distilled water. If diluted any more than this it begins to act like AZ-1350 developer and if the dilution factor is less the etch rate becomes uncontrollable.

To experimentally verify the theory of this section, thin layers ($\sim 10\ \mu m$) of Shipley AZ-1350 photoresist were exposed with 441.6-nm radiation from a He–Cd laser. A number of exposures were made on each sample. After development a layer of aluminum was evaporated on the samples to provide reflectivity necessary in order to measure the layer thickness by the Tolansky interference technique [7.33].

Figure 7.1 is a plot of the results for two different development techniques. The upper curve represents development with AZ-303 (4:1 dilution) for 20 s at 25 °C while the lower one represents development with AZ-1350 (no dilution) for 20 s at 25 °C. It is immediately obvious from these two curves that the AZ-303 developed curve shows a higher degree of linearity for a greater exposure range than does the AZ-1350 developed curve.

The "speed" or sensitivity of a photoresist is always of practical interest [7.9, 11]. Comparing the two curves for a constant value of $\Delta d$, the AZ-303 developer provides an apparent 2.5 to 1 increase in sensitivity over the AZ-1350 developer. The shapes of both curves conform well to the predictions of (7.10). However, although both curves conform well to (7.10), the higher degree of linearity provided by the AZ-303 developed curve makes it more desirable for holographic recording. To investigate this further, consider Fig. 7.2, which is a plot of the AZ-303 developed photoresist data over the linear range. From this

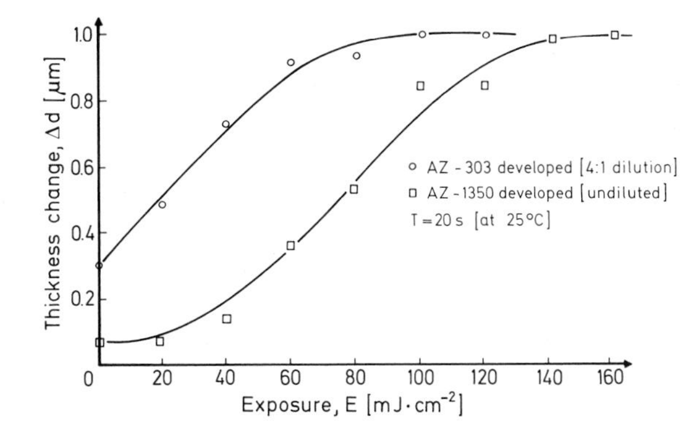


Fig. 7.1. Thickness change $\Delta d$ as a function of exposure $E$ for two different development techniques

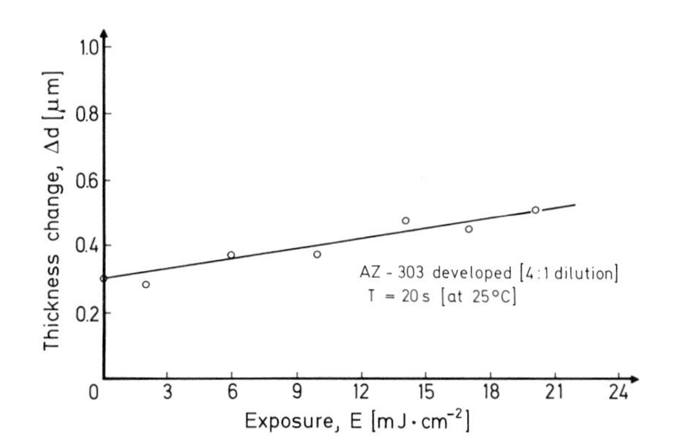


Fig. 7.2. Thickness change $\Delta d$ as a function of exposure $E$ using Shipley AZ-303 developer

curve we can attempt to obtain some of the photoresist parameters. At $E=0$, $\Delta d = r_2 T$, and thus, for $T=20$ s, $r_2 = 0.015\ \mu\text{m}\cdot\text{s}^{-1}$.

Figure 7.3 is a plot of $\Delta d$ vs. $T$ for $E=0$ and $E=10\ \text{mJ}\cdot\text{cm}^{-2}$. As expected from (7.10) the relationship is linear. Using (7.10) and the slopes of Fig. 7.3, we have

$$r_2 = 0.015\ \mu\text{m}\cdot\text{s}^{-1}$$

and

$$\Delta r \alpha_0 = 0.5 \times 10^{-3}\ \mu\text{m}\cdot\text{cm}^2\cdot\text{mJ}^{-1}\cdot\text{s}^{-1}\,.$$

This value of $r_2$ is the same as obtained from Fig. 7.2. To determine $\Delta r$ a value for $\alpha_0$ is necessary. From (7.6) $\alpha_0 = \eta_q \alpha / hf$ where $h = 6.23 \times 10^{-31}$ mJ·s and $v = 6.8 \times 10^{14}\ \text{s}^{-1}$. *Broyde* [7.18] has found that for AZ-1350 photoresist $\eta_q = 3.2\%$ and $\alpha = 6.25 \times 10^{-17}$ cm$^2$. Therefore, $\alpha_0 = 0.5 \times 10^{-2}$ cm$^2\cdot$mJ$^{-2}$. Using this value for $\alpha_0$ we have

$$\Delta r = r_1 - r_2 = 0.1\ \mu\text{m}\cdot\text{s}^{-1}\,.$$

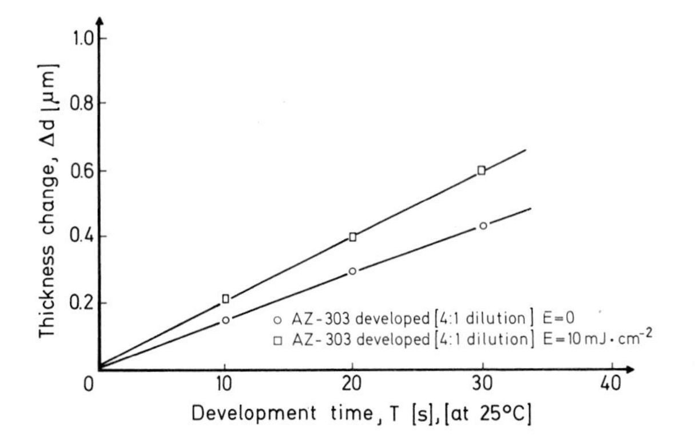


Fig. 7.3. Thickness change $\Delta d$ as a function of development time $T$ for two different values of exposure $E$

Because $r_2 = 0.015\,\mu\text{m}\cdot\text{s}^{-1}$ then

$$r_1 = 0.115\,\mu\text{m}\cdot\text{s}^{-1}.$$

We have thus shown in this section that the use of AZ-303 developer with AZ-1350 photoresist relieves the material nonlinearity usually associated with photoresist for a very usable range of exposures, thereby allowing for higher holographic efficiencies. It was also shown that the use of this developer provides an apparent increase in material sensitivity.

### 7.5.2 Shipley AZ-1350 Resolution Capability

Because a hologram itself looks nothing like the image it stores but rather is an interference pattern associated with the image, we cannot look at the hologram *directly* and attempt to determine material resolution in terms of commonly used criteria. Looking for an appropriate criterion for resolution in this case we note that the efficiency (power diffracted into image divided by incident readout beam power) of a phase hologram is a monotonically increasing function of the depth of modulation (i.e., the depth of the corrugations in the developed photoresist surface). But the depth of modulation is related to the quality of recorded phase variation which in turn is related to the fineness of detail in the formation being recorded. Thus, it is considered appropriate to measure resolution capability in terms of efficiency.

Since the simplest hologram is formed by two plane waves coming together at a recording plate, the experimental setup of Fig. 7.4 was used to determine photoresist resolution capability. The recollimated He–Cd laser beam (wavelength 441.6 nm and approximately 1 cm in diameter) in Fig. 7.4 is separated into two paths which are later brought together again by means of pivotal mirrors. These pivotal mirrors and the mobile slide holder allow adjustment of the angle between the two beams at the recording plate while maintaining one of the beams normal to the recording plane.

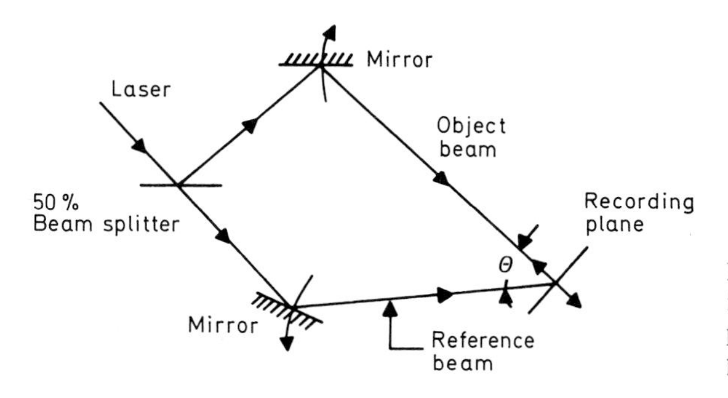


Fig. 7.4. Experimental setup for determining resolution capability of Shipley AZ-1350 photoresist

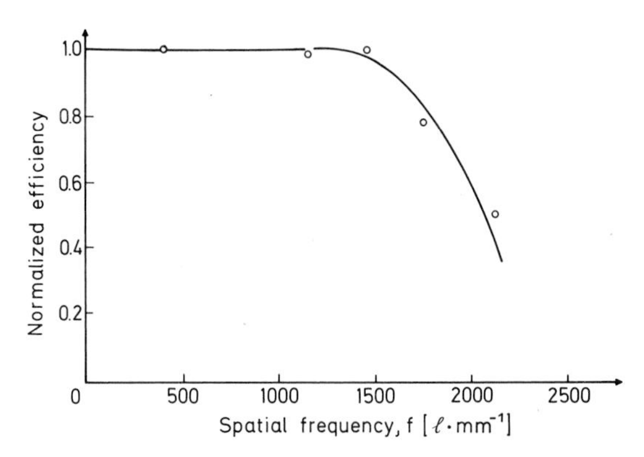


Fig. 7.5. Spatial frequency response of Shipley AZ-1350 photoresist

It is generally true that a definite relationship exists between coating thickness and photographic resolution, although this relationship is sometimes complicated by other factors. Items such as light source, developing technique, and material surface condition all affect the resolution capabilities of photoresist layers. Four different thicknesses (all less than 2 μm) were evaluated but no definite resolution dependence on thickness was observed.

The result of this experiment for Shipley AZ-1350 photoresist (using AZ-303 developer) is shown in Fig. 7.5. This spatial frequency response curve indicates that the maximum useful recording frequency is 1500 $1 \cdot mm^{-1}$.

### 7.5.3 Shipley AZ-1350 Relief Phase Hologram Design Procedure

In this section, the design procedure for recording useful (i.e., acceptable signal-to-noise ratio >25 dB, high efficiency ~3–5%, exposure sensitivity ~a few $mJ \cdot cm^{-2}$) relief phase holograms in AZ-1350 photoresist are presented. Experimental recording examples resulting from the application of this design process are compared with the theory of the previous sections.

There are three principal types of noise that plague hologram images: cosmetic defects, material scattering, and intermodulation noise. Let us examine these in turn.

In order to obtain all the benefits of holography, it is necessary to use highly coherent light. Unfortunately, coherent light tends to exaggerate the noise caused by cosmetic defects, arising from scratches, dimples, bubbles, thickness variations, reflections, and dust or dirt in and on the recording system components. These cosmetic defects are usually indiscernible with noncoherent illumination but they can be and usually are a very serious problem with coherent illumination.

As pointed out elsewhere [7.34], the solution to the cosmetic noise problem is redundancy. Redundancy is achieved by illuminating the object with a multiplicity of beams generated by a phase grating, mirror array, or pinhole array. The holograms described in this chapter were recorded using such a pinhole array redundancy device (to essentially eliminate the cosmetic noise problem) since it produces the best results to date [7.34].

Scattered light is not a serious problem with photoresist materials since these materials do not have the grain-like structure of photographic film [7.12]. Thus, we have not considered this aspect of image reproduction in this chapter.

Intermodulation distortion is not random in nature as other forms of noise generally are, and thus, its effect is best described as a degradation in image quality rather than noise in the true sense. Nevertheless, it is conveniently treated as a noise component for numerical calculations. As pointed out early in this chapter, the origin of IM noise is the object signal itself; because most objects are quite complex, a rigorous analysis of IM noise would be extremely difficult and possibly of little practical value in most cases. The theoretical attempts at determining $s/n_{IM}$ presented previously were for the simplest cases (such as 2 object points). This simple case analysis is valuable for helping us understand the basic process involved. With this knowledge as groundwork, the questions that really needs answering in the general case include:

1) What test pattern does one use to measure $s/n_{IM}$?
2) What is an acceptable $s/n_{IM}$?
3) How does one measure $s/n_{IM}$?

The best pattern chosen to measure IM noise is shown in Fig. 7.6. It is a white square with a black background, representing a worst-case object. Figures 7.6a–c are hologram images having increasing efficiency and thus increasing IM noise photographed from the face of a TV monitor. Figure 7.6a is a 1.5% efficient hologram with an $s/n_{IM}$ of 32 dB; Fig. 7.6b is a 5.4% efficient hologram with $s/n_{IM}$ of 23 dB; and Fig. 7.6c is a 10.8% efficient hologram with $s/n_{IM} < 20$ dB.

*Budrikis* [7.35] in an article on fidelity measures and criteria for visual communications stated:

> "One way of approaching our subject would be to launch immediately into a review of all that is known about luminance vision. But our purpose will be served better if we first determine what it is that we want to know. To this end we will first examine the practical view of visual fidelity and how it is assessed. We will find that the basis for it is an overall *subjective evaluation* of achievement within an agreed context" (emphasis added).

With this viewpoint in mind and referring to Fig. 7.6, it is subjectively clear that an $s/n_{IM}$ of greater than 25 dB would be acceptable to most observers.

Fig. 7.6a—c. Examples of images produced from holograms as a function of average exposure $E_0$. (a) $s/n_{IM}=32$ dB, $K=10$, $E_0=5.5$ mJ·cm$^{-2}$, $\eta=1.5$%. (b) $s/n_{IM}=23$ dB, $K=10$, $E_0=10$ mJ·cm$^{-2}$, $\eta=5.4$%. (c) $s/n_{IM}<20$ dB, $K=10$, $E_0=16.8$ mJ·cm$^{-2}$, $\eta=10.8$%

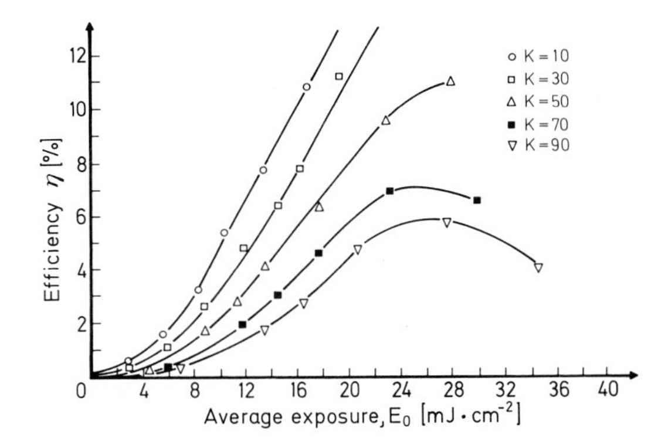


Fig. 7.7. Efficiency $\eta$ as a function of average exposure $E_0$ for a variety of reference-to-object beam ratios $K$

Holograms were recorded with a He–Cd laser ($\lambda_0=441.6$ nm). The recording material was approximately a 1-μm layer of Shipley AZ-1350 positive photoresist. The object transparency was the one shown in Fig. 7.6. The hologram size was 6 × 10 mm. The holograms were recorded with different exposures $E_0$ and *a* variety of beam ratios $K$. They were read out with a He-Ne laser ($\lambda_R=632.8$ nm) and their efficiencies measured. Figure 7.7 presents the results from these holograms. As predicted by (7.23) we see that as $K$ increases, $\eta$ decreases for a given value of $E_0$. Also, as $E_0$ increases, $\eta$ increases for a given value of $K$ until,

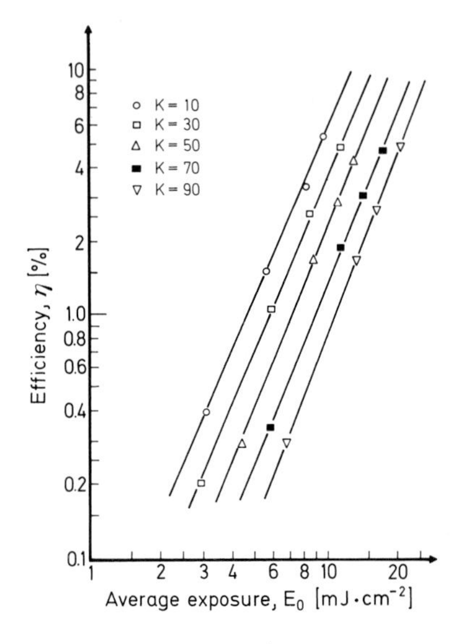


Fig. 7.8. Efficiency $\eta$ as a function of average exposure $E_0$ for a variety of reference-to-object beam ratios $K$

beyond a certain value of $E_0$, the efficiencies begin to decrease. This results from the fact that a certain high values of $E_0$, the exposure has begun to burn through the thin photoresist layer; i.e., the photoresist can no longer support the hologram pattern. Of course, as $K$ increases, burn-through occurs at lower values of $E_0$ since the reference beam power is much higher than the signal power. *Vikram* and *Siroki* [7.36] have considered this problem of insufficient photoresist thickness and recommend the following pre-exposure calculation for the minimum layer thickness ($d_{\min}$) needed:

$$d_{\min} = mE_0\,, \tag{7.33}$$

where $m = \Delta r \alpha_0 T$ for our case. Since the lower exposures are more important, a display of the lower portion of the data of Fig. 7.7 on log–log paper in Fig. 7.8 reveals that, indeed,

$$\eta = \text{constant}\, E_0^2\,. \tag{7.34}$$

Thus, for efficiencies less than 5% we can expect useful results.

The $s/n_{IM}$ measurements of these holograms were carried out by first viewing the holographic images with a vidicon TV camera and displaying the images on a TV monitor; next the actual signal (white square) and noise (value of IM degradation surrounding the white square) values were measured electronically. Figures 7.9 and 7.10 display the results of those measurements. Figure 7.9 is a plot of $s/n_{IM}$ vs $K$ for a variety of $\eta$'s. Both figures consider only holograms in the acceptable efficiency range; viz., $\eta$ approximately 5% and less.

These results differ from any of the theoretical results of the previous section, as expected, because of the different objects used in each case and the method

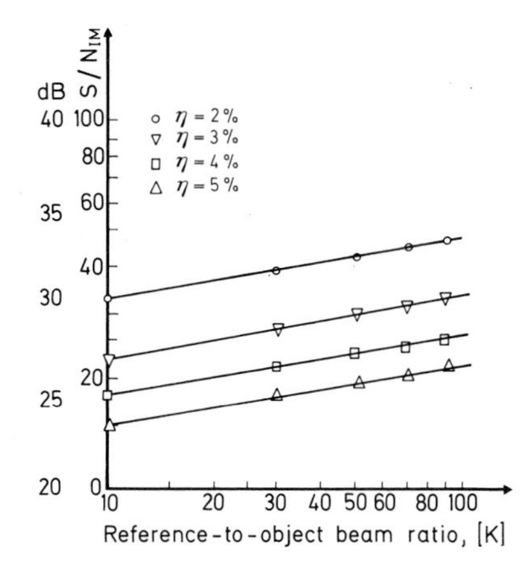


Fig. 7.9. Signal-to-IM noise ratio $s/n_{IM}$ as a function of reference-to-object beam ratio $K$ for a variety of efficiencies $\eta$

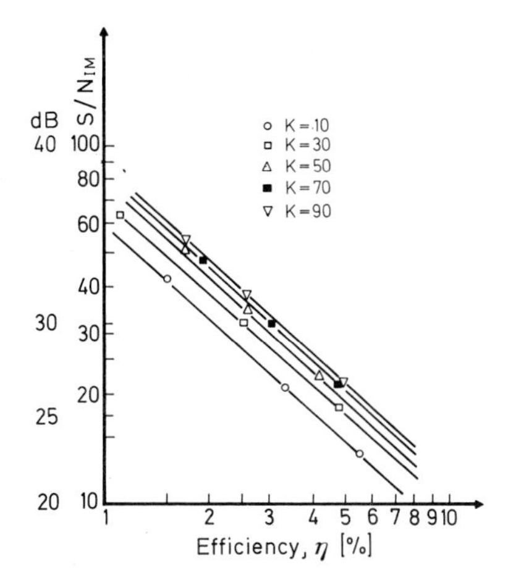


Fig. 7.10. Signal-to-IM noise ratio $s/n_{IM}$ as a function of efficiency $\eta$ for a variety of reference-to-object beam ratios $K$

used to measure $s/n_{IM}$ in this chapter [7.6]. Obtaining an accurate measure of $s/n_{IM}$ is a very difficult task because the measurement technique must be qualified as to object and technique used.

The attempt here was not to empirically derive an $s/n_{IM}$ relationship to conform with a given theory but rather to show what relationship results in practice. To this end we used a "worst case" test pattern as the object and for the measurement techniques we included the entire effect of IM as noise. The acceptable quality range was a subjective determination and was shown, using Fig. 7.6, to be greater than 25 dB. A final vivid example of the tradeoff between IM distortion and hologram efficiency for a test pattern object is shown in Fig. 7.11.

Thus, thin relief phase holograms recorded in Shipley AZ-1350 positive photoresist have useful efficiency and signal-to-noise values (limited by the intrinsic nonlinearity) up to 5% and greater than 25 dB, respectively, for a wide variety of objects. Also, the useful exposure range is 5 to 20 $mJ \cdot cm^{-2}$. These

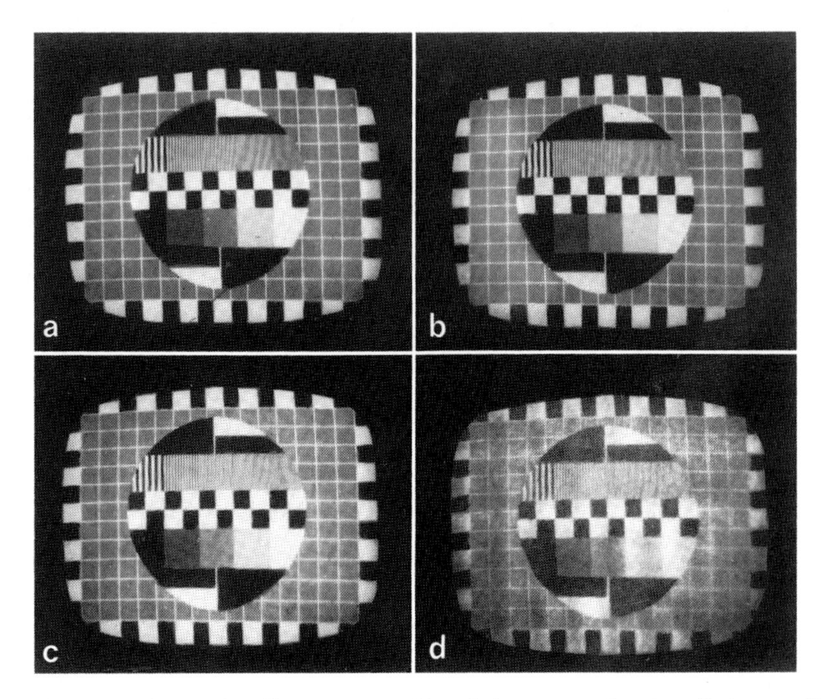


Fig. 7.11a—d. Examples of images produced from holograms as a function of hologram efficiency $\eta$. (a) $\eta=1.3\%$; (b) $\eta=5\%$; (c) $\eta=10\%$; (d) $\eta=20\%$

values are all dependent on the value of reference-to-object beam ratio $K$ chosen. Thus, using Figs. 7.8–10 one can determine the desired and/or necessary recording parameters to achieve particular tradeoff requirements for a given application. This design procedure technique can also be applied to other photoresists.

## References

7.1 R.A.Bartolini, W.J.Hannan, D.Karlsons, M.J.Lurie: Appl. Opt. **9**, 2283 (1970)
7.2 R.A.Bartolini, N.Feldstein, R.J.Ryan: J. Electrochem. Soc. **120**, 1408 (1973)
7.3 K.R.Dunham: Solid State Tech. **14**, 41 (1971)
7.4 K.S.Pennington: *Handbook of Lasers* (Chemical Rubber Co. Press, 1971) p. 563
7.5 Shipley Co., Newton, Massachusetts, Technical Data Bulletin
7.6 R.A.Bartolini: Appl. Opt. **11**, 1275 (1972); and Appl. Opt. **13**, 129 (1974)
7.7 M.J.Beesley, J.G.Castledine: Appl. Opt. **9**, 2720 (1970)
7.8 S.L.Norman, M.P.Singh: Appl. Opt. **14**, 818 (1975)
7.9 F.J.Loprest, E.A.Fitzgerald: Photo. Sci. Eng. **15**, 260 (1971)
7.10 Eastman Kodak Co., Rochester, New York, Technical Data Bulletin
7.11 M.S.Htoo: Photo. Sci. Eng. **12**, 169 (1968)

7.12 M.J. Beesley, J.G. Castledine, D.P. Cooper: Elec. Lett. **5**, 257 (1969)
7.13 GAF Co., New York City, New York, Technical Data Bulletin
7.14 R.G. Zech, J.C. Dwyer, F. Fichter, M. Lewis: Appl. Opt. **12**, 2822 (1973)
7.15 P.A. Hunt Corp., Palisades Park, New Jersey, Technical Data Bulletin
7.16 R.G. Brandes, R.K. Curran: Appl. Opt. **10**, 2101 (1971)
7.17 K.G. Clark: Solid State Tech. **14**, 48 (1971)
7.18 B. Broyde: J. Electrochem. Soc. **117**, 1555 (1970)
7.19 D.B. Ostrowsky, A. Jacques: Appl. Phys. Lett. **18**, 556 (1971)
7.20 D.G. Dalgoutte: Opt. Commun. **8**, 124 (1973)
7.21 C.V. Shank, R.V. Schmidt: Appl. Phys. Lett. **23**, 154 (1973)
7.22 W. Tsang, S. Wang: Appl. Phys. Lett. **24**, 196 (1974)
7.23 S. Austin, F.T. Stone: Appl. Opt. **15**, 1071 (1976)
7.24 V. Balzani, V. Carassiti: *Photochemistry of Coordination Compounds* (Academic Press, New York 1970)
7.25 J.C. Urbach, R.W. Meier: Appl. Opt. **8**, 2269 (1969)
7.26 R.A. Bartolini, J. Bordogna, D. Karlsons: RCA Rev. **33**, 170 (1972)
7.27 L.H. Lin: J. Opt. Soc. Am. **61**, 203 (1971)
7.28 H. Dammann: J. Opt. Soc. Am. **60**, 1635 (1970)
7.29 W.T. Cathey: J. Opt. Soc. Am. **56**, 1167 (1966)
7.30 W.H. Lee, M.O. Greer: J. Opt. Soc. Am. **61**, 402 (1971)
7.31 J.A. Jenney: Appl. Opt. **11**, 1371 (1972)
7.32 L.J. Fried, R. Flachbart, D.E. Itlen, J.W. Raiseski, F.W. Anderson, K.V. Dalel: J. Electrochem. Soc. **117**, 1079 (1970)
7.33 W.L. Bond, F.M. Smits: Bell System. Tech. J. **35**, 1209 (1956)
7.34 A.H. Firester, E.C. Fox, T. Gayeski, W.J. Hannan, M.J. Lurie: RCA Rev. **33**, 131 (1972)
7.35 Z.L. Budrikis: Proc. IEEE **60**, 1635 (1970)
7.36 C.S. Vikram, R.S. Siroki: Appl. Opt. **10**, 2790 (1971)

# 8. Other Materials and Devices

J. Bordogna and S. A. Keneman

With 1 Figure

In this final chapter we discuss a number of additional optical recording materials and devices which, taken together with those discussed above, cover the principal media of current interest in holographic recording. Of course, referring to such recording media as "holographic" implies at least that they are capable of supporting micrometer resolution. Several of the materials and devices presented here do not as yet, but are included because of the likelihood that further interest and research will soon improve their resolution capabilities.

The media discussed are characterized as *materials* or *devices* depending on their fabrication and use. The former possess a single constituent that is directly light sensitive. The latter are composite structures containing a light-sensitive element, but requiring more than light alone to operate properly. The *materials* discussed include magneto-optic films, chalcogenide glasses, and metal films. The *devices* include ferroelectric-photoconductor devices, elastomer devices [1], liquid-crystal-photoconductor devices, and photoconductor-Pockel's-effect devices—referred to in the text as *layered devices with photoconductor write-in.* The background for this presentation has its beginnings in [8.1]. Subsequent review articles by *Chen* and *Zook* [8.2] and *Bartolini* et al. [8.3] are of complementary general interest.

To begin our discussion, it is convenient to define a number of terms used repeatedly by workers in the field to describe the operational effectiveness of a particular material or device: record or write energy density (quantity of energy per unit area required for recording—frequently referred to as "sensitivity"); record time (time required to record a hologram); erase time (time to clear storage medium for next hologram); diffraction efficiency (percentage of light energy in reconstructed image compared to incident energy of readout light); linearity (measure of attainable contrast or "gray scale"); resolution capability (number of resolvable lines per linear distance); cycle lifetime (number of times medium can be recorded and erased without deterioration of performance); and natural decay time (length of time hologram can be stored with no sustaining power). These parameters are also frequently useful when comparing the relative merits of available materials and devices for a given holographic recording application.

---

[1] The elastomer devices, although not light sensitive, are included for generality, because they may be useful for holographic imaging.

## 8.1 Magneto-Optic Films

Magnetic holograms may be stored by Curie-point writing in ferromagnetic materials. Although there are many magneto-optic materials [8.4] that would be candidates for holographic recording, manganese-bismuth (MnBi) [8.5–10] has received considerable attention; hence, this section will concentrate on MnBi.

Holograms are written in MnBi by Curie-point writing [8.11–13], a technique studied prior to the advent of magnetic holography. Sufficient heat is applied through absorption of the incident light beam to raise the MnBi film above its Curie temperature (360 °C), but not to its decomposition temperature (450 °C). This carries the film to a paramagnetic state. When the light is removed, the film cools back through the Curie point and the local film regions take on a magnetization determined by the magnetic fields produced by surrounding MnBi film regions or by an applied external magnetic field.

Since Curie-point writing is a thermal process, any wavelength strongly absorbed by MnBi could, in principle, be used to store holograms. The absorption coefficient for MnBi is $\alpha \gtrsim 3 \cdot 10^5\ \mathrm{cm}^{-1}$ for wavelength $\lambda < 700\ \mathrm{nm}$ [8.13]; however, thermal diffusion through the film and substrate occurs and constrains one to use short light pulses, on the order of 10 ns [8.6], such as those produced by a pulsed ruby laser ($\lambda = 694.3\ \mathrm{nm}$). The write energy density required to store a hologram is approximately $0.3\ \mathrm{mJ/mm^2}$ [8.5], and erasure of the hologram may be performed by application of a magnetic field (without illumination)—a 4000-Oe field is suitable [8.5].

Readout is performed by using the Faraday effect in transmission or the Kerr effect in reflection. Hologram efficiencies in the range 0.01% to 0.05% have been achieved. Although a polarizing filter is not required for readout, it has been used to greatly improve signal-to-noise ratio since the first order output is polarized at 90° relative to the incident beam and zeroth order output [8.6].

MnBi films for holography are fabricated in the thickness range 300 to 700 nm by evaporating Mn and Bi separately onto a mica substrate and annealing for 72 h in vacuum at approximately 300 °C [8.14, 15]. The use of mica substrates has permitted the fabrication of single crystal MnBi films [8.15], whereas earlier films [8.14] had been polycrystalline and were not reproducible in their characteristics [8.16].

Magnetic holography has been performed in EuO [8.17] and, of course, MnBi. Nonholographic Curie-point writing has been demonstrated in a number of magneto-optic materials: GdIG [8.18–20], PtCo [8.21], GdCo [8.22], $CrO_2$ [8.23], Co-P [8.24], and MnAlGe [8.25]. Many of these, summarized and compared elsewhere [8.3], are not capable of hologram storage because of insufficient resolution [only several hundred line pairs per millimeter (lp/mm)]. As noted by *Cohen* and *Mezrich* [8.26], the achievable resolution is determined by the minimum domain size. For the class of materials desirable for magnetic holography (where the one axis of easy magnetization is perpendicular to the surface), the domain size is dependent on the magnetic parameters and cannot be controlled arbitrarily.

The interested reader is directed to [8.26] and [8.27] for more of the details, theory, and systems analysis of magnetic holography.

## 8.2 Metal Films

Experiments by *Gerritsen* and *Heller* in 1967 demonstrated that simple holograms could be stored by vaporization of matter [8.28]. Specifically, evidence was presented that patterns on the order of 0.1 μm could be engraved directly in electrolytic copper and high purity silicon (polished and several millimeters thick) by the interfering energy of two laser beams. Attempts at recording the hologram of a fine wire screen in copper yielded recognizable real and virtual images on readout.

Subsequent experiments by *Amodei* and *Mezrich* [8.29] and *Fourney* and *Barker* [8.30] have shown that thermal-induced visible light holograms can be recorded effectively in thin films of bismuth (Bi). (*Chen* and *Zook* [8.2] report storage of optical information by vaporization of rhodium, but there is no mention of hologram storage.) Additionally, *Decker* et al. [8.31] have conducted infrared holographic storage experiments with 10 to 80 nm thick films of bismuth, antimony (Sb), cadmium (Cd), and paraffin. Corollary work by *Mayden* [8.32] has provided detailed theoretical and experimental data on general micromachining and image recording on thin metallic films by laser beams. (*Dalisa* et al. [8.33] have reported holographic recording on the surface of silicon but by photoanodic reaction rather than thermal induction.)

Holographic recording by vaporization is based on the principle that energy absorbed by a thin film ($\sim$ 10-20 nm) during very brief exposure to a laser beam diffuses away relatively slowly. Consequently, the material vaporizes from regions of the surface in amounts that are proportional to the integrated intensity of light absorbed over that region. The result is a surface relief pattern corresponding to the spatial intensity variations of the recording wavefront. Because the film is thin, the relief pattern does not significantly affect the phase of the light traversing it on readout, thus yielding an amplitude hologram, which can be read out by reflection or transmission with comparable efficiency. For example, in tests with 7.5 to 20 nm thick Bi films vacuum deposited on glass substrates [8.29], transmission and reflection readout produced efficiencies of 6%, which are close to the theoretical maximum for amplitude holograms.

The sensitivity of vaporization holograms depends to some extent on film thickness. *Amodei* and *Mezrich* found experimentally that for films approximately 10 nm thick, the energy density required to record is less than 50 mJ/cm$^2$, which compares favorably with materials such as photochromics and magneto-optic films.

A distinctive feature of vaporization holograms, of course, is the fact that lasers of any wavelength can be used for recording since the recording technique is heat dependent rather than wavelength dependent. Furthermore, these holograms may be operated in a linear region because the energy required to reach boiling temperature is much lower than the vaporization energy. Other operating parameter data include a 5 to 20 ns record time and resolution of 1000 lp/mm.

## 8.3 Layered Devices with Photoconductor Write-In

The layered devices described in this section derive their image or hologram write-in capability from the photoconductive layer. Storage, as well as optical readout, is provided by the other active layer. Transparent electrodes on one or both surfaces provide the required electric fields and access for the readout illumination.

### 8.3.1 Ferroelectric-Photoconductor Devices

The idea of a ferroelectric-photoconductor (FE-PC) device for image storage is not particularly new [8.34]; however, it has been only relatively recently that means of achieving optical readout were suggested [8.35, 36]. Early devices incorporated optical storage and electrical readout—clearly not suitable for hologram storage.

Image storage has been demonstrated in a number of FE-PC device types but, to date, hologram storage has been performed only in those devices based on $Bi_4Ti_3O_{12}$. For this reason, our attention will focus on this ferroelectric. However, devices based on other ferroelectric materials will be described as a means of introducing other relevant physical phenomena. The general schematic construction for these devices is shown in Fig. 8.1a.

Storing an image or hologram in an FE-PC device requires simultaneous illumination and applied voltage. Figure 8.1a illustrates storage of a simple pattern. Prior to storage, it is assumed that the ferroelectric's remanent polarization perpendicular to the plane of the device has been switched down. Next the light pattern is displayed on the photoconductor (PC) layer and a voltage of the *Write* polarity is applied. The applied voltage must, of course, be in excess of the coercive voltage of the ferroelectric (FE) and several factors (including PC response, field/light pulse width, and FE switching characteristic) will determine the degree of switching. (On the basis of this phenomenological process, an RC circuit model of the FE-PC device has been described [8.37].) Of course, the device may be tailored to a particular laser source by appropriate photoconductor selection. Further, the light energy required for hologram storage depends on the PC gain. More will be said on this subject in Section 8.3.5.

Ideally, either the width of the light pulse or the voltage pulse may be used to control the exposure. As a practical matter, however, controlling the voltage

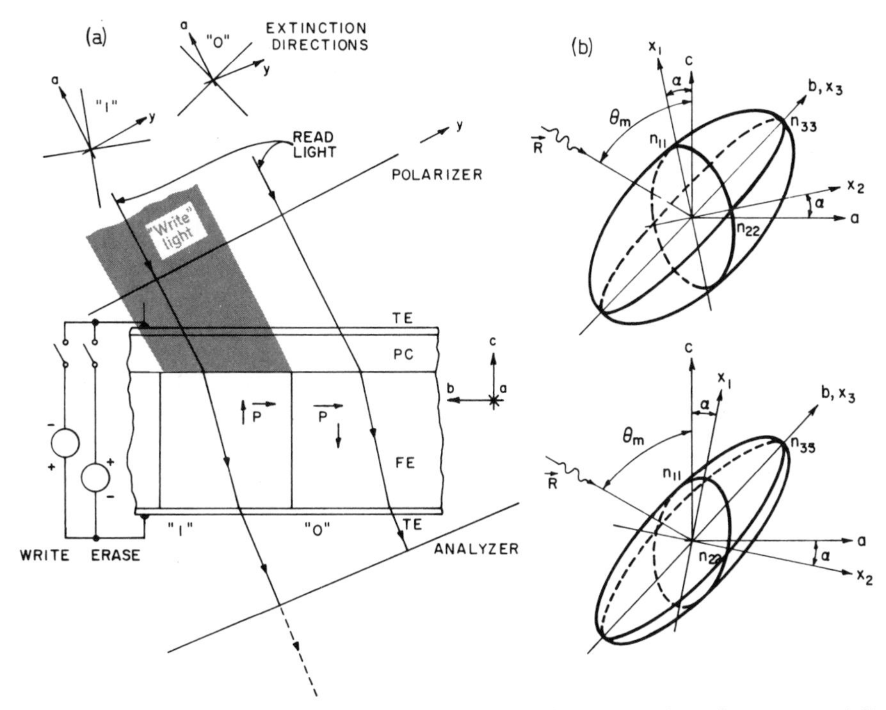


Fig. 8.1a and b. FE-PC (a) $Bi_4Ti_3O_{12}$ device geometry showing *Write* and *Read* processes and (b) orientation of optical indicatrix for $Bi_4Ti_3O_{12}$ for the two states of $c$-axis remanent polarization $P_{Rc}$ (with fixed $P_{Ra}$)

provides the advantage of decreasing unwanted switching caused by room illumination and dark conductivity of the PC.

An FE-PC hologram may be erased in two ways. First, severe overexposure will leave the FE in a uniform state due to the nonzero conductivity throughout the photoconductor. (This ERASE operation leaves the device in the opposite state from which it started. Hence, a voltage polarity reversal will be needed for the next *Write* operation.) More conventionally, the voltage is reversed and the device is illuminated uniformly. (Note that the simultaneous application of light and electric field for erasure differentiates this device from the magneto-optic films of Sec. 8.1; e.g., in MnBi, field alone is required for erasure.)

FE-PC readout depends on the characteristics of the ferroelectric layer. First, it must possess a switchable remanent polarization (or at least a switchable polarization component) perpendicular to the plane of the device. Second, the optical properties of these two states must be different. To date, several such differences have been exploited:

1) optical indicatrix reorientation—$Bi_4Ti_3O_{12}$,
2) birefringence change—PLZT,
3) change in optical scattering state—PLZT,
4) change in axis of optical rotation—$Pb_5Ge_3O_{11}$.

The optical indicatrix reorientation is, perhaps, best explained in terms of the bismuth titanate ($Bi_4Ti_3O_{12}$) FE-PC device. $Bi_4Ti_3O_{12}$, a member of crystal class *m* [8.38], is prepared in single crystal platelets by flux growth [8.39]. It possesses four stable remanent polarization states (having both *c*-axis and *a*-axis components) in a detwinned crystal. If $P_{Ra}$ is first poled [8.40] (in the plane of the crystal) uniformly, then switching $P_{Rc}$ between its two states leads to non-180° switching. As shown in Fig. 8.1b, the orientations of the optical indicatrix for the two states differ by a rotation of $2\alpha$ (roughly 50°) around the crystalline $b(x_3)$-axis. Since $Bi_4Ti_3O_{12}$ is biaxial, $n_{11} \neq n_{22} \neq n_{33}$. Two techniques for observing the change in optical indicatrix orientation (or equivalently, reading out a stored pattern in an FE-PC device) are possible: 1) *birefringence mode:* tilting the crystal about the *a*-axis to achieve a difference in extinction directions for the two states [8.35, 36, 41–43] illustrated in Fig. 8.1a, and 2) *phase mode:* tilting the crystal about the *b*-axis to achieve a phase difference for light polarized perpendicular to *b* [8.42, 44]. The reader may easily convince himself that light normal to the *ab* plane (i.e., parallel to the *c*-axis) will not be adequate to distinguish the two states. Both techniques have been used for reconstruction [8.41, 44] and to display direct images [8.42, 45] in $Bi_4Ti_3O_{12}$-based FE-PC devices.

"Strain-biasing" of ceramic PLZT (lanthanum-doped lead zirconate-lead titanate) ferroelectrics allows, effectively, non-180° switching of the remanent polarization of $P_R$ [8.46–48], as described for $Bi_4Ti_3O_{12}$ above. With these ferroelectric materials, FE–PC devices capable of optical readout are possible. PLZT platelets [8.47], coated with a photoconductor layer and transparent electrodes, are held in tension or compression, causing ferroelectric domains to line up parallel and anti-parallel to the strain axis (in the platelet) and resulting in zero net remanent polarization. Switching the device with fields normal to the platelet produces a normal component of remanent polarization. The platelet birefringence is relating to this remanent polarization, allowing readout of a stored pattern. Holographic resolution has yet to be demonstrated.

PLZT FE-PC devices may also be made which utilize a scattering readout mode [8.49, 50]. Again, the resolution of 40 lp/mm [8.49] falls far short of holographic requirements. A different scattering FE-PC device based on field-induced transitions between FE and anti-FE states in PLZT has also been suggested [8.51].

Lead germanate ($Pb_5Ge_3O_{11}$), an optically active ferroelectric, has been used for image storage in an FE-PC device [8.52]; in this material, the sign of the optical rotation depends on the polarization and, hence, may be switched between two distinguishable states [8.53, 54]. Polarizing optics are used for direct image readout but would not be needed for hologram reconstruction as was the case in MnBi magnetic holograms (Sect. 8.1). Resolution of only 57 lp/mm for 500-μm crystal thickness has been demonstrated [8.52], but hologram storage has not been reported.

As mentioned above, holograms have been stored in FE-PC with $Bi_4Ti_3O_{12}$ FE and ZnSe PC layers [8.41]. Write energy density was 10 $\mu J/mm^2$ at

$\lambda = 514.5$ nm and 1.26-μm gratings were stored in FE-PC devices with 50-μm-thick FE and 2 μm-thick PC. (Electron-beam-writing experiments suggest that the ultimate resolution achievable in $Bi_4Ti_3O_{12}$-FE-PC-device holography may depend on crystal orientation with respect to the interfering beams [8.55].) Grating readout efficiencies (at $\lambda = 632.8$ nm for decreased photoconductor absorption) of 0.01% were achieved in the phase mode; however, in the birefringence mode, only $10^{-3}$% to $10^{-4}$% was possible [8.44].

Hologram storage in FE-PC devices depends on resolution in the FE layer. The $Bi_4Ti_3O_{12}$ results may well be attributable to triangular-shaped domains (observed in so-called display-mode bars) rather than domains passing through the entire crystal thickness [8.45]. If this is true, perhaps thin film layers of $Bi_4Ti_3O_{12}$ [8.56–58] or other ferroelectrics should permit improved holographic performance of FE-PC devices.

FE-PC devices are not without their problems. The bismuth titanate device involves difficult fabrication and processing techniques because of its single-crystal nature. Large-area devices are difficult to achieve. Furthermore, an optical degradation has been observed after numerous switching cycles with metal electrodes directly on the $Bi_4Ti_3O_{12}$ surfaces [8.59]. However, such degradation was not observed in FE-PC devices with photoconductive layers [8.45]. The PLZT devices are simpler to fabricate because of the ceramic FE layer. However, ferroelectric ceramics are notorious for their electrical and optical fatigue [8.60], which would restrict their useful lifetime. And, of course, adequate resolution has yet to be demonstrated.

Perhaps the ultimate FE-PC for holographic applications would utilize thin films of $Bi_4Ti_3O_{12}$ mentioned above [8.56, 57]. This particular structure might represent the optimum compromise between the difficulties of single-crystal technologies and long-term ceramic fatigue.

### 8.3.2 Liquid-Crystal-Photoconductor Devices

The ability to store optical images in liquid crystals, at both visible [8.61] and infrared [8.62] wavelengths, has prompted their use in electrically and photoconductively controllable real-time imaging devices. *Kiemle* and *Wolff* [8.63] reported recyclable holographic storage in an electrically controllable device consisting of mixed (90% nematic, 10% cholesteric) liquid-crystal sandwich cells having matrix-shaped electrodes. Holographic recording was achieved as a result of dynamic scattering induced by application of a DC or low-frequency AC electric field across the electrodes. While initially the nematic ordering is intact and the nematic-cholesteric mixture is clear, the dynamic scattering caused by the applied field emulsifies the cholesteric material. Because this colloidal-like suspension can influence the coherence of an incident laser-light wavefront, holographic recording is feasible. Experiments [8.63] indicate that this optical scattering state can persist for weeks without a sustaining field or can be erased by an AC field in 20 to 500 ms. A typical restoration field strength is 5 V/μm (rms) at 3 kHz.

The matrix-shaped electrode structure required for the electrically controllable device is complicated to construct and, worse, limits resolution capability. The implementation of the photoconductively controlled liquid-crystal device overcomes, in principle, both of these problems. Such a device is achieved simply by replacing the electrode matrix structure by a photoconductive layer. Of the photoconductive materials tested, zinc sulfide (ultraviolet recording) [8.64, 65] and cadmium sulfide (visible light recording) [8.66] have given the best results. Holographic recording has been reported by *Poisson* [8.65] with resolution of 500 lp/mm.

In structure, the photoconductively controlled liquid-crystal device consists of a layer of liquid crystal and a layer of photoconductor sandwiched between two electrode layers of conductive material (e.g., tin oxide). The photoconductive layer is designed to have a "nonilluminated" resistance that is higher than that of the liquid-crystal layer but an "illuminated" resistance that is lower. In the nonilluminated condition, an applied DC voltage across the electrodes of the devices lies almost entirely across the photoconductive layer. Thus, the low voltage across the liquid-crystal layer is not sufficient to produce scattering effects, and the liquid crystal remains transparent. However, when a wavefront illuminates the photoconductor, its resistance is lowered, causing more voltage to appear across the liquid crystal. This higher voltage produces sufficient current to produce dynamic scattering and allow image storage.

The liquid-crystal-photoconductor device permits image recording and readout at a rate controllable by the magnitude of the applied voltage and composition of the materials used. The sensitivity appears to be better than that of other materials, being several orders of magnitude greater than that of photochromic films, for example.

### 8.3.3 Photoconductor-Pockel's Effect Devices

These devices consist of a layer of photoconductive, electro-optic single-crystal material covered by insulating dielectric layers on one or both faces and sandwiched between transparent electrodes (or one electrode and a simple ohmic contact). Experiments using highly resistive photoconductive ZnS, $Bi_{12}SiO_{20}$, and ZnSe as the crystal materials have been reported in the literature [8.67–72].

In operation, a DC voltage applied across the material is modulated by an incident optical wavefront in accordance with the photoconductive properties of the crystal, thereby storing the information in the wavefront structure. Since the material is simultaneously electro-optic, readout of the stored electrostatic pattern can be accomplished by the Pockel's effect, i.e., the pattern is reconverted to an optical wavefront by local phase retardations resulting from the transmission of uniformly polarized light through the material. The combination of both photoconductive and electro-optic functions in a single material has an advantage over a device using separate laminated photoconductive and electro-optic films because of the fabrication problems inherent in interfacing a multilayered structure of this kind. On the other hand, the complexities involved in fabricating a uniform single material which optimally exhibits both functions is actually more difficult than the interfacing problem at the present state of the art.

The ZnS and $Bi_{12}SiO_{20}$ materials used in the reported devices are films on the order of 30-μm and 150-μm thickness, respectively. At the present state of the art, they appear to offer comparable resolution capability: 10 μm and approximately 12 μm. Images have been recorded and read out in both devices but there have been no reported experiments of holographic storage.

### 8.3.4 Elastomer Devices

If the thermoplastic material in the thermoplastic-deformation devices discussed in Chapter 6 is replaced by an elastomer, a useful device for holographic imaging (but not permanent storage) is created. Such a device, termed the "ruticon" [2], has been investigated by *Sheridon* [8.73].

The ruticon device has a layered structure consisting of a conductive transparent substrate, a thin photoconductor, a thin deformable elastomer layer, and a means (either through corona discharge or by using a deformable electrode) of placing a voltage across the elastomer and photoconductive layers. If a deformable electrode is used, it can take three forms: 1) a conductive liquid, 2) a conductive gas, or 3) a thin optically opaque flexible metal; the separate devices resulting from these three different deformable electrodes are identified, respectively, as α-ruticon, β-ruticon, and γ-ruticon. The most interesting of these is the γ-ruticon because it shows most promise as a practical device.

*Kermisch* [8.74] has developed a detailed theory of the image formation mechanism in the γ-ruticon. In operation, a DC electric field ($<600$ V) between the thin flexible metallic electrode and the transparent electrode of the γ-ruticon establishes the initial uniform charge. Holographic information impinging on the device through the transparent substrate causes changes in the electric field across the photoconductor and, hence, across the elastomer. These field changes create mechanical forces that cause the elastomer, and consequently the thin metal layer, to deform and form a phase hologram.

In readout, light is reflected with high efficiency from the metal layer to reconstruct the original object wavefront. If the metal layer is nontransparent, readout light will not affect the conductivity modulation of the photoconductor layer. Erasure is accomplished by simply removing the DC field.

The operating parameters of the γ-ruticon include write energy density (or sensitivity) on the order of 0.1 $mJ/cm^2$, millisecond record and erase times, readout efficiency of 15%, good linearity, and 1000 lp/mm resolution.

---

[2] From the Greek "rutis" for wrinkle and "icon" for image.

### 8.3.5 Advantages and Limitations of Photoconductor-Based Devices

Even the novice can quickly appreciate the sought-for advantages in PC-based holographic devices: the light does not produce the optical change, it merely triggers it. The work is done not by relatively inefficient laser illumination but, rather, by batteries. One envisions photoconductors carefully selected for their spectral response and their high gain (the number of electrons generated per photon). A situation analogous to photographic film, where chemistry supplies the gain, has been created. Further, the medium is recyclable.

While all the above is true to some degree, there are limitations. A device with three components (photoconductor, active layer, transparent electrode) and three interfaces is inherently more difficult to produce reliably than is a single material. But such considerations are beyond the scope of this chapter. What is relevant to the present discussion, however, is the limitation in write energy which will be experienced as higher gain photoconductors are used [8.75]. *Rose* argues that increases in photoconductor gain above unity will not bring about a proportionate decrease in write energy. In this limit, the required exposure $W_r$ is given by

$$W_r = (0.5 \times 10^7 E) \quad \text{photons/cm}^2, \tag{8.1}$$

where $E$ is the electric field (in V/cm) developed across the photoconductor. A comparison between (8.1) and observed exposures $W_0$ for a variety of devices shows that $W_0 \gtrsim W_r$.

## 8.4 Chalcogenide Glasses

The chalcogenide glasses, and amorphous materials in general, have received considerable attention since the late 1960's when *Ovshinsky* described reversible electrical switching in evaporated films of $As_{30}Ge_{10}Si_{12}Te_{48}$ [8.76]. The chalcogenide family includes, of course, all compounds containing an element from group VI of the periodic table. Since the materials of interest are amorphous, classical stoichiometry does not apply and, hence, the number of family members is essentially unlimited.

A large number of chalcogenide glasses undergo long-lasting optical changes when illuminated. In some cases the light energy is converted to heat and the heat performs the storage. In other cases, the light produces bond breaking and free carriers which initiate a more complicated process. Spot storage, image storage, and hologram storage have been demonstrated.

Any one following the literature might conclude that the physical mechanisms to which optical storage is ascribed are as numerous as the methods of fabrication and the materials themselves. Describing all of them would be cumbersome if not impossible. Instead, we will describe those groups of materials which have received the most attention and, hence, have the greatest current technological importance:

1) Materials exhibiting an amorphous-crystalline phase change.
2) $As_2S_3$ and related materials.
3) Photodoping layer devices.

### 8.4.1 Materials Exhibiting an Amorphous-Crystalline Phase Change

The optical properties of the amorphous state of material differ from those of the crystalline. Bulk pieces of a Te-based glass (81 %) with As and Ge exhibited a 30% increase in reflectivity in going from the amorphous to the crystalline state [8.77]. This experiment, with bulk pieces heated in an oven, suggests the feasibility of laser-induced heating to cause an optical change.

Laser-induced amorphous-crystalline phase changes were produced in sputtered $Te_{81}Ge_{15}Sb_2S_2$ films with focused argon-laser ($\lambda = 514.5$ nm) illumination [8.78]. Microsecond switching with application of 25 mJ/mm$^2$ write energy density was observed. A model explaining both amorphous → crystalline and crystalline → amorphous changes was suggested [8.78]. Simply, to end up with the amorphous state, one must heat the material above the melting point $T_m$ and cool between $T_m$ and the glass transition temperature $T_g$ quickly. To end up in the crystalline state, one must heat the film to the interval from $T_g$ to $T_m$ and hold it there long enough to form crystallites before cooling.

Separating thermal and nonthermal effects of the illumination is often difficult; however, in the above case, the authors suggest that nonthermal effects are important [8.78]. Strong photocrystallization has been found in other materials also, including pure Se [8.79].

The resolution and high write power density may make holography with this class of chalcogenide glass difficult. Light-induced crystallites of 0.5-μm diameter have lead to projections of 500 to 1000 lp/mm resolution [8.80].

Improved sensitivity to light has been achieved by simultaneous application of light and voltage to a chalcogenide device similar to the layered photoconductor devices of Section 8.3: *Terao* et al. [8.81] stored crystalline spots in amorphous $Te_{60}Ge_{30}As_{10}$ films with a write energy density of approximately 15 μJ/mm$^2$.

Continued research on spot-storage [8.82, 83] and image storage [8.84, 85] may lead to materials suitable for amorphous → crystalline holography.

### 8.4.2 $As_2S_3$ and Related Materials

$As_2S_3$ has received considerable attention as an optical storage medium [8.86–92] and several investigators [8.86, 93] have performed hologram storage. Perhaps its apparent simplicity (two components in a stoichiometric combination), its ease of fabrication (simple vacuum evaporation of films from the bulk), and its sensitivity to argon-laser illumination have contributed to its popularity. Ever since its useful infrared transmission was reported [8.94], investigators have studied its photoconductivity [8.95, 96], its softening properties [8.97], its high index of refraction [8.98], and other characteristics unrelated to our topic of interest.

Although light-induced optical changes in orpiment (naturally occurring crystalline $As_2S_3$) and realgar (naturally occurring crystalline $As_4S_4$) were long known to mineralogists, the recent interest in optical storage in $As_2S_3$ and related materials dates from the research of *Kostyshin* et al. [8.99] on photodoping in $As_2S_3$ (see Sect. 8.4.3).

For the purpose of this discussion, we group many other materials with $As_2S_3$: other members of the As–S family; $As_2Se_3$ and other As–Se glasses; and, perhaps, the entire As–Ge–S–Se family. As illustrated below, these materials respond to light by a variety of mechanisms, but they are related in at least several ways: 1) they are primarily used as evaporated single-material films, not multi-layer devices; and 2) the amorphous-crystalline transition, although it may occur, is not of prime importance for the holography application.

$As_2S_3$ is well suited to hologram storage with an argon laser and essentially "permanent" readout with a He–Ne laser [8.86, 90]. Resolution is quite good ($>2860$ lp/mm) and high hologram efficiency ($\sim 80\%$ for 632.8-nm-readout of a grating in a 10-μm-thick film) has been obtained [8.86]. The high efficiency for such a relatively thin storage medium is made possible by the large light-induced index of refraction change, $\Delta n$. $\Delta n$ depends on film deposition conditions [8.90] and has been measured to be between 0.05 and 0.10 [8.89, 90, 93].

Hologram storage has been performed in a number of chalcogenide glasses in both film and bulk form. Further, considerable discussion of the physical mechanisms for the observed optical storage phenomena has taken place. The interested reader is referred to the literature summarized below for more information in both areas:

| Material | Study of optical storage and discussion of physical mechanisms | Hologram storage results |
|---|---|---|
| $As_2S_3$ | [8.87, 89, 90, 92] | [8.86, 93] |
| $As_xS_{100-x}(x<40)$ | [8.100] | [8.93, 101, 102, 103] |
| $As_xS_{100-x}(x>40)$ | [8.88, 90, 104, 105, 106] | — |
| $As_xSe_{100-x}$ | [8.87, 89, 90, 107] | — |
| As-S-Se-Ge | [8.108, 109] | [8.93, 110] |

In the holographic storage discussed above, a hologram through the volume of the material was produced. Such holograms may be converted to surface-relief holograms by etching in basic solution. This has been demonstrated in $As_2S_3$ [8.111] and various As–S–Se–Ge compositions [8.112]. Sensitivity is enhanced (or, equivalently, efficiency is increased for comparable exposure) because the index difference between film and ambient is much larger than the original light-induced $\Delta n$ of the material. Efficiency increases of an order of magnitude have been observed [8.90, 112]. Overcoating with a suitable metal (i.e., one which will not thermally diffuse into the chalcogenide film [8.113, 114] gives holograms a new degree of permanence [8.90].

### 8.4.3 Photodoping Devices

Photodoping refers to the light-induced diffusion of certain metals into an adjacent chalcogenide glass. As already indicated in Section 8.4.2, the recent study of optical storage effects in $As_2S_3$ and related materials was initiated by *Kostyshin* et al. [8.99, 115, 116] in their research on photodoping. The presence of a metallic layer in contact with the film was found to increase sensitivity [8.99], which obeyed an *Urbach* [8.117] law [8.115]. Hologram storage was performed [8.116].

Further studies of metal-$As_2S_x$ structures were performed by *Shimizu* et al. [8.118–121] and photoengraving processes were derived from the photodoping.

*Kikuchi* et al. [8.122–124] performed photodoping with Ag and Cu into bulk $As_2S_3$ glass [8.123] and derived an energy band model for the Ag–$As_2S_3$ interface [8.122].

Holographic performance parameters for the photodoping system $As_2S_3$–Ag have been reported by *Kostyshin* et al. [8.116] as:

Resolution $>2000$ lp/mm$^2$.
Approx. write energy $= 10$ mJ/mm$^2$ ($\lambda = 488$ nm).
Approx. grating efficiency $= 20\%$ ($\lambda = 632.8$ nm).

### 8.4.4 Summary

In the short time the chalcogenide glasses have been studied, they have come to provide several convenient alternatives to holographers. Either as simple evaporated films or in conjunction with a metal layer, holograms may be efficiently stored with a cw argon laser; efficient readout can be obtained in the spectral region of low absorption; and permanence can be achieved by etching and overcoating with appropriate metals.

## References


8.1 J. Bordogna, S. A. Keneman, J. J. Amodei: RCA Rev. **33**, 227 (1972)
8.2 D. Chen, J. D. Zook: Proc. IEEE **63**, 1207 (1975)
8.3 R. A. Bartolini, H. A. Weakliem, B. F. Williams: Opt. Eng. **15**, 99 (1976)
8.4 See, for instance, D. Chen: Magneto-Optic Materials. In *Handbook of Lasers*, ed. by R. J. Pressley (CRC Press, Cleveland, Ohio 1971), Chapt. 16, pp. 460—477

8.5 R.S.Mezrich: Appl. Phys. Lett. **14**, 132 (1969)
8.6 R.S.Mezrich: Appl. Opt. **9**, 2275 (1970)
8.7 T.C.Lee, D.Chen: Appl. Phys. Lett. **19**, 62 (1971)
8.8 W.K.Unger: Appl. Opt. **10**, 2788 (1971)
8.9 M.Tanaka, T.Ito, Y.Nishimura: IEEE Trans. MAG-**8**, 523 (1972)
8.10 H.Haskal, G.E.Bernal, D.Chen: Appl. Opt. **13**, 866 (1974)
8.11 L.Mayer: J. Appl. Phys. **29**, 1003 (1958)
8.12 L.Mayer: J. Appl. Phys. **29**, 1454 (1958)
8.13 D.Chen, J.F.Ready, G.E.Bernal: J. Appl. Phys. **39**, 3916 (1968)
8.14 H.J.Williams, R.C.Sherwood, F.G.Foster, E.M.Kelley: J. Appl. Phys. **28**, 1181 (1957)
8.15 D.Chen: J. Appl. Phys. **37**, 1486 (1966)
8.16 L.Mayer: J. Appl. Phys. **31**, 384S (1960)
8.17 G.Fan, K.Pennington, J.H.Greiner: J. Appl. Phys. **40**, 974 (1969)
8.18 J.T.Chang, J.F.Dillon,Jr., U.F.Gianola: J. Appl. Phys. **36**, 1110 (1965)
8.19 N.Goldberg: IEEE Trans. MAG-**3**, 605 (1967)
8.20 R.E.MacDonald, J.W.Beck: J. Appl. Phys. **40**, 1429 (1969)
8.21 D.Treves, J.T.Jacobs, E.Sawatzky: J. Appl. Phys. **46**, 2760 (1975)
8.22 P.Chaudhari, J.J.Cuomo, R.J.Gambino: Appl. Phys. Lett. **22**, 337 (1973)
8.23 R.K.Waring,Jr.: J. Appl. Phys. **42**, 1763 (1971)
8.24 D.Treves, R.P.Hunt, B.Dickey: J. Appl. Phys. **40**, 972 (1969)
8.25 R.C.Sherwood, E.A.Nesbitt, J.H.Wernick, D.D.Bacon, A.J.Kurtzig, R.Wolfe: J. Appl. Phys. **42**, 1704 (1971)
8.26 R.W.Cohen, R.S.Mezrich: RCA Rev. **33**, 54 (1970)
8.27 D.Chen: Appl. Opt. **13**, 767 (1974)
8.28 H.J.Gerritsen, M.E.Heller,Jr.: J. Appl. Phys. **38**, 2054 (1967)
8.29 J.J.Amodei, R.S.Mezrich: Appl. Phys. Lett. **15**, 45 (1969)
8.30 M.E.Fourney, D.B.Barker: Appl. Phys. Lett. **21**, 21 (1972)
8.31 G.Decker, H.Herold, R.Röhr: Appl. Phys. Lett. **20**, 490 (1972)
8.32 D.Mayden: Bell System. Tech. J. **50**, 1761 (1971)
8.33 A.L.Dalisa, W.K.Zwicker, D.J.Debitetto, P.Harnack: Appl. Phys. Lett. **17**, 208 (1970)
8.34 R.M.Schaffert: U.S. Patent 3,148,354, issued September 8, 1964
8.35 S.E.Cummins: Proc. IEEE **55**, 1536 (1967)
8.36 S.E.Cummins: U.S. Patent 3,374,473, issued March 19, 1968
8.37 T.E.Luke: IEEE Trans. ED-**16**, 576 (1969)
8.38 S.E.Cummins, L.E.Cross: J. Appl. Phys. **39**, 2268 (1968)
8.39 A.D.Morrison, F.A.Lewis, A.Miller: Ferroelectrics **1**, 75 (1970)
8.40 M.M.Hopkins, A.Miller: Ferroelectrics **1**, 37 (1970)
8.41 S.A.Keneman, G.W.Taylor, A.Miller, W.H.Fonger: Appl. Phys. Lett. **17**, 173 (1970)
8.42 S.E.Cummins, T.E.Luke: IEEE Trans. ED-**18**, 761 (1971)
8.43 S.A.Keneman, G.W.Taylor, A.Miller: Ferroelectrics **1**, 227 (1970)
8.44 S.A.Keneman, A.Miller, G.W.Taylor: Appl. Opt. **9**, 2279 (1970)
8.45 S.A.Keneman, A.Miller, G.W.Taylor: Ferroelectrics **3**, 131 (1972)
8.46 A.H.Meitzler, J.R.Maldonado, D.B.Fraser: Bell System. Tech. J. **49**, 953 (1970)
8.47 J.R.Maldonado, A.L.Meitzler: Proc. IEEE **59**, 368 (1971)
8.48 J.R.Maldonado, L.K.Anderson: IEEE Trans. ED-**18**, 774 (1971)
8.49 W.D.Smith, C.E.Land: Appl. Phys. Lett. **20**, 169 (1972)
8.50 G.H.Haertling, D.B.McCampbell: Proc. IEEE **60**, 450 (1972)
8.51 A.Kumada, G.Toda, Y.Otomo: Proc. 5th Conf. on Solid State Devices, Tokyo, 1973;—Supp. J. Japan Soc. Appl. Phys. **43**, 150 (1974)
8.52 S.E.Cummins, T.E.Luke: Proc. IEEE **61**, 1039 (1973)
8.53 H.Iwasaki, K.Sugii: Appl. Phys. Lett. **19**, 92 (1971)
8.54 T.Yamada, H.Iwasaki, N.Niizeki: J. Appl. Phys. **43**, 771 (1972)
8.55 S.E.Cummins, B.S.Hill: Proc. IEEE **58**, 938 (1970)
8.56 W.J.Takei, N.P.Formigoni, M.H.Francombe: Appl. Phys. Lett. **15**, 256 (1969)

8.57 S.Y.Wu, W.J.Takei, M.H.Francombe, S.E.Cummins: Ferroelectrics **3**, 217 (1972)
8.58 S.Y.Wu, W.J.Takei, M.H.Francombe: Appl. Phys. Lett. **26**, 22 (1973)
8.59 G.W.Taylor, S.A.Keneman, A.Miller: Ferroelectrics **2**, 11 (1971)
8.60 W.C.Stewart, L.S.Cosentino: Ferroelectrics **1**, 149 (1970)
8.61 G.H.Heilmeier, J.E.Goldmacher: Proc. IEEE **57**, 34 (1969)
8.62 T.Sakusabe, S.Kobayashi: Japan. J. Appl. Phys. **10**, 758 (1971)
8.63 H.Kiemle, U.Wolff: 1970 Annual Meeting of Optical Society of America 28 Sept.–2 Oct. 1970
8.64 J.D.Margerum, J.Nimoy, S.Y.Wong: Appl. Phys. Lett. **17**, 51 (1970)
8.65 F.Poisson: Opt. Commun. **6**, 43 (1972)
8.66 G.Assouline, M.Hareng, E.Leiba: Proc. IEEE **59**, 1355 (1971)
8.67 D.S.Oliver, P.Vohl, R.E.Aldrich, M.E.Behrndt, W.R.Buchan, R.C.Ellis, J.E.Genthe, J.R.Goff, S.L.Hou, G.McDaniel: Appl. Phys. Lett. **17**, 416 (1970)
8.68 D.S.Oliver, W.R.Buchan: IEEE Trans. ED-**18**, 769 (1971)
8.69 S.L.Hou, D.S.Oliver: Appl. Phys. Lett. **18**, 325 (1971)
8.70 J.Feinleib, D.S.Oliver: Appl. Opt. **11**, 2752 (1972)
8.71 P.Nisenson, S.Iwasa: Appl. Opt. **11**, 2760 (1972)
8.72 S.G.Lipson, P.Nisenson: Appl. Opt. **13**, 2052 (1974)
8.73 N.K.Sheridon: IEEE Trans. ED-**19**, 1003 (1972)
8.74 D.Kermisch: Appl. Opt. **15**, 1775 (1976)
8.75 A.Rose: IEEE Trans. ED-**19**, 430 (1972)
8.76 S.R.Ovshinsky: Phys. Rev. Lett. **21**, 1450 (1968)
8.77 J.Feinleib, S.R.Ovshinsky: J. Non-Cryst. Solids **4**, 564 (1970)
8.78 J.Feinleib, J.Deneufville, S.C.Moss, S.R.Ovshinsky: Appl. Phys. Lett. **18**, 254 (1971)
8.79 J.Dresner, G.B.Stringfellow: J. Phys. Chem. Solids **29**, 303 (1968)
8.80 S. Asai, E.Maruyama: Proc. 2nd Conf. Solid State Devices, Tokyo, 1970;—Supp. J. Japan Soc. Appl. Phys. **40**, 172 (1971)
8.81 M.Terao, H.Yamamoto, E.Maruyama: Proc. 4th Conf. Solid State Devices, Tokyo 1972;—Supp. J. Japan Soc. Appl. Phys. **42**, 233 (1973)
8.82 M.Terao, H.Yamamoto, S.Asai, E.Maruyama: Proc. 3rd Conf. Solid State Devices, Tokyo, 1971;—Supp. J. Japan Soc. Appl. Phys. **41**, 68 (1972)
8.83 A.W.Smith: Appl. Opt. **13**, 795 (1974)
8.84 S.R.Ovshinsky, P.H.Klose: J. Non-Cryst. Solids **8–10**, 892 (1972)
8.85 S.R.Ovshinsky, H.Fritzsche: IEEE Trans. ED-**20**, (1973)
8.86 S.A.Keneman: Appl. Phys. Lett. **19**, 205 (1971)
8.87 J.S.Berkes, S.W.Ing,Jr., W.J.Hillegas: J. Appl. Phys. **42**, 5908 (1971)
8.88 K.Tanaka, M.Kikuchi: Solid State Comm. **11**, 1311 (1972)
8.89 J.P.Deneufville, S.C.Moss, S.R.Ovshinsky: J. Non-Cryst. Solids **13**, 191 (1973/74)
8.90 S.A.Keneman: *Optical Storage Effects in Arsenic Trisulfide and Related Materials* (Dissertation, University of Pennsylvania, Philadelphia, Pennsylvania 1974)
8.91 K.Tanaka: Appl. Phys. Lett. **26**, 243 (1975)
8.92 E.Ruske: Phys. Stat. Sol. **28**a, K151 (1975)
8.93 Y.Ohmachi, T.Igo: Appl. Phys. Lett. **20**, 506 (1972)
8.94 R.Frerichs: J. Opt. Soc. Am. **43**, 1153 (1953)
8.95 H.Schlosser: J. Appl. Phys. **28**, 512 (1957)
8.96 S.W.Ing,Jr., J.H.Neyhart, F.Schmidlin: J. Appl. Phys. **42**, 696 (1971)
8.97 S.S.Flaschen, A.D.Pearson, W.R.Northover: J.Am. Ceram. Soc. **43**, 274 (1960)
8.98 W.S.Rodney, I.H.Malitson, T.A.King: J. Opt. Soc. Am. **48**, 633 (1958)
8.99 M.T.Kostyshin, E.V.Mikhailovskaya, P.F.Romanenko: Fiz. Tverd. Tela **8**, 571 (1966) [English transl.: Sov. Phys.—Sol. St. **8**, 451 (1966)]
8.100 A.D.Pearson, B.G.Bagley: Mater. Res. Bull. **6**, 1041 (1971)
8.101 R.G.Brandes, F.P.Laming, A.D.Pearson: Appl. Opt. **9**, 1712 (1970)
8.102 V.I.Mandrosov, E.I.Pik, G.A.Sobolev: Opt. Spectrosk. **34**, 1198 (1973) [English transl.: Opt. Spectrosc. **34**, 695 (1973)]

8.103 V.I.Mandrosov, E.I.Pik, G.A.Sobolev: Opt. Spectrosk. (USSR) **35**, 131 (1973) [English transl.: Opt. Spectrosc. **35**, 75 (1973)]
8.104 K.Tanaka, M.Kikuchi: Solid State Commun. **12**, 195 (1973)
8.105 K.Tanaka, M.Kikuchi: Solid State Commun. **13**, 669 (1973)
8.106 A.Matsuda, M.Kikuchi: Solid State Commun. **13**, 285 (1973)
8.107 Y.Asahara, T.Izumitani: Phys. Chem. Glasses **16**, 29 (1975)
8.108 T.Igo, Y.Toyoshima: J. Non-Cryst. Solids **11**, 304 (1973)
8.109 A.Hamada, M.Saito, M.Kikuchi: Solid State Commun. **11**, 1409 (1972)
8.110 T.Igo, Y.Toyoshima: Proc. 5th Conf. Solid State Devices, Tokyo, 1973—Supp. J. Japan Soc. Appl. Phys. **43**, 106 (1974)
8.111 S.A.Keneman: Thin Solid Films **21**, 281 (1974)
8.112 Y.Utsugi, S.Zembutsu: Appl. Phys. Lett. **27**, 508 (1975)
8.113 L.A.Freeman, R.F.Shaw, A.D.Yoffe: Thin Solid Films **3**, 367 (1969)
8.114 S.Maruno, T.Yamada, M.Noda, Y.Kondo: Japan. J. Appl. Phys. **10**, 653 (1971)
8.115 M.T.Kostyshin, P.F.Romanenko, E.P.Krasnozhenov: Fiz. Tekh. Polup. **2**, 1164 (1968) [English transl.: Sov. Phys.—Semi. **2**, 973 (1969)]
8.116 M.I.Kostyshin, E.P.Krasnojonov, V.A.Makeev, G.A.Sobolev: The Use of Light Sensitive Semiconductor—Metal Systems for Holographic Applications. In *Applications of Holography, Proceedings of the International Symposium of Holography*, ed. by J.-Ch. Vienot, J.Bulabois, J.Pasteur (Univ. Besancon, France 1970) paper 11.7
8.117 F.Urbach: Phys. Rev. **92**, 1324 (1953)
8.118 H.Sakuma, I.Shimizu, H.Kokado, E.Inoue: Proc. 3rd Conf. Solid State Devices, Tokyo, 1971;—Supp. J. Japan Soc. Appl. Phys. **41**, 76 (1972)
8.119 I.Shimizu, H.Sakuma, H.Kokado, E.Inoue: Bull. Chem. Soc. Japan **44**, 1173 (1971)
8.120 I.Shimizu, H.Sakuma, H.Kokado, E.Inoue: Photo. Sci. Eng. **16**, 291 (1972)
8.121 E.Inoue, H.Kokado, I.Shimizu: Proc. 5th Conf. Solid State Devices, Tokyo, 1973;—Supp. J. Japan Soc. Appl. Phys. **43**, 101 (1974)
8.122 A.Matsuda, M.Kikuchi: Solid State Commun. **13**, 401 (1973)
8.123 H.Mizuno, K.Tanaka, M.Kikuchi: Solid State Commun. **12**, 999 (1973)
8.124 A.Matsuda, M.Kikuchi: Proc. 4th Conf. Solid State Devices, Tokyo, 1972;—Supp. J. Japan Soc. Appl. Phys. **42**, 239 (1973)

# Additional References with Titles

## Chapter 4

J.W.Burgess, R.J.Hurditch, C.J.Kirby, G.E.Scrivener: Holographic storage and photoconductivity in PLZT ceramic materials. Appl. Opt. **15**, 1550 (1976)
Dae M.Kim, Rajiv R.Shak, T.A.Ralison, F.K.Tittel: Nonlinear dynamic theory for photorefractive phase hologram formation. Appl. Phys. Lett. **28**, 338 (1976)
C.M.Verber, N.F.Hartman, A.M.Glass: Formation of integrated optics components by multiphoton photorefractive processes. Appl. Phys. Lett. **30**, 272 (1977)
W. Bollman, H.J.Stohr: Incorporation and mobility of $OH^-$ ions in $LiNbO_3$ crystals. Phys. Stat. Sol. **39**, 477 (1977)
W.Bollman: The origin of photoelectrons and the concentration of point defects in $LiNbO_3$ crystals. Phys. Stat. Sol. **40**, 83 (1977)
R.Orlowski, E.Kratzig, H.Kurz: Photorefractive effects in $LiNbO_3$:Fe under external fields. Opt. Comm. **20**, 171 (1977)
Y.Ohmori, M.Yamaguchi, K.Yoshino, Y.Inuishi: Optical damage in $LiNbO_3$ induced by X-ray irradiation. Japan J. Appl. Phys. **16**, 181 (1977)
I.B.Barkan, S.I.Marennikov, M.V.Entin: Holographic storage in $LiNbO_3$ crystal at high temperatures. Phys. Stat. Sol. (a) **38**, K139 (1976)
J.M.Spinhirne, D.Ang, C.S.Joiner, T.L.Estle: Simultaneous holographic and photocurrent studies of the photorefractive effect in $LiTaO_3$ and $LiNbO_3$. Appl. Phys. Lett. **30**, 89 (1977)
G.Chaunussot, A.M.Glass: A bulk photovoltaic effect due to electron-phonon coupling in polar crystals. Phys. Lett. **59A**, 405 (1976)
U.Ohmori, M.Yamaguchi, K.Yoshino, Y.Inuishi: Electron hall mobility in reduced $LiNbO_3$. Japan J. Appl. Phys. **15**, 2263 (1976)
A.Zylbersztejn: Thermally activated trapping in Fe-doped $LiNbO_3$. Appl. Phys. Lett. **29**, 778 (1976)
W.D.Cornish, M.C.Moharam, L.Young: Effects of applied voltage on hologram writing in lithium niobate. J. Appl. Phys. **47**, 1479 (1976)
W.D.Cornish, M.G.Moharam, L.Young: Ellipsometric investigation of optical damage in lithium niobate. Ferroelectrics **10**, 153 (1976)
D. Von der Linde, A.M. Glass: Multiphoton processes for optical storage in pyroelectrics. Ferroelectrics **10**, 5 (1976)
E.Kratzig, H.Kurz: Spectral dependence of the photorefractive recording and erasure process in doped $LiNbO_3$. Ferroelectrics **10**, 159 (1976)
A.M.Glass, D.Von der Linde: Photovoltaic photoconductive and excited state dipole mechanisms for optical storage in pyroelectrics. Ferroelectrics **10**, 163 (1976)
M.G.Moharam, L.Young: Hologram writing by the photorefractive effect with Gaussian beams at constant applied voltage. J. Appl. Phys. **47**, 4048 (1976)
S.F.Su, T.K.Gaylord: Refractive-index profile and physical process determination in thick gratings in electro-optic crystals. Appl. Opt. **15**, 1947 (1976)
D.M.Kim, R.R.Shak, T.A.Robson, F.K.Tittel: Study of the equivalent electron drift field characteristics in $LiNbO_3$ by phase holography. Appl. Phys. Lett. **29**, 84 (1976)
S.F.Su, T.K.Gaylord: Determination of physical parameters and processes in hologram formation in ferroelectrics. J. Appl. Phys. **47**, 2757 (1976)

W. D. Cornish, L. Young, M. L. W. Thewalt: Space charge fields: A fringe technique for observing such fields during hologram writing in $LiNbO_3$. Appl. Opt. **15**, 1258 (1976)
H. Kurz: Photorefractive recording dynamics and multiple storage of volume holograms in photorefractive $LiNbO_3$. Optica Acta **24**, 463 (1977)
E. Kratzig, H. Kurz: Photorefractive and photovoltaic effects in doped $LiNbO_3$. Optica Acta **24**, 475 (1977)

## Chapter 5

D. Casasent, F. Caimi: Adaptive photodichroic matched spatial filter. Appl. Opt. **15**, 2631 (1976)
W. C. Collins, M. J. Marrone: Photodichroic materials as adaptive spatial filters in real-time spectral analysis. Appl. Phys. Lett. **28**, 260 (1976)
D. Casasent, F. Caimi: Photodichroic recording and storage material. Appl. Opt. **15**, 815 (1976)

## Chapter 6

T. C. Lee, J. W. Lin, O. N. Tufte: Thermoplastic—Photoconductor for optical recording and storage—New developments. Proc. of SPIE Seminar on Optical Storage Materials and Methods, Paper 11, Vol. 123 (1977), to be published
D. S. Lo: Infrared laser heating for thermoplastic recording. Proc. of SPIE Seminar on Optical Storage Materials and Methods, Paper 12, Vol. 123 (1977), to be published
J. Gaynor: Laser sensitive deformable films. Proc. of SPIE Seminar on Optical Storage Materials and Methods, Paper 02, Vol. 123 (1977), to be published

# Author Index


Aftergut, S. 154, 190
Afterness, R. 93
Alphonse, G. A. 104, 120
Amodei, J. J. 104, 108, 111, 118, 231
Anderson, H. R. 179, 181
Ashcheulov, Y. V. 152

Barker, D. B. 231
Bartolini, R. A. 229
Beauchamp, H. L. 185
Blume, H. 159
Brandes, R. G. 87, 90
Brown, G. H. 133
Broyde, B. 219
Budd, H. F. 170
Budrikis, Z. L. 222
Burke, W. 130

Carlsen, W. S. 123
Case, S. K. 93, 97
Cathey, W. T. 216
Chang, M. 91, 93
Chen, D. 229, 231
Chen, F. S. 101
Clark, M. G. 106
Cohen, R. W. 230
Colburn, W. S. 166, 168, 174–176, 180, 186, 199
Collier, R. J. 151
Credelle, T. L. 188, 190, 191, 196, 198, 199
Curran, R. K. 83, 84, 93–95

Dalisa, A. L. 231
D'amia, L. 120
Dammann, H. 216, 217
Decker, G. 231
Denisyak, Y. N. 266
Dubow, J. B. 166, 175, 180, 186

Fillmore, G. L. 94, 95
Fourney, M. E. 231
Friesem, A. A. 156
Frieser, H. 34, 37, 40

Gabor, D. 1, 22, 50
Gaynor, J. 164, 172, 190
Gerritsen, H. J. 231
Glass, A. M. 112
Greer, M. O. 216

Heller, M. E. 231

Jenney, J. A. 216

Kakuchy, M. 241
Kermisch, D. 237
Kessler, H. C. 144, 145
Kestigian, M. 122
Kiemle, H. 235
Kogelnik, H. 10–12, 18, 152, 171
Kostyshin, M. T. 240, 241
Kurtz, C. N. 199

La Macchia, J. T. 118
Lamberts, R. L. 199
Lanzl, F. 156
Lee, T. C. 187
Lee, W. H. 216
Leith, E. N. 1
Leonard, C. D. 96
Lin, L. H. 85, 87, 90, 94, 95, 185, 216
Lo, D. S. 187, 201
Luty, F. 158

Mathies, D. L. 169–171, 199
Mayden, D. 231
McCauley, D. G. 86, 89, 97
Meier, R. W. 216
Meyerhofer, D. 79, 85, 94, 96
Mezrich, R. S. 230, 231
Micheron, F. 119

O'Brian, B. 78
Ovshinsky, S. R. 238

Pennington, K. S. 98
Phillips, W. 120
Poisson, F. 236
Przibram, K. 133

Roder, U. 158
Rose, A. 238

Schaffert, R. M. 190
Schmidlin, F. W. 169, 190
Schneider, I. 156
Scrivener, G. E. 154
Sewell, G. S. 172
Shankoff, T. A. 75, 76, 81–84 93–95
Shaw, W. G. 133
Sheridon, N. K. 168, 178, 237
Shimizu, I. 241
Siroki, R. S. 224
Smith, H. M. 31
Spong, F. W. 188, 190, 191, 196, 198, 199
Staebler, D. L. 111, 121, 144
Stark, H. M. 190
Stewart, W. C. 97
Sukhanov, V. I. 152

Terao, M. 239
Thaxter, J. B. 117, 122
Tomlinson, W. J. 152–155
Tompkins, E. N. 168, 180, 199
Townsend, P. L. 118
Tubbs, M. R. 154
Tynan, R. F. 94, 95

Upatnieks, J. 1, 96
Urbach, J. C. 216, 241

Van Heerden, P. J. 2
Vikram, C. S. 224
Von der Linde, D. 112, 121

Walker, J. L. 156
Wolff, U. 235

Young, L. 107, 123

Zook, J. D. 229, 231

# Subject Index

Absorption 13, 14, 75
 constant 4, 13, 117
 spectrum 81, 83, 135–139
Adjacency effect 29, 38, 44, 53
Amplitude 4, 10
 hologram 3, 4, 6, 14–16, 21, 28, 31, 34, 61, 68
 transmittance 2, 4, 5, 18
Antihalation coating 26
Antireflection 26
Anomalous dispersion 158, 159

Bandwidth 7, 186, 187
Beam ratio 8, 28, 40, 51, 57
Bessel function 9, 10
Birefringence 234
Bleached
 emulsions 5, 28
 holograms 59, 69
Bleaching 59–62, 69
 gaseous 63
 reversal 65
Bragg
 angle 10, 11, 16, 67
 condition 11–15, 17, 33, 52, 66, 69
 effect 68

Capacity, storage 96, 102, 118, 119, 122, 125, 126, 130, 133, 151, 155, 158, 206, 209, 234, 235, 238
Carrier frequency 22, 44, 50
Chalcogenide glass 229, 238–241
Charge density 166, 167, 169, 175
Checkerboard 48
Circular grain model 48
Coercive field 113
Coherence 1, 26
Coherent
 waves 10
 light 37
Color hologram 21, 26
Computer hologram 31, 34
Conjugate image 5, 8
 wave 10
Contrast ratio 141
Corona 164, 166, 167, 169, 180
Coupled wave theory 10, 11, 16
Crosstalk 159
Curie temperature 114, 115, 117–119, 230
Cycle lifetime 229
Czochralski method 115, 118

D-log E curve 24, 28
Dark decay 173, 202, 229
Dephasing measure 12
Developer 29, 31
 monobath 31, 55
 rapid 33
Dichroism 158
Dichromated gelatin 5, 62, 75
Dielectric relaxation time 110
Diffraction efficiency 5–8, 10, 13–18, 26, 28, 29, 31, 34, 35, 39–41, 46, 48, 50–52, 54–68, 75, 82, 85, 86, 92, 94, 95, 98, 110, 111, 123, 152, 154, 155, 158, 176, 180, 186, 188–197, 200, 201, 205, 215, 218, 220, 223–225, 229, 230, 237, 240
Diffuse illumination 1
Diffusion 102, 103, 105, 107, 111, 124, 127
 mottle 37, 47
Dispersion, anomalous 158, 159
Dopant 120, 121, 135
Drift 102–108, 111, 124
Drying 30
Duplicate hologram 98
Dynamic range 54, 126–128, 130

Eidophor 171
Elastomer 237
Emulsion stress 56
Erase mode sensitivity 141–149
Erasure 102, 108–111, 113, 114, 116, 117, 120–122, 125–128, 130, 133–136, 138, 144, 147, 148, 159, 164, 166, 167, 174–176, 178, 200, 201, 204, 206, 229, 233, 237
Exposure 2, 4, 18
Exposure latitude 28

Faraday effect 230
Ferricyanide 63
Ferroelectric crystal 5, 101, 232, 234
Fixing, 30, 33, 109–114, 163, 164, 166, 211
Foral 105 181
Fourier
  analysis 42
  coefficient 42
  hologram 3, 58, 186
  transform 37
Fraunhofer
  diffraction 3, 39
  hologram 3
Fresnel
  diffraction 3
  hologram 3
Frequency, carrier 22, 44, 50
  response 187, 189, 190
  spatial 6, 27, 34, 38–43, 53, 54, 62, 67
Frost 162, 169–171, 190, 198, 199

Gabor
  condition 22
  hologram 3
Gelatin 77
  dichromated 5, 62, 75
Ghost image 174–176, 202, 214
Glass, chalcogenide 229, 238–241
Granularity 18
  noise 43

Herschel effect 49
Hologram
  amplitude 3, 4, 6, 8, 14–16, 21, 31, 34, 61, 68, 78
  bleached phase 59
  bleached reflection 69
  color 21, 66
  computer generated 31, 34
  duplicate 98
  Fresnel 3
  Fourier Transform 3, 58, 186
  Fraunhofer 3
  Gabor 3
  image plane 27, 44
  in-line 3
  Leith-Upatnieks 3
  lensless Fourier Transform 3
  Lippmann 3, 21, 22, 67, 68
  magnetic 230
  off-axis 3
  phase 3–6, 9, 10, 13, 16, 18, 21, 28, 59, 61, 65, 66, 68, 75, 129, 159, 171, 199, 209, 220, 221, 225, 237, 240
  plane 3, 6
  reflection 3, 12, 16, 17, 68, 69, 75
  relief 5, 75, 79, 171, 240
  thick 3, 5, 18, 66, 68, 80, 87, 109, 133
  thin 3, 5
  transmission 12, 13, 68, 69, 75
  vaporization 231, 232
  volume 3, 6, 10
Holographic
  exposure index 48, 49
  interferograms 32
  interferometry 5, 26, 32, 57, 133
  non-destructive testing 27
  optical elements 97
Hypersensitization 33, 49, 57

Illuminating beam 1, 4
Image plane hologram 27, 44
In-line hologram 3
Index of refraction 4, 15, 22, 23, 31, 33, 60, 65, 68, 75, 79, 82, 83, 85, 92, 102, 110, 114, 119, 120, 123–129, 152, 154, 158, 178, 192, 213, 214
Infra red 49, 50
*In situ* development 32, 33
Intermodulation 32, 52, 53, 57, 64, 65, 67, 69, 198, 199, 205, 213–215, 222
Interferometer, real-time 32
Interferometry, holographic 5, 26, 32, 57, 133

K 28, 40, 51, 57
Kerr effect 230
Kossel lines 70

Latent image 23, 24, 49, 70, 166, 167
  electrostatic 202
  fading 28
Leith-Upatnieks hologram 3
Lensless Fourier Transform hologram 3
Lippmann, G. 66
  hologram 3, 21, 22, 67, 68
Liquid crystal 235, 236

Magneto-optic films 229, 230
Magnetic hologram 230
Metal films 231
Microdensitometer 38, 39, 42
  multiple-sine-slit 41, 55
Mie theory 23
Modulation 6, 8, 104
  transfer function (see MTF)
Moire 42

Monobath developer 31, 55
MTF 6, 23, 27, 29, 34–44, 47, 49, 51–56, 63, 65, 67
 apparent 44, 53
 as-if 41, 55

Noise 5, 10, 18, 45, 50, 75, 89, 95, 129, 159, 170, 174, 197, 199, 213, 214, 217, 218, 222, 224, 225
 Wiener 130
Nonlinearity 19
 intrinsic 198, 199, 213, 214, 218

Off-axis hologram 3
Optical
 indicatrix 234
 memory 21
 path 4

Parallax 1
Phase 4, 9, 10, 12, 24, 42, 171, 231
 grating 64
 hologram 3–6, 9, 10, 13, 16–18, 21, 28, 59, 61, 65, 66, 68, 69, 75, 79, 129, 159, 171, 199, 209, 220, 221, 225, 237, 240
 modulation 31, 55, 59
 noise 19
 shift 91, 192, 213, 236
Photochromic 4, 18, 19, 133
 absorption spectrum 136, 139
 switching 138, 139, 141
Photoconductivity 111
Photoconductor 19, 171, 176
 fabrication 182
Photodichroic 156
Photodoping 241
Photoionization 106
Photopolymerization 79
Photoplastic 164
Photoresist 5, 209
Photovoltaic 103, 107–109, 116
Piccopale 179, 184
Plane hologram 3, 6
Pockels effect 236
POTA 54
Processing 28, 82, 86, 87, 89, 164, 166

Q
 parameter 3, 6, 66
 switching 23
 values 10

Rapid developers 33
Ratio
 beam 8, 57
 contrast 141
 signal-to-noise 5, 41, 44, 57, 61, 64–66, 96, 155, 180, 198, 199–202, 215
Rayleigh theory 23
Real-time interferometer 32
Reciprocity failure 23
Redundancy 222
Reflection hologram 3, 12, 16, 17, 68, 69, 75
Relief 4, 18, 24, 31, 38, 53, 55, 56, 59, 61, 62, 64, 65, 79, 80, 193, 196, 213, 231
 hologram 5, 75, 79, 171, 240
Remeika method 118
Resolution 95, 129, 133, 155, 173, 186, 190, 191, 205, 209, 220, 230, 234–240
Reversal
 bleach 65
 development 31
 process 64
Ronchi ruling 39
Ruticon 168, 178, 237

Scattered flux 19, 26, 48–50, 63, 75, 129, 130, 222
 spectrum 44–47
Sensitivity 93, 123–127, 139, 152–155, 158, 176, 177, 182, 191, 194, 196, 203, 205, 209, 218, 221, 229, 237, 240
 erase mode 141–149
 spectral 24, 238
Sensitizing dye 30
Signal-to-noise ratio 5, 41, 44, 57, 61, 64–66, 96, 155, 180, 198, 199
Signal, bandwidth 7, 186, 187
Spatial
 frequency 6, 27, 34, 38–43, 53, 54, 62, 67,
 filter 24, 25, 29, 31, 55, 63
 filtering 54
Spectrum
 absorption 81, 83, 135, 137
 angular 198
 scattered flux 44–47
 Wiener 45
Spread function 47
 line 36, 37
Spectral
 sensitivity 24, 238
 sensitization 23
Speckle 6, 19, 32, 44, 47, 50, 67
Staybelite ester 178, 179, 181, 184, 201, 202, 204
Stop bath 30

Storage capacity 96, 102, 118, 119, 122, 125, 126, 128, 130, 133, 155, 158, 206, 209, 234, 235, 238
Stress relief 57
Surface relief 4, 15, 18, 24, 31, 38, 53, 55, 56, 59, 64, 65, 193, 231
Switching 232, 233, 234, 238, 239

Tanning 55, 59, 61, 62, 77, 80
Temperature, Curie 114–119, 230
Thermoplastic 5, 18, 19, 171
 fabrication 189
Thick hologram 3, 5, 18, 66, 68, 80, 87, 109, 133
Thin hologram 3, 5
Time, dielectric relaxation 110
Transform plane 41
Transmission hologram 12, 13, 68, 69, 75
Traps 102, 103, 107, 109

Vaporization holograms 231, 232
Volume hologram 3, 6, 10

Wetting agent 30
Wiener spectrum 45

# Springer Series in Optical Sciences

Editor: D. L. MacAdam

Volume 1
W. Koechner
## Solid-State Laser Engineering

287 figures, 38 tables. XI, 620 pages. 1976
ISBN 3-540-90167-1
*Contents:* Optical Amplification. – Properties of Solid-State Laser Materials. – Laser Oscillator. – Laser Amplifier. – Optical Resonator. – Optical Pump Systems. – Heat Removal. – Q-Switches and External Switching Devices. – Mode-Locking. – Nonlinear Devices. – Design of Lasers in Relation to their Application. – Damage of Optical Elements. – Appendices: Laser Safety. Conversion Factors and Constants.

Volume 2
R. Beck, W. Englisch, K. Gürs
## Table of Laser Lines in Gases and Vapors

IV, 130 pages. 1976
ISBN 3-540-07808-8

Volume 3
## Tunable Lasers and Applications

Proceedings of the Loen Conference, Norway 1976
Editors: A. Mooradian, T. Jaeger, P. Stokseth
238 figures. VIII, 404 pages. 1976
ISBN 3-540-07968-8
*Contents:* Tunable and High Energy UV-Visible Lasers. – Tunable IR Laser Systems. – Isotope Separation and Laser Driven Chemical Reactions. – Nonlinear Excitation of Molecules. – Laser Photokinetics. – Atmospheric Photochemistry and Diagnostics. – Photobiology. – Spectroscopic Applications of Tunable Lasers.

Volume 4
V. S. Letokhov, V. P. Chebotayev
## Nonlinear Laser Spectroscopy

193 figures, 22 tables. XVI, 466 pages. 1977
ISBN 3-540-08044-9
*Contents:* Introduction. – Elements of the Theory of Resonant Interaction of a Laser Field and Gas. – Narrow Saturation Resonances on Doppler-Broadened Transition. – Narrow Resonances of Two-Photon Transitions Without Doppler Broadening. – Nonlinear Resonances on Coupled Doppler-Broadened Transitions. – Narrow Nonlinear Resonances in Spectroscopy. – Nonlinear Atomic Laser Spectroscopy. – Nonlinear Molecular Laser Spectroscopy. – Nonlinear Narrow Resonances in Quantum Electronics. – Narrow Nonlinear Resonances in Experimental Physics.

Volume 5
M. Young
## Optics and Lasers

An Engineering Physics Approach
122 figures, approx. 6 tables. XIV, 208 pages. 1977
ISBN 3-540-08126-7
*Contents:* Ray Optics. – Optical Instruments. – Light Sources and Detectors. – Wave Optics. – Interferometry and Related Areas. – Holography and Fourier Optics. – Lasers. – Electromagnetic and Polarization Effects.

Volume 6
B. Saleh
## Photoelectron Statistics

With Applications to Spectroscopy and Optical Communication
85 figures, 8 tables. Approx. 420 pages. 1977
ISBN 3-540-08295-6
*Contents:* Tools from Mathematical Statistics: Statistical Description of Random Variables and Stochastic Processes. Point Processes. – Theory: The Optical Field: A Stochastic Vector Field or Classical Theory of Optical Coherence. Photoelectron Events: A Doubly Stochastic Poisson Process. – Applications: Applications to Optical Communication. Applications to Spectroscopy.

Volume 7
## Laser Spectroscopy III

Proceedings of the Third International Conference, Jackson Lake Lodge, Wyoming, USA, July 4–8, 1977
Editors: J. L. Hall; J. L. Carlsten
296 figures. XII, 468 pages. 1977
ISBN 3-540-08543-2
*Contents:* Fundamental Physical Applications of Laser Spectroscopy. – Multiple Photon Dissociation. – New Sub-Doppler Interaction Techniques. – Highly Excited States, Ionization, and High Intensity Interactions. – Optical Transients. – High Resolution and Double Resonance. – Laser Spectroscopic Applications. – Laser Sources. – Laser Wavelength Measurements. – Postdeadline Papers.

Springer-Verlag
Berlin
Heidelberg
New York